GROWTH, STABILITY AND EQUITY IN AGRICULTURE
A Quantitative Analysis

About the Author

Arup Chattopadhyay obtained his M.A. in Economics with Statistics & Econometrics from the University of Burdwan and stood first in first class. He received his Ph.D. degree from Vidyasagar University. The author contributes articles and papers regularly to professional journals, seminars and conferences. Besides the academic assignments the author occupies quite a few important positions in both administrative and academic fields within and outside the University. He is also the Associate Editor of the V.U. Journal of Commerce, Midnapore, West Bengal. Recently he joined the post of Reader in the Department of Commerce, the University of Burdwan, West Bengal.

GROWTH, STABILITY AND EQUITY IN AGRICULTURE
A Quantitative Analysis

ARUP CHATTOPADHYAY
Department of Commerce
The University of Burdwan
West Bengal

CONCEPT PUBLISHING COMPANY, NEW DELHI-110059

ISBN 81-7022-923-5

First Published 2001

Printed and Published by
Ashok Kumar Mittal
Concept Publishing Company
A/15-16, Commercial Block, Mohan Garden,
New Delhi-110059 (India)

Phones: 5648039, 5649024

Fax: 091–(11)–5648053

E.mail: publishing@conceptpub.com

Preface

In the era of world-wide fading out of communism and the advent of globalisation and market economy a common curiosity has arisen among the scholars to probe the clues behind the longest lived Left Front Government (LFG) in West Bengal in general and its strong hold in rural West Bengal in particular. Different scholars have tackled this issue from different angles, but the most prominent thinking runs in terms of an *expost* analysis by enquiring into the performance of West Bengal agriculture. The performance of agriculture has so far been examined mainly from the measurement of growth and stability in agriculture. But the existing studies, as observed, are mostly suffering from methodological weaknesses in almost every phase of measurement of growth and stability, which probably also lead to differences of opinion among the scholars in this field. Further the equity aspect of growth and stability has not been properly taken care of by the scholars in West Bengal agriculture. An attempt is made in this book to make up these gaps in the earlier studies as far as practicable.

Apart from rigorous statistical examination of growth and stability and their joint effect on equity with the help of newly established improved techniques, the book covers some new issues, like: (a) determination of relative contributions of locational factors and individual crops towards overall agricultural growth, (b) determination of growth contributing factors through factor analysis, (c) statistical examination of causes of instability, (d) application of sophisticated inequality measures using primary date etc. The general period chosen for the analysis is from the crop year 1957-58 to 1994-95 which is again divided into two sub-periods, namely, from 1957-58 to 1976-77 (i.e., Pre-LFG regime) and 1977-78 to 1994-95 (i.e., LFG regime), to facilitate the comparative analysis of performance of West Bengal agriculture before and during the LFG rule. The findings of this study, based on improved methodology, are quite interesting and different from those of the existing studies.

In the course of this study I have been helped by many persons, both directly and indirectly. My greatest debt is to Prof. Purnendu Sekhar

Das who stimulated my interest in the subject and guided me in my work at almost every step. In spite of his various preoccupations, he spent his valuable time in going through critically each and every word of the manuscript. Actually my words fail me in expressing my feelings and gratitude towards him.

I am also indebted to Prof. S. Srinivasan, Department of Industrial Engineering and Management, I.I.T., Kharagpur, who helped me a lot in learning the computer operations and also in using computer facilities from his institution. I am thankful to my colleagues, Prof. A.Jana, Dept. of Political Science, Dr. S.Majumder, Dept. of Economics and Dr. K.C.Paul, Head , Dept. of Commerce, Vidyasagar University, for their advice and assistance in solving various problems encountered by me during the study period. I am grateful to the U.G.C., Govt. of India for the award of teacher fellowship.

I must also profusely thank my students, Debashis Gangopadhyay and Jayanta Debnath, who helped me in data collection and processing of data. I cannot acknowledge by name many employees of the Govt. of West Bengal who provided me with opportunities to go through the necessary government publications and periodicals, but I wish to record my debt to them. I shall be failing in my duty, if I do not express my heartiest thanks to my informants and respondents who supplied me the Primary data required for the study.

Finally, I would like to thank my wife, Arunima, who during my difficult period of this study not only took the responsibility of all household affairs which relieved me from many anxieties but also helped me in various activities of the present work. I feel guilty to my daughter Payel and my son Anik who were deprived of my attention and affection during the research work. Hopefully they will forgive me when they will grow up and see this volume of work that I produced for my Ph. D. Degree.

Of course, I am alone responsible for any omissions and commissions in this study.

Arup Chattopadhyay

24 April, 2000

Contents

List of Tables

1

Introduction

1.1 Importance of the Study

The debate in development economics is unending and continuing on the choice of appropriate strategy for the development of 'Less Developed Countries'(LDCs) which are mostly suffering from the twin problems of unemployment/underemployment and poverty/ malnutrition as the consequences and causes of massive population growth, low productivity, inadequate capital formation, low employment opportunities, high debt burden, chronic inflation, acute inter-regional and inter-personal disparities etc. In the early writings, dated back to Second World War the future development of the LDCs was viewed mainly as a recapitulation of the past historical experience of the developed countries; more specifically, at that time, industrialisation was thought to be synonymous with development and agriculture as a catalytic agent on it on the consideration of its 'product contribution', 'factor contribution' and 'market contribution' [Kuznets (1961)]. This pattern was also highlighted in the so-called "Dualistic Models" of growth [Lewis (1954), Ranis-Fei (1961), Jorgenson (1966), etc.]. But from the growth pattern of the LDCs it was commonly observed that inspite of the rapid industrial growth in some of the LDCs (even higher than those in the then developed countries) the basic problems of unemployment and poverty remained there as, apart from other reasons [Todaro (1977), Ch.4], the LDCs were mostly dependent on agriculture whose neglect might have thwarted the real development process of the LDCs. Consequently, in the economic literature [Coale and Hoover (1958), Dovring (1959), Johnston and Mellor (1961), Nicholls (1963), Mellor (1966)] from early sixties, the role of agriculture in promoting economic development was sufficiently underlined considering the potential contribution of agriculture as (i) to provide food and fiber for an expanding population with rising purchasing power, (ii) to provide capital including foreign exchange for the economic transformation, (iii) to provide a direct increase in rural welfare and (iv) to protect the prospects of the business cycles of foreign trade and industrial production. Now the debate centres not around the industry *vs.* agriculture but around the alternative strategies of agricultural development, adopted by the LDCs with different structures and political power bases and with different sets of objectives [Griffin (1974)]. These alternative strategies may be broadly classified and termed in three ways, viz., (1) improvement (or 'minimalist policy package') and transformation (adopted extensively in Africa and Asia and radically by China, Cuba, Tanzania, etc.), (2) extensive margin and intensive margin (popularly known as 'Green Revolution' in Asia) and (3) uni-model (Japanese-Taiwanese development patterns) and bimodel (followed by Mexico and Colombia) [Johnston and Kilby (1975)]. Irrespective of the strategies followed by the LDCs, they have two common features. These are : (a) the agriculture has been practised and nourished under the restrictions and regulations of the government, and (b) agricultural growth in output has got its momentum with different degrees with and without a 'shared growth' (i.e., equitable distribution of growth) [Johl (1975), Neale (1978), Chinn (1979), Vyas (1983), Shand (1987), etc.]. But the debate on the development strategy in agriculture does not end here. Rather in the present context of globalisation and liberalisation this debate has been renewed with much warmth. The supporters of "free market approach" [Johl (1995)] argue that large-scale corporate, capitalist agriculture is more efficient than the individual peasant farming as the former leads to better allocative efficiency, induces higher private investment in agriculture and results in higher output. For this they (mainly the large landholders, capitalists and the policy makers getting support from the world organizations) demand for the removal of all sorts of regulations and restrictions operative on agriculture, by which price distortions would be avoided and they believe that 'setting prices right' sets everything else right. But this free-market approach gets stiff resistance from the notable agricultural economists, like V.S. Vyas (1994), C.H. Hanumantha Rao (1995),

K. Parikh and S. Narayana (1995), S.N. Mishra (1997), etc. and their common perception is that agricultural policy reform issues in LDCs should be considered in the light of the specific goals like 'growth, equity, employment, poverty alleviation and food security'. "For certain social purposes and in the case of market failure, such as national food security and food security for the poor, market intervention is necessary and desirable. Similarly, there is no case for removal of ceilings on land ownership if maximization of agricultural output and employment remains the social goals of development." [Mishra (1997), p A25].

From the foregoing brief historical analysis it is evident that there is no consensus strategy which may be followed by the LDCs for the development of agriculture. Rather different countries depending on different socio-economic, political and institutional bases have adopted different strategies and within a country also different strategies have been adopted over different time periods. For instance, in India during initial phase of plan period transformation approach (through institutional reforms) was pursued; later on after the mid-sixties that was shifted to intensive margin (known as green revolution) and at present the policy of deregulation is coming forward. Therefore, the crucial issue before the policy makers of a country is to select and frame the most appropriate strategy for agricultural development. This can be resolved in two ways : either through *ex-ante* analysis by evaluating the established theoretical arguments in relation to the particular situation or through *ex-post* analysis by examining the past performance of agriculture systematically. But agricultural activity is so diverse and peculiar that theoretical reasoning applicable for a country may fail to bring desired result for others. On the contrary, past performance gives better guidance not only to take corrective measures but also to frame the future policies successfully. In the direction of second line of thinking we have selected three main functions [out of five* as described by SUAN, (the Southeast Asian Universities Agroecosystem Network), Marten (1988)] of an agroecosystem, viz., "Growth, Stability and Equity" for the proper and judicious assessment of agricultural performance.

---

* Five Functions of an agroecosystem are productivity, stability, viability, sustainability and equity. Here, productivity, however, directly implies the growth in agricultural production.

Growth measurement through trend fitting exercise is a convenient way of summarizing a long time series without going into the complexities of functional interplay of different economic and non-economic variables and that's why perhaps the technique has been widely used in Indian agricultural scenario. From the measured output growth rates we can compare the production performance of agriculture over different crops and regions for different time periods. Moreover, through growth decomposition we can find out the contributions of different growth components (say, area, yield, etc.), different crops and different regions towards overall crop-output growth rate. Growth measurement exercise can also be extended to search out the relative effects of the growth contributing factors (say, institutional, technological, etc.). In short, from growth analysis we can empirically examine the record of agricultural performance related to the adopted particular strategy and by doing so we can get idea what has been done, what ought to be done, how much falls short of the required result and how that required result can be achieved.

Stability concerns fluctuations in production that result from numerous fluctuations in an agroecosystem's physical, technical and social environment, like, variations in rainfall, temperature and humidity, periodic attacks of diseases, insects and weeds, use of high yielding varieties (which are more vulnerable than local varieties to fluctuating environmental stresses), extension of cultivation to the marginal lands and soil degradation (due to salinity, water-logging, deforestation, excessive use of chemical fertilizer, etc.), fluctuations in input and output prices etc. Stability, closely related to growth, is assessed in terms of the fluctuation of production around a long-term trend, and sources of production instability can be measured from its components' fluctuations. An analysis of stability gives the understanding on the agricultural products' and incomes' security that determines the liquidity, viability and general well beings (i.e., the ability to meet up financial obligations) at the household level and the general prospects of business cycle at the country level. Therefore, stability is as much of importance as growth because although there is high rate of growth, instability in agricultural production will create hardship to the inhabitants as well as the policy makers and in consideration of the economic criteria the effectiveness of an agricultural strategy should be assessed from the angle of both growth and stability in agricultural production.

Apart from economic consideration, the agriculture, specially in the LDCs, has also a social dimension in regard to employment generation and poverty alleviation. The real development of agriculture should be measured from the equity consideration, i.e., from the consideration of its growth percolation to different sections of rural people, specially to the small and marginal farmers and agricultural labourers which are the most sufferers from the twin problems of poverty and unemployment. Equity is most commonly measured in terms of the evenness of distribution of agricultural income (i.e., power of spending) or of rural consumption pattern (i.e., actual spending) or also with respect to households' access to inputs, such as, land, capital, etc. This can also be measured from the changes in the indicators of socio-economic well-beings, like, literacy, infant mortality, poverty, etc. All of these are mainly the end results of the implementation of a particular agricultural strategy and these results are likely to differ for different adopted strategies on agriculture even if the hierarchical and situational characters of the system properties [G.G. Marten (1988)] remain unchanged. For instance, even if the households vary in their access to input and output markets, equitability for cash production may be lower than that for food production.

But in applying these three parameters, viz., growth, stability and equity, for assessing the performance of agriculture, one may face difficulty as these three parameters do not play the desired roles independently; rather they are often logically inter-related. For example, higher growth may be attained by increasing harvests in bad years (i.e., by reducing the impact of drought through irrigation or by reducing the impact of pest attacks etc. through the use of pesticides, insecticides, etc.), thereby making harvests more even from year to year, and increasing stability. Higher growth may also be associated with lower stability if the higher production is achieved by means of high-yielding varieties that are more vulnerable than local varieties to fluctuating environmental stresses such as droughts and pest attacks. Likewise, growth and stability affect the equitability in numerous ways. For example, if agriculture is rain-fed, losses are more severe for farmers whose land is poor in moisture retention and therefore vulnerable to drought. On the contrary, with an improved water supply (through irrigation) the growth will increase with increased yield and for everyone if water is distributed according to needs, the stability will be greater; all these lead to greater equitability because crop losses often do not afflict farmers at random. However, higher growth and greater stability can also lead to lower equitability (i) if the overall supply of water is not sufficient to provide reliable irrigation to all farmers in the area (only some of the farmers may receive irrigation service), (ii) if some farmers are lucky enough to harvest when prices are high and others when prices are low, (iii) if new strategy in agriculture may induce social changes that lead to unequal land holdings, etc.

Due to inter-relationships among these parameters one cannot avoid the question of value judgment, specially when they are of conflicting nature (i.e., negatively associated). For example, if the social consideration (i.e., equity) conflicts with the economic consideration (i.e., growth and stability), some may attach more importance to the former than to the latter while others may take the opposite view. However, the domain of the value judgment (which is more prominent in the theoretical justification of a particular agricultural strategy over others) has been minimized and objectivity has been re-enforced in our study. The study, being an *expost* analysis, empirically evaluates the performance of the agroecosystem for a particular region (having, more or less, constant environmental and social setting) over the periods through the measurements of growth, stability and equity in agriculture; their specific types of observed inter-relationships also help us to specify the pattern and state of agricultural development of the region. Further, the desired state of agricultural development (with higher growth, more stability and greater equity) for the region over the periods can be searched out from these empirical verifications along with the examination of growth and stability contributing factors.

1.2 Choice of the Region

In the era of world-wide fading out of communism and the advent of globalisation and market economy, a common curiosity has arisen among the academicians to probe the clues behind the longest lived

Left Front Government (LFG) in West Bengal in general and its strong hold in rural West Bengal in particular. The scholars [Kohli (1987), Bandyopadhay (1988), Dreze and Sen (1989), Lieten (1990), Webster (1990), Swaminathan (1990), Harriss (1983,1992,1993), etc.] have put forward different reasons for uninterrupted regime of the LFG in West Bengal, like (i) successful implementation of land reforms along with 'operation barga', (ii) well developed 'Panchayat Raj System' which has led to greater participation of common men in the socio-economic and socio-political institutions, (iii) implementation of government's beneficiary-oriented programmes such as IRDP, JRY, etc. at the minimum possible leakage, (iv) setting up of village-level agricultural labourers' organizations through *Krishak Sabha* movement, (v) reduction of money-lenders' activities through the expanding coverage of institutional finances, (vi) restoration of social dignity of the oppressed people, etc. These factors have significantly contributed to the retention of the LFG's hold in the rural sector of the State. These and similar other findings around the positive view have, however, also been opposed and contested by some scholars [Bhattacharya, Chattopadhay and Rudra (1987), Mallick (1992), Gazder (1992), Khasnabis (1982,1994), etc.] who are of the opinion that (i) there is no remarkable improvement in the level of living of the rural households in West Bengal under LFG regime, (ii) development literature in the State is the manipulation and mis-interpretation of the figures at best and illusion at worst, (iii) the rural development is skewed towards CPI(M) sympathizers and party workers, (iv) the CPI(M) in power has effectively sold out to dominant class interest, (v) if land redistribution was thought to be a panacea to the problem of rural poverty, then it has clearly failed in West Bengal, (vi) "the tenancy is still operating [inspite of operation barga under LFG rule] under the general rule of semi-feudalism. The reforms do not seem to have changed the existing mode of production in a radical way........... the informal credit agencies are fast making a come back in rural West Bengal" [Khasnabis (1994) P. A198], etc.

But all these studies, mainly based on micro level village survey, are so diverse and partial in character that no definite general conclusions can be drawn from these studies. Considering this, perhaps, a different line of thinking has been evolved to evaluate the performance of LFG in the rural sector from the assessment of the growth performance of agriculture which is still the most important dominant sector in West Bengal as nearly two-thirds of the population being engaged in farming contribute almost 35 percent of the 'State Domestic Product'. In this line of thinking it has been observed from the recent studies [CMIE(1993), Saha and Swaminathan (1994), Sawant and Achuthan (1995), Sen and Sengupta (1995), Bhalla and Singh (1997), etc.], that the situation of backwardness and low productivity of West Bengal agriculture, as depicted in the earlier writings of Das (1978), Sen Committee (1983), Boyce (1987), Vaidyanathan (1988), Bhalla and Tyagi (1989), etc. has changed towards continuing rapid growth in production since the early 1980s. For instance, A. Sen and R. Sengupta [Rogaly (1995)] estimated that the growth rate of rice production in West Bengal was 7.6 per cent in the 1980s in comparison to the growth rate of 1.9 per cent during the period 1950-51 to 1973-74 [as estimated by Rudra (1978)]. However, Saha and Swaminathan (1994) estimated growth in rice production as 6.9 per cent during the period 1981-82 to 1990-91. According to CMIE's (1993) estimates the average annual growth in output of foodgrains in West Bengal increased from 0.6 per cent between 1969-70 and 1981-82 to 6.5 per cent between 1981-82 and 1991-92. During the more or less same period Saha and Swaminathan (1994) have shown that the overall growth in agricultural production in West Bengal increased from 2.74 per cent to 6.4 per cent; but according to the estimate of Bhalla and Singh (1997) this growth rate is 5.39 per cent during the period 1980-83 to 1992-95. Taking the same data base we have seen that Saha-Swaminathan's estimated growth rate (6.4 per cent) reduces to 5.31 per cent only through the inclusion of few years more (i.e., for the period 1981-82 to 1994-95). Actually the scholars have conflicting views on the exact nature of agricultural growth in West Bengal. "Boyce speculated that the true rate of agricultural growth in West Bengal in 1980s may have around 4 per cent per year" [Rogaly *et al.* (1995), P.1862]. Boyce has further pointed out that this growth rate (specially in rice production) was pushed up by abnormally low outputs in 1981-82 and 1982-83. Considering the maximum targeted growth rate of 4 to 4.5 per cent in agricultural production for our country in the present condition [Mishra (1997)], this high growth in West Bengal agriculture may be termed as abnormal and Dutta Roy (1994) has cast doubt on the reliability of official data, published for West Bengal agricultural production, used to estimate the growth rates. The concluding remark in this regard was, perhaps, given by Rogaly *et al.* (1995) while saying that, "there is strong evidence from West Bengal to indicate rapid growth in agricultural output since the mid-1980s (even if the average annual rate was 4 to 5 per cent per year rather than 6 to 7 per cent)" [p. 1864].

From the discussion it is clear that the equity aspect has been totally ignored in the agricultural growth analyses and, on the other hand, in the diversified micro studies main emphasis has been placed on the equity consideration. Moreover, in both sorts of analyses there are conflicting views among the scholars. To resolve these issues from proper perspective and also from an integrated approach (i.e., the consideration of both equity and growth along with stability) in assessing the performance of agriculture we have taken West Bengal as our case study. Another chief reason of choosing this State of India as our case study is that West Bengal possesses some unique features like (i) apart from having different political power base, decentralised planning has been implemented in the State through local participation and well developed Panchayat Raj system and (ii) as the strategy for agricultural development, technological improvements [as depicted in Table 5, P. A-8, Bhalla and Singh (1997)] have taken place along with the successful implementation of institutional reforms [as explained by Lieten (1990)]. These have created much interest among the notable world economists as a development model to be followed for an underdeveloped region.

1.3 Coverage of the Time Period

The general period chosen for the analysis of growth and stability in West Bengal agriculture is from the crop year [crop year consists of the Bhadoi (i.e., July) crops of one year to the rabi harvest (i.e., June) of the next] 1957-58 to 1994-95 which is again divided into two sub-periods, namely, from 1957-58 to 1976-77 (i.e., Pre-LFG regime) and from 1977-78 to 1994-95 (i.e., LFG regime), to facilitate the comparative analysis of the performance of West Bengal agriculture before and during the LFG rule. But in case of equity measurement we are forced to consider small sample periods depending on the availability of data. Likewise, due to non-availability of uninterrupted time series data at the district level we have taken cross sectional study at three points of time, viz., 1970-71, 1980-81 and 1990-91, to determine the effects of growth and stability contributing factors.

Regarding the selection of time period one question may arise: why has 1957-58 been chosen as the starting year inspite of the availability of official data on West Bengal agriculture since independence ? 1957-58 has been chosen as the starting year on the following ground. Because of the implementation of the "States Re-organization Act of 1956" which resulted in the transfer of Purulia district and Islampur Sub-division (of West Dinajpur district) from Bihar to West Bengal and also transfer of fringe zones (of Midnapore district) from West Bengal to Bihar or from Orissa to West Bengal, the data prior to 1957-58 are not geographically comparable with the post-1957-58 data.

One may also raise objection to the choice of 1977 as the cut-off point for the sub-periods on the ground that the effect of the policies of the LFG on the agricultural sector has not been readily reflected from the very date when it assumed power. Though this is apparently true yet in growth measurement exercise this is not so as the initial years act only as base years on which the later years are included as assessment years of agricultural performance. Further in trend analysis the terminal years should be the normal cropped years which will not happen if one chooses the early years of 1980s as cut-off year (as was erroneously chosen in the recent studies).

1.4 Hypotheses of the Study

There is only one broad-based hypothesis for undertaking this study:

Under the adopted strategies of both transformation (i.e., institutional reforms) and improvement (i.e., technological changes), the performance of West Bengal agriculture in the light of growth, stability and equity is much better during LFG regime in comparison to the earlier period.

From this general hypothesis the following main testable hypotheses may be formulated:

(i) The rate of growth in agriculture has been significantly higher and more evenly distributed among crops and districts in West Bengal during the period from the late seventies to the mid-nineties (Sub-period II) than those during its immediately preceding period from mid-fifties to late seventies (Sub-period I).

(ii) The growth in agricultural production in West Bengal during the Sub-period II has taken place mainly because of growth in productivity.

(iii) The contributions of technological factors to the agricultural growth in the State have been significantly higher than those of non-technological factors [including institutional factors during the early period (i.e., early seventies)]; but during the sub-period II both these two sets of factors affect West Bengal agriculture with equal strength.

(iv) The sub-period II has experienced relatively less fluctuations in the agricultural production in the State than in the sub-period I and that has been possible due to extension of irrigation and good rainfall.

(v) Lower instability in agricultural production in the State during sub-period II (in comparison to sub-period I) is due to lower instabilities both in area and productivity.

(vi) The benefits of agricultural growth, measured in terms of wages, land distributions, level of consumption, etc., in the State have increasingly accrued to the poorers (i.e. small and marginal farmers and agricultural labourers) during sub-period II and consequently the rural poverty has declined during this period.

(vii) Growth and stability in West Bengal agriculture are positively associated and they are also jointly positively associated with the equity i.e., the 'trickle-down theory' is operating in West Bengal during sub-period II.

1.5 Plan of Work

The study has been organized in the following way:

The first three chapters have been devoted respectively to the introductory analysis (Chapter 1), literature survey (Chapter 2), and data base and methodology of the study (Chapter 3).

Chapter 4 deals with the selections of trend equation, data

adjustments and the estimation of growth,rates (for the whole period and for the sub-periods) of agricultural production in the State as well as the districts. For the State we have taken all the major crops and crop groups (altogether 22 in number) and for the districts only four major crop groups, namely, total rice, total foodgrains, total non-foodgrains and all crops combined. The decomposition of crop-output growth rate into its components (such as area and productivity) in all cases and the contributions of the districts as well as different crops to the overall output growth rate in the State have also been estimated and presented in this chapter.

Chapter 5 discusses the issues like the measurement of nature and extent of instability in agricultural production with crop-wise and district-wise disaggregation, decomposition of output instability and the measurement of association between growth and instability.

We have examined the causes of agricultural growth and instability from the host of technological, institutional, demographical and weather determined factors in West Bengal with both cross section and time series data in Chapter 6.

Chapter 7 deals with the measurement of equity and also examines the association among growth, stability and equity. As pointed out earlier, the limitations of the available data as well as non-availability of reliable data on certain specific variables have forced us to restrict our study to measurement of equity only from the consideration of wage rates of agricultural labourers, the distributions of land and consumption among rural inhabitants and incidence of rural poverty, so far as secondary data are concerned and also from the distribution of income (along with the decomposition of income inequality into its different sources of income) as available from the primary survey, conducted by the researcher himself.

In the last concluding Chapter 8 the broad findings of the study have been summarized.

2

Literature Survey

In this chapter an attempt is made to make a review of the existing literature on the methods of measurement of growth, stability and equity, particularly those used in Indian agriculture and also on the major empirical findings related to these three issues in West Bengal agriculture.

2.1 Existing Methods for Growth Analysis in Agriculture

The subject of growth analysis in agriculture is a tricky area, especially because of the high degree of volatile fluctuations in agricultural production data and the dependence on a number of factors such as the choice of the period, cut-off points, the statistical techniques employed, etc. For these reasons, there is a need for a systematic inquiry into these aspects of the problem. Also there is a need for caution in interpreting the results. These points have been taken up in a number of studies made earlier [Minhas and Vaidyanathan (1965), Parikh (1966), Minhas (1966), Sen (1967), Rudra (1970,1982), Dharm Narain (1977), Reddy (1978), Das (1978), Bhalla and Alagh (1979), Vaidyanathan (1979), Srinivasan (1979), Dandekar (1980), Ray (1983), Sawant (1983), Mahendradev (1987), Boyce (1987), Hanumantha Rao (1992), Saha and Swaminathan (1994), Sawant and Achuthan (1995), Bhalla and Singh (1997), etc.]. Research activity related to the measurement of agricultural growth is largely a post-independence phenomenon in India, mainly because of significant improvements in data base and researchers' urge to identify the factors lying behind the agricultural stagnation and development. Apart from the earlier growth calculations either by comparing two arbitrarily chosen end points or by comparing two averages of some arbitrarily chosen periods at the beginning and end of the series [which were highly criticised by Vidya Sagar (1980)], the normal statistical procedure, used in most of the studies, to obtain measure of growth rate, is to postulate a hypothetical trend function and to estimate its parameters (by ordinary least square method) which offer the measure of growth rate over the period of a series. In this normal statistical procedure three problems crop up. These are : (i) choice of trend equation, (ii) avoidance of volatile fluctuations in the data set and (iii) estimation of sub-periods, growth rates. The other two problems, related to the growth measurement exercise, (on which most of the scholars have paid their due attentions) are: (i) decomposition of agricultural output growth into its components and (ii) the identification of growth contributing factors. We discuss below the contributions of the researchers to each of these problem areas.

Empirically it has been observed that the popular forms of trend equation for growth estimation are linear, parabolic, exponential, log-quadratic, higher degree polynomial and S-type growth curves (Gompertz, Logistic etc.). But each of these trend equations has its own interpretation and restriction upon the character of the growth process. For example, linear trend implies constant absolute growth [which can be divided by the harmonic mean of the series, as argued by Minhas (1966), to get average compound growth rate]; similarly exponential or log-linear trend [$\log_e Y_t = a+bt$] implies constant compound growth rate, b [however, the correct formulation is e^b-1, as pointed out by Dandekar (1980)]. Although parabolic and log-quadratic trends are more flexible than linear and exponential in that they permit non-zero second derivative with respect to time (implying acceleration or deceleration in the growth process) they nevertheless impose restriction that the second order derivatives are constant [In this connection it is to be noted that in the case of second degree polynomial equation as t and t^2 are correlated, time has to be normalised (shifting origin to the mid-point of the period) for the avoidance of multicollinearity problem, as argued by Reddy (1978)]. One way to allow a non-constant second derivative is to add a cubic term of time in the trend equation [as used by Ray (1983)]; but this process is unending. Further, by doing so, the very objective of growth measurement would either be violated (having more than one turning points in the growth curve of polynomial of degree three or more) or be meaninglessly laborious. Like log-quadratic trend, S-type growth curves [applied by Rudra (1970), Dey (1975), Reddy (1978), Das (1978), Chattopadhyay and Bhattacharya (1986), etc.] are also open to the criticisms regarding the behaviour of the acceleration/deceleration in the growth process. Further, the results of non-linear estimation procedure [either by 'Group Average Method' or by transformation approach (as prescribed by Reddy and Chattopadhyay *et al.*) or by search technique] in case of S-type growth curves are non-comparable with the results of the OLS method, applied in the cases of other trend equations, as some of these procedures are arbitrary and defective and for others the relative merits of the estimation method (*vis-à-vis* others) have not been established by the Monte-Carlo experiment.

So in growth measurement exercise the choice of trend equation from amongst the available alternatives is very crucial. The most commonly used statistical criterion as the 'test of goodness of fit' for choosing one particular equation out of all available trend equations is R^2 or $\bar{R}^2$ (i.e., coefficient of determination adjusted by the degrees of freedom). But many scholars have raised several questions against this criterion. Reddy (1978) and Rudra (1982) demonstrated that different functional forms (having different implications on growth process) provided comparably equal good fits when applied to a given set of time series data. Das (1978) has rightly pointed out that, "...... if we consider some mis-specified equations, one of them must have the highest R^2. Moreover, when the variables are .... each serially correlated, there is possibility that multiple determinant coefficient is spurious and no valid conclusions can be drawn on the basis of this R^2" (p. 1742). This will also happen when the basic econometric criteria for regression (viz., homoscedastic, non-autocorrelated, random disturbance terms, etc.) are not satisfied. Further, R^2 or $\bar{R}^2$ is non-comparable among alternative trend equations having different forms of dependent variable (such as Y_t, $\log Y_t$, $1/Y_t$, etc.) [Maddala (1988)]. This limitation of R^2 or $\bar{R}^2$ as a criterion for choosing an equation from amongst a number of equations has not always been taken into consideration by many researchers.

Reddy (1978) prescribed two other criteria to test the goodness of fit, namely, (i) "accuracy of growth rates" (defined as minimum standard error of estimated growth rate) and (ii) "accuracy of prediction". Boyce (1987) made the following comments on Reddy's suggestions: (i) "Here the problem arises that the ranking of alternative specifications are not necessarily consistent over time, since these are annual measures which (except in the exponential case) change through time", and (ii) "this means a preference .... for specification in which the standard error and coefficient of variation steadily diminish towards the end of the time series, rather than remaining constant or increasing. Since the behaviour over time are imposed by functional form, this procedure also involves an element of arbitrariness". [p. 267]. In this regard Rudra (1982) advocated the non-parametric tests (based on rank correlation coefficient) in preference to parametric tests. The former, according to him, provides a more rigorous measure of the statistical significance of changes in growth rates through time. These non-parametric tests were also highly criticised by Boyce (1987) from his empirical investigations. As there is no decisive statistical ground for model specification among alternatives, recent studies are arbitrarily using R^2 or $\bar{R}^2$ as the criterion for test of goodness of fit, leaving arguments and counter-arguments aside. In recent studies the widely used trend equations are exponential and log-quadratic. The reason for this can be advanced on the lines of arguments forwarded by Dandekar (1980) and Boyce (1987): "The exponential (log linear) form is chosen as a starting point, rather than the simple linear form, owing to the *a priori* expectation that change in output in a given year is more likely to be a constant percentage of output in the preceding year than a constant diminishing percentage of it. Moreover, it would seem reasonable to assume that larger errors—in the statistical sense of deviations of actual from trend values—are associated with larger output and hence that a multiplicative error term is more appropriate than an additive one" [Boyce (1987) p. 258]. But these arguments do not seem either plausible or convincing to us particularly against the possibility of diminishing returns in agricultural production and the existence of various forms of growth process. Further, without non-linearity tests [Maddala (1988) pp. 177-180] one should not establish the effectiveness of logarithmic functions over the functions in the natural scale for a particular series.

The presence of volatile fluctuations in the data is also a cause of concern to those interested in fitting trend lines to these data, since it may distort the trend growth rate by introducing upward or downward bias in it. Pindyck and Rubinfeld (1981) have argued that "it is desirable to smooth a time series and thereby eliminate some of the more volatile short-term fluctuations. Smoothing might be done before making forecast or simply to make the time series easier to analyze and interpret" [p. 473]. Several suggestions have been made to correct for the presence of volatile fluctuations in the data set. First technique, adopted by Srinivasan (1979), Rath (1980), Rao *et al.* (1980), etc., is to eliminate extreme observations having exceptionally high or low values (to be determined in probabilistic term, i.e., the outliers of 5 per cent or 10 per cent band limit around trend line) and to fit the trend equation through the remaining normal observations for the estimation of trend growth rate. But in this technique the element of subjective judgment is unavoidable so far as the formation of the band limit is concerned. Moreover, in a small sample time series data (where population is infinite) such type of elimination is untenable.

Mahendradev (1987), Saha and Swaminathan (1994), etc. have used another technique in which rainfall has been included in the trend equation as a proxy explanatory variable for weather to minimize the effects of volatile fluctuations in weather on the estimated trend growth rate in agricultural output. But one has to accept that rainfall is the only one among a large number of factors that can explain the wide fluctuations in the agricultural production data. Therefore, this procedure may not have the desired effect. Further, using rainfall as a proxy variable for weather and the process of estimating total rainfall index are likely to introduce measurement error in the variable. In this case the straight forward application of OLS method to estimate the coefficient of rainfall yields inconsistent estimate. Sawant (1983), Sawant and Achuthan (1995), etc. have considered three year centered moving averages of the original series of the dependent variable for the trend equation to minimize the influence of inherent fluctuations in the data set. But, as we see, this technique is also not free from limitations. Due to smoothing of the original series the trend pattern may be distorted and further due to smoothing the disturbance term may be serially correlated about which nothing has been said in these studies.

In the growth measurement exercise another subject of concern among the scholars is measuring sub-period growth rates for the comparison of period-wise (dividing the whole period into a few sub-periods) performance of the variable concerned. To measure sub-period growth rates we get, so far, three approaches. One of the approaches, used by the early scholars [Das (1978), Alagh and Sharma (1980), Rath (1980), etc.], is to fit separate growth functions for each sub-period. As this method involves estimation of a large number of parameters, it results in the loss of degrees of freedom [Srinivasan (1979)]. An alternative approach, most commonly used by the scholars, is to fit a single trend equation with dummy variable. For example, the form of the log-linear trend equation with dummy variable for two sub-periods is :

$$\log Y_t = a + b_1 t + b_2(t.D)$$

where D=0 for the first sub-period and D=1 for the second sub-period. Here the respective growth rates in the first and second sub-periods are b_1 and $(b_1 + b_2)$.

To test the significant difference between two sub-periods' growth rates, here (like first approach) the usual joint t test is carried out (also the chow test in the first approach). But the underlying assumptions for these tests are hardly fulfilled in reality. Further, Boyce (1987) has shown that this approach, along with the first one, for the measurement and comparison of sub-periods' growth rates suffers from "discontinuity bias" which implies that each of the estimated sub-periods' growth rates is higher or lower than the estimated growth rate for the whole-period. To overcome this difficulty Boyce (1987) has introduced another approach in which "kinked exponential model" (pp. 267-271) is fitted for estimating sub-period growth rates. Of late this approach is being used by most of the recent studies. However, we can raise questions against this approach also : (a) why is the exponential specification chosen over others and (b) why should we place arbitrary kinked restriction at the break point? Moreover, as we have observed, in the face of volatile fluctuations in the data set kinked exponential model leads to over-estimation or under-estimation in the sub-periods' growth rates, specially if the outliers of opposite natures (extreme low value and extreme high value) lie on two opposite sides of the break point.

Another problem related to the growth measurement exercise is to decompose the agricultural output growth into its various components, like area, yield, cropping pattern [shifts from relatively low-value (per hectare) crops to high-value crops or *vice versa*], etc. From an in-depth literature survey we get three distinct approaches in the decomposition of growth in agricultural output, viz., (i) 'additive scheme', (ii) 'multiplicative scheme' and (iii) trend approach.

The first systematic study in the field was made by Minhas and Vaidyanathan (1965) in which, by 'additive scheme', they decomposed agricultural output growth into growths in area, yield and cropping pattern and the interaction between yield and cropping pattern.

$$[(Q_t - Q_o) = (A_t - A_o)\Sigma_c\, a_{co} Y_{co} P_{co} + A_t \Sigma_c a_{co}(Y_{ct} - Y_{co}) P_{co} + A_t \Sigma_c (a_{ct} - a_{co}) Y_{ct} P_{co} + A_t \Sigma_c (a_{ct} - a_{co})(Y_{ct} - Y_{co}) P_{co}$$

where Q, A, a, Y and P denote respectively output, area, proportion of area under crop c, yield and price; t denotes terminal period and o denotes base period]. Later on the scholars have expanded this decomposition model mainly to take into account the additional components. For example, Sondhi and Singh (1975) added three more interaction terms [namely, (i) area and yield, (ii) area and cropping pattern and (iii) area, cropping pattern and yield] in the 'Minhas-Vaidyanathan Model' replacing current period area (A_t) by base period area (A_0) for each component other than area. Decomposing gross agricultural output at prevalent prices [i.e., Q_t at P_t and Q_o at P_0] Vidya Sagar (1977) further expanded the decomposition into a seven component version, viz., area, yield, price and their interactions (Area-price, area-yield, yield-price and Area-price-yield). Dharm Narain (1977) incorporated an additional component in the 'Minhas-Vaiyanathan Model', viz., the locational component (shift in crop locations from low productivity regions to high productivity regions and *vice versa*). These 'additive scheme' of decomposition, however, in general, suffer from severe limitations. These models decompose absolute growth [dQ] in output in contrast to the usual practice of reporting the comparable agricultural performance through relative growth [dQ/Q]. Further, the inaccuracies in these decomposition models have been covered up through the interaction terms which are more technical than economically meaningful.

So far as the multiplicative decomposition models are concerned, Parikh's study (1966) appears to be the first in a series of such studies made later on. Parikh (1966) expressed the output index as the product of the indices of area, cropping pattern and yield:

$$Q_t / Q_o = (A_t / A_o).(\Sigma_c a_{ct} Y_{ct} P_{co} / \Sigma_c a_{co} Y_{ct} P_{co}).(\Sigma_c a_{co} Y_{ct} P_{co} / \Sigma_c a_{co} Y_{co} P_{co})$$

In an independent study a similar exercise was done by Dayal (1966). But 'Parikh-Dayal' scheme remains inconsistent in regard to the uses of Laspeyre's method in the index of yield and Paasche's formulation in the index of cropping pattern. Using only Laspeyre's index for each component, Minhas [see Vidya Sagar (1980), P. 51] improved the Parikh-Dayal version through the inclusion of a residual component (similar to the interaction term in the additive scheme). Realising the discomfortable features of the residual term, Jamal and Zaman (1992) further tried to improve this multiplicative decomposition model by using Fisher's index (i.e., the geometric mean of Laspeyre's and Paasche's indices) in place of Laspeyre's index for each component. In their model, actually, the residual term has been distributed among the area effect, price effect, yield effect and crop-mix (i.e., cropping pattern) effect. But, like additive scheme, the multiplicative decomposition models also might give erroneous and biased results as the components are estimated, in general, by comparing only the end points or some averages of the end points of the series.

To minimize the effect of volatile fluctuations in the data set in decomposition exercise Dandekar (1980) and Sagar (1980) argued that this exercise should be based on the entire series of observation. Accordingly, we get trend approach of decomposition [Ray (1983), Boyce (1987), Saha and Swaminathan (1994), etc.] in which, taking all the observations into account, separate exponential trends were fitted for each of the components of the Parikh-Dayal model and then output growth rate was expressed as the sum of the component growth rates. This approach is similar to the production function approach as generalised Cobb-Douglas production function is fitted, replacing inputs by components. But, like Parikh-Dayal version, in this approach also components' effects, are likely to be overestimated or underestimated. More specially, as used by Boyce, the yield effect (being Laspeyre's index) would be overestimated and the crop-mix effect (being Paasche's index) would be underestimated. To avoid these difficulties, perhaps, Saha and Swaminathan (1994) have used a very simplified but meaningful two factor decomposition model in which output growth rate is expressed as summation of area and productivity growth rates (without considering the intricacies of index numbers, as they are definitionally related i.e., output = area × productivity). But their method of trend fitting exercise, like others, may be criticized generally on different grounds, as pointed out earlier.

Lastly, along with the growth estimation, a few studies [Das (unpublished research note, 1981), Boyce (1987) etc.] also considered the determination of the effects of leading inputs on the growth of output/yield by using simple regression analysis (taking one input in the production function at a time) or multiple regression analysis (taking a few inputs at a time in the production function). In these studies, multicollinearity problem was either avoided through model simplification or simply ignored. Further, these analyses are mostly partial in consideration to the non-consideration of the interactions and joint effects of inputs, taken together, on output/yield.

2.2 Existing Methods of Instability Measures in Agriculture

Instability as a counterpart of stability measures the extent and nature of fluctuations in the agricultural production. Measurement of instability has mainly been evolved as a related issue to the growth measurement exercise in agriculture. The notable scholars who gave their close attention to this particular topic are Sen (1967), Das (1978), Rao *et al.* (1980), Mehra (1981), Hazell (1982,1984), Ray (1983), Dhawan (1983), Mahendradev (1987), Boyce (1987), Hanumantha Rao *et al.* (1988), Bandyopadhyay (1989), Saha and Swaminathan (1994), etc. From an in-depth study of these works it is seen that altogether there are four aspects of instability measurement exercise, namely, (i) measurement of instability, (ii) association between growth and instability, (iii) decomposition of output instability and (iv) causes of instability.

In regard to the measurement of instability in agricultural production we get broadly two techniques: Summary measure and Trend measure. The prevalent summary measures of instability, adopted by Barker *et al.* (1981), Mehra (1981), Hazell (1982, 1984), Ray (1983), Dhawan (1983), Hanumantha Rao *et al.* (1988), etc., are standard deviation (also the standard error of the estimated growth rate), variance and coefficient of variation either measured from the dispersion in production around trend line (fitted for growth estimation) or from the year-to-year fluctuations in production. But, in general, these summary measures have been highly criticised on the following grounds: (i) These measures could not fully grasp the necessary information on the change of instability through time. (ii) Summary measures around trend line are suspected when fluctuations are large and frequent (that lead to distortion in trend estimation) or when the trend is itself mis-specified. (iii) Measures based on year-to-year fluctuations are more influenced by the short-term variations than the long-term variations and one can change the conclusion on instability simply by adding or removing one or two years having strong outliers in a period. (iv) Like growth rates, standard errors (and the residual variances) may be strongly affected by the discontinuities between the separate estimates for sub-periods.

As an improvement over the summary measures the scholars presented the trend measures of instability. We get two broad categories of trend measure. The first one is to estimate separate trends (linear or exponential) for 'peak' and 'trough' years and to see whether these two trends are converging, meaning thereby declining instability overtime, or diverging, signifying thereby increasing instability overtime. For selecting peaks and troughs there are mainly three criteria. (a) The usual local maxima and minima definitions, used by Sen (1967), Das (1978), etc. This criterion is, however, open to the criticism that very small output differences can cause certain years to be included or excluded. To avoid this limitation Rao *et al.* (1980) suggested two alternative criteria, namely, (b) identifying local maxima and minima by a certain percent margin (say 5% or 10%) and (c) identifying maxima and minima by a certain percent (5% or 10%) deviation from the estimated trend. These two criteria, however, introduce arbitrariness in identifying maxima and minima. The criterion (c) further introduces the additional problem of trend specification. Further, in all these criteria instability is measured by using only a few selected data points, instead of using full information. Moreover, if selected peak or trough points happen to be small in number, the estimates of peak or trough trends would be noticeably affected. These different criteria may also be inconsistent in regard to the conclusion finally derived on the nature of instability. For instance, empirically Boyce (1987) has found that for a given data set (West Bengal agriculture during the period 1949-80) "the year-to-year criteria indicate divergence, and hence rising instability, while the trend criteria indicate convergence, and hence declining instability, underscoring the problem of arbitrariness in criteria for selection of peak and trough points" [p. 273].

To overcome these difficulties, in the recent studies [Boyce (1987), Mahendradev (1987), Saha and Swaminathan (1994), etc.], another trend measure of instability has been used by fitting the time trend (linear or exponential) on the 'statistic' of instability and observing the sign of the slope parameter with statistical significance (negative sign implies declining instability over time and *vice versa*). Mahendradev (1987) has used this 'statistic' of instability as 'a nine-year moving standard deviation in year-to-year changes in production'. On the other hand, Boyce (1987) and Saha and Swaminathan (1994) have used this 'statistic' of instability as the squared or absolute values of the detrend output data (similar to Glejser's test for heteroscedasticity). In comparison to Mahendradev's statistic [which is based on standard deviation and nine-year (chosen arbitrarily) moving average period], the Boyce's statistic is more reliable as, being the detrend statistic, it captures all sorts of variations (namely seasonal, cyclical and irregular variations) in the agricultural production. However, this method of instability measurement also, like most of the others, highly depends on the correct specification of the growth trend; otherwise specification error will cover up the true instability.

Some studies considered the nature of association between growth and instability, along with their measurements, either from the simple visual inspection of the movements of the estimated results on growth rates and summary measures/nature (i.e., converging or diverging) of instability over different sub-periods [Sen (1967), Mehra

(1981), Hazell (1982), Ray (1983), etc.] or by regressing instability (levels and trends) against growth rates cross sectionally over different crops and regions. However, the regression method appears to be better than the other one since it is amenable to statistical analysis and inferences.

Like growth decomposition, instability in agricultural output has also been decomposed into its components by some scholars [Hazell (1982), Ray (1983), Hanumantha Rao *et al.* (1988), etc.] to determine its sources. Perhaps, the first systematic study on the decomposition of instability (measured by the variance around trend) was made by Hazell [1982, 1984]. Through mathematical and statistical identities he developed a seven factor (viz. changes in mean, variance-covariance of area and yield and their interactions, etc.) decomposition model for individual crops. For crop-groups these seven factors had been added over five factors [viz., variances within crops, among crops, within states, among states, etc.]. Hazell's model was meaninglessly labourious and could not avoid the residual term which was difficult to define economically. Later on, Ray simplified this decomposition model as outlined below (for individual crops):

Since production is equal to area multiplied by yield [i.e., $Q_t = A_t.Y_t$], it can be written that

$$G_q = G_a + G_y \text{ and } Var(G_q) = Var(G_a) + Var(G_y) + 2Cov(G_a, G_y)$$

[Here G implies growth and Var and Cov imply respectively variance and covariance].

Therefore, changes in the variability of annual output growth rates [as a measure of instability], calculated from the original series, can be expressed as the corresponding changes in area and yield and their synchronous movements. This decomposition model can easily be extended by taking cropping pattern as a separate factor.

But Hanumantha Rao *et al.* (1988) show that Ray's model give erroneous result if the components have trend-like movements. More specifically, in case of rising trend (declining trend) instability would be overestimated (underestimated). To eliminate this difficulty, they prescribed for the calculations of variances and covariance around respective trend growth rates, instead of mean of annual growth rates. Even with this improvement one cannot avoid the difficulties if each of the components does not precisely follow exponential trend and if some strong outliers are present in the series.

A few studies have also considered the factors (such as rainfall, technology, ecological degradation etc.) behind the causes of instability in agricultural output. But their studies are more descriptive than analytical.

2.3 Empirical Findings of Earlier Researches on Growth and Stability in West Bengal Agriculture

In consideration to growth and stability in West Bengal agriculture we get a good number of earlier studies in which this State of West Bengal was taken either exclusively or as a part of the particular region (say, eastern India or Bengal region) or simply as a constituent State of India. If we summarize the broad findings of all these studies, it would not only be voluminous but also repetitive in nature. Therefore, here we have mentioned only those findings of the studies which either maintain chronology in regard to time periods and crops or give contrasting views, if any, in relation to others. In this connection special emphasis has been given to the studies of Boyce (1987) and Saha and Swaminathan (1994) which are not only exhaustive and compact in this field but also consider the whole period since independence.

From the estimate of the Ministry of Food and Agriculture (1966) it was observed that the growth rates in agricultural production in West Bengal were 0.89 percent during 1952-53 to 1961-62 and 1.94 per cent during 1952-53 to 1964-65, in contrast to 3.23 per cent and 3.42 per cent growth rates in Indian agricultural production during the same periods. Das (1978, 1981) estimated that agricultural growth rate in crop output in West Bengal was 2.07% during the period 1950-51 to 1973-74 and taking 1965 as cut off year he further showed that in the post green revolution period (after 1965) the agricultural production growth rate sharply increased to 2.15 per cent from 1.04 per cent growth rate of pre-green revolution period. This was also corroborated by Boyce (1987), extending each sub-period by a few years. For instance, according to Boyce's estimate the growth rate in agricultural production in West Bengal sharply increased from 1.20 per cent during 1949-64 to 2.27 per cent during the period 1965-80. According to Srinivasan's (1979) estimate during the more or less same period [1949-50 to 1977-78], on the other hand, the national trend growth rate of all crops' output was over 3 percent. From all these estimates it is clear that agricultural growth rate in West Bengal was very low prior to mid-sixties and after 1965 this improved significantly (with trend break 1.07 percent) but was still much lower than the national average growth rate.

Boyce's (1987) study was further extended to crop-wise and district-wise analyses. "At the district level, however, the changes between the sub-periods [before and after mid-sixties] were quite mixed, with declines in the estimated growth rates in the four northern districts, namely, Jalpaiguri, Darjeeling, West Dinajpur and Cooch behar, and in the Western districts of Purulia, all of which surpassed the State's average growth rate in the first sub-period and fell below it in the second" [p. 69]. Between two sub-periods the remarkable improvement occurred in six districts, viz. Hooghly, Malda, Howrah, Murshidabad, Midnapore and Burdwan [arranged rank-wise]. Regarding crops, the significant improvement in growth rates between two sub-periods took place in the productions of boro rice, wheat, oilseeds, potatoes and aus rice; in all other cases growth rates were constant or declining. Though area was the main component behind the output growth of sesamum, boro rice, wheat and aus rice, the overall output growth in the State during second sub-period was mainly contributed by the growth in productivity [61.2%]. During the first sub-period, on the other hand, area was the main component (its share was 67.5%) behind output growth. We get the same observation from the study of Das (1981).

In respects of output instability, however, we get contrasting views. Boyce observed that output instability in West Bengal did not change over the 1949-80 period. Das (1978) observed that instability in both agricultural production and foodgrains production declined significantly during the period 1950-51 to 1973-74. On the other hand, Mahendradev (1987) established that instability in foodgrains production in West Bengal was increasing during the period 1960-61 to 1984-85. These contrasting observations might be due to the adoption of different techniques and different time periods of measuring instability. Boyce further observed that the factors behind the slow growth rate of agricultural output in West Bengal [compared to the national level] were low levels of irrigation development, including poor water management and control and the elements of the agrarian structure, including inequality of land-holdings.

Since the early 1980s the situation in West Bengal agricultural scenario changed dramatically from backwardness and low productivity to high and rapid growth, as was observed by the scholars, like Hanumantha Rao (1992), Saha and Swaminathan (1994), Sawant and Achuthan (1995), Rogaly *et al.* (1995), Bhalla and Singh (1997), etc. The broad findings of these studies, also with conflicting views among the scholars, have already been mentioned in our introductory analysis (section 1.2, Chapter 1). Here we shall only consider the study of Saha and Swaminathan (1994) to highlight some other important features of their findings.

According to Saha and Swaminathan, break-through in rice production (particularly in aman rice through improvements in yield) was mainly responsible for the overall good agricultural growth performance in West Bengal from the early 1980s. The other crops whose growth rates improved very significantly were potatoes and mustard (mainly through the expansion of area). On the other hand, output growth was slow in case of wheat, stagnant in case of Jute. In the case of pulses the growth in production was negative; all these happened because of decline in the cultivated area. As regards to the district-wise pattern of growth during 1981-82 to 1990-91 the authors observed that, "with the exception of Darjeeling and Jalpaiguri, the index of aggregate crop production grew by over 5 per cent annually in all the districts of West Bengal. .... Bankura and Birbhum were ranked at the top, with growth rates of 9.5 and 9.3 per cent respectively. Medinipur and Puruliya followed with growth rates of 8.5 and 8.4 per cent. Next in rank were Nadia and Haora. In all these six districts, growth rates were higher than the state average" [p. A-3]. However, one can suspect the growth figures of Birbhum and Purulia where after the inclusion of rainfall index in the trend equation (as done by the authors themselves), the growth rates declined by more than 1.5 per cent; this might happen due to wide fluctuation in the data of two districts.

Saha and Swaminathan, further, observed that the high growth in agricultural output in West Bengal (during 1980-81 to 1990-91) was mainly attributed to the growth in productivity [81%] and the growth performance of the 1980s could not be explained in terms of the good rainfall, except in few districts [like Burdwan, etc.]. This observation is very much startling. Burdwan being relatively endowed with assured irrigation depends on the vagaries of weather whereas in the rain-fed districts [like Purulia, etc.] agriculture does not depend on rainfall. Regarding instability, the authors found that the period of high growth was not associated with greater instability in the levels of production. Finally, the authors speculated that the factors behind West Bengal's good agricultural production performance from early 1980s were the 'landmark programme of land reform' and 'the establishment of new democratic Panchayat institutions'. However, applying statistical tools (estimating trans-log cost function) a recent study established that "institutional factors played a major role in raising production and productivity in the post panchayat years" [Mukherjee and Mukhopadhyay (1994), p. 14]. But, however, from a non-random and small sample village survey (in rice crop only) of such study, no general conclusion should be derived.

2.4 Existing Literature on the Equity Measures

Equity aspects of growth is considered in terms of the nature of the distribution of benefits, arising out of growth process, among different sections of the population. The nature of the distribution of benefits is customarily measured in terms of inequality in the distribution of either assets or income or consumption or simply from the incidence of poverty in a developing country. Measures of inequality are of two types: positive measures (such as coefficient of variation, standard deviation on logarithms, Theil's entrophy measure, Gini coefficient, etc.), which make no explicit use of any concept of social welfare, and normative measures (such as Dalton's measure, Atkinson's measure, Sen's measure, etc.), which make explicit formulation of social welfare and loss incurred from unequal distribution. Inspite of the appealing features [Sen (1973)], the normative measures are of little significance in the empirical investigations as, in practice, it is very difficult to assign a particular form of social welfare function without arbitrary consideration or without the question of value judgment and interpersonal comparison of utility. That's why, perhaps, the most widely used measure of inequality in the empirical studies is Gini coefficient which is, in consideration of its relative merits and demerits (as shown by Sen), better than all other positive measures. The most widely used formulation of the Gini coefficient [Yntema (1933)] is:

$$\text{Gini coefficient, } G = 1 - \sum_{T=1}^{T+1} f_t (q_t + q_{t+1})$$

where T+1 is the number of income classes, f_t is the relative frequency of the t th income class and q_t is the cumulative proportion of income (or consumption, etc.) received by the units (say households) earning income less than or equal to the upper limit of the t th income class.

Later on, however, this formulation of Gini coefficient has been improved by Kakwani (1980) and Lerman and Yitzhaki (1985) to avoid the bias in inequality measure [as Yntema's formulation considers the inequality only among classes, not within classes] and to measure the contributions and marginal effects of each income components on the overall measured inequality.

Likewise, for the measurement of poverty (after defining the appropriate poverty line on the basis of either minimum calorie in-take or minimum power of spending) we have two types: absolute measures (head count and family count ratios) and relative measures (poverty gap ratio, Sen's index, Kakwani index, etc.) [Kakwani (1980)]. Though distribution-sensitive relative measures are better than the absolute measures, yet in all the official publications the incidence of poverty in a country is represented mainly through head count ratio (i.e., the ratio of number of persons below poverty line to the total number of persons).

To know the nature of association between growth and equity, more specially, to know how much equity is restored with the growth process [the reverse causation has also been examined theoretically by Kuznet (1955) and empirically in India by Datta (1986), which is not our consideration here], the most of the existing studies have followed a very simple technique in which the movements of growth rate and appropriately defined equity index are considered either over arbitrarily defined two points of time or over two sub-periods or cross sectionally at a point of time. But this technique is very naive in the sense that the association between growth and equity cannot be tested statistically and also arbitrary in the sense that results may be erroneous because of improper selection of time period. However, a few studies have considered this aspect statistically. To the best of our knowledge there are three such approaches: (i) application of Logit-Probit models [Bardhan (1986)], (ii) production function approach [Shah (1976)] and (iii) regression approach [Narain (1976), Ahluwalia (1978, 1986), Ninan (1994), etc.]. Through cross sectional study Logit-Probit models determine the probability that a person lies below the poverty line with the growth process. But Srinivasan (1986) has pointed out that trickle-down theory has a dynamic aspect, which should be tested by applying time series analysis, rather than cross section analysis. In the production function approach, first the elasticity of substitution is estimated from the empirical estimation of production function and then applying "Hicksian Laws" of factor share the future movements in the shares of factor incomes can be predicted. For instance, in a two factor case if elasticity of substitution is greater than unity, the relative share of the factor whose relative price is falling would go up. But this approach is based on some theoretical restrictive assumptions (like competitive factor market, factor-neutral technology, etc.) which are rarely satisfied in reality. As an easy escape route the economists [Jose (1988), Sengupta and Gazdar (1995), etc.] have simply examined the movements of wage rates over time to get an idea of the operation of 'trickle down theory' in agricultural sector. In the regression approach, poverty as a negative index of equity has been regressed over time taking per capita agricultural production (along with others) as the explanatory variable to determine the existence of 'trickle down theory'. In this regression analysis Narain (1979) used price index as another explanatory variable which was highly criticised by Ahluwalia (1986) as price variable also entered in the calculation of poverty. On the same ground we can recommend the use of total agricultural production instead of per capita agricultural production specially when head count ratio is taken as the measure of poverty. It might be one of the reasons why Hanumantha Rao *et al.* (1986) used gross value of crop output in their regression analysis. Further, the inclusion of time as a separate explanatory variable may be criticized on the ground of multicollinearity problem because over time agricultural output has a trend effect. On the same grounds [inclusion of price and population as explanatory variable, apart from the restrictive simplification of the some explanatory variables and non-consideration of multicollinearity problem (very likely to arise between agricultural production and consumption inequality] one can criticize the study of Ninan (1994).

With the advent of new strategy in agriculture (from 1960s) in the LDCs a common curiosity arose among both the national and international scholars to examine the validity of 'growth percolation' or 'trickle down' theory [i.e., the 'Kuznets U-hypothesis']. Presenting some theoretical arguments Rudra (1992, pp.190-191) opined that "the New Technology has had the effect of aggravating the already acute inequality of distribution of income and wealth among the population dependent on agriculture" [p. 190]. This pessimistic view was supported by the results of the empirical investigation in India and abroad by many scholars, like Frankel (1971), Cleaver (1972), Griffin (1974), Rajaraman (1975), Shah (1976), Grabowski (1979), Griffin and Ghosh (1979), Bardhan (1973, 1986), Dantwala (1987), etc. The contrary view was also held by those [Johl (1975), Neale (1978), Ahluwalia (1978, 1986), Narain (1979), Blyn (1980), Pray (1981), Leaf (1983), Hanumantha Rao *et al.* (1986), Paul (1989), etc.] who were impressed by the gains of green revolution accruing to the small farmers. There were some others [Chinn (1979), Hayami (1980), Bhalla and Chadha (1982), Vyas (1983), Kada (1983), Shand (1987), etc.] who argued that new technology was itself scale neutral and the effects on the distributional equity were dependent on the access of farmers to the necessary inputs and on the other sources of non-farm incomes. However, these differences in observation of the earlier studies arise mainly due to differences in techniques applied to different samples covering different countries and different time periods. From these studies, however, one may come to the general conclusion that new strategy in agriculture gives main emphasis on agricultural growth and along with that equity in the distribution of income can be achieved more and more by facilitating (through institutional reforms) more and more access of the farmers to the input markets (including land).

To enquire into the relation between growth and equity (i.e., the existence of 'trickle down theory') in West Bengal agriculture Bardhan (1986) made a cross section study for the year 1977-78, using NSS data at the disaggregated household level (covering 550 sample villages of West Bengal) and used both 'Logit' model and 'Linear Regression' model. In the Logit model the probability of an agricultural labour household falling below the poverty line was the dependent variable while in the Linear Regression model per capita monthly expenditure (at 1960 prices) of agricultural labour households was the dependent variable. The rate of growth in agricultural production in the district (where the sample village was located) was the explanatory variable in both the models. His finding was that the 'trickle-down theory' did not operate in West Bengal [i.e., agricultural growth was significantly positively related with the Logit-model's dependent variable and significantly negatively related with the Linear Regression model's dependent variable]. The author derived the same conclusion taking 'primarily-cultivator households' in place of 'agricultural-labour households'. In the author's own words, "other things remaining the same, the probability of an agricultural-labour household's sliding below the poverty line seems to be higher if it is in a district where agricultural production has grown at a faster rate" [p. 84] and "the variable GROWTH has a highly significant association with a lower level of living in the cultivator household, other things remaining the same" [p. 87]. The author, however, admitted the limitations of such type of cross section study in place of time-series study.

After Bardhan's work we do not get any study in West Bengal, in which the association between these two economic indicators (viz., growth and equity) during the Left Front Government regime has been examined critically. However, some studies which had the titles as "growth and distribution", "agricultural growth and poverty", etc. may be connected to this particular issue. These works may be classified in terms of their areas of studies, such as, (i) land reforms, (ii) wage rates and (iii) poverty and level of living.

A number of studies have been made considering an effective land reforms as means of bringing down rural poverty with agricultural growth. These studies [Rudra (1981), Sengupta (1981), Khasnabis and Chakraborty (1982), Bandhyopadyay (1988), Webster (1989), Lieten (1990), Mallick (1992), Khasnabis (1994), etc.] examined the implementation aspect (along with real beneficiaries) of land reforms in West Bengal during LFG regime. In consideration of their findings we get disagreements, as indicated earlier [Chapter

1, section 1.2]. However, these disagreements mainly arose (from village level surveys) in regard to the qualitative aspects of land reforms. As regards the success of land reforms in quantitative terms, Lieten observed, "In the early seventies West Bengal took the twelfth position among all India states in terms of the quantity of estimated surplus, or less than 1.8 percent of the national figure. Ten years later, it occupied the first place in terms of declared land surplus, with 16.1 percent of the national achievement. At that time the state accounted for 1.4 million of a total of 3.2 million beneficiaries" [Lieten (1990), p. 2265].

Some studies considered the movement of wage rates overtime in West Bengal as an indicator of the well-beings of agricultural labourers. Bardhan (1986) observed from the analysis of Logit model that increase in wage rate was negatively associated with the incidence of rural poverty in West Bengal. Jose (1988) showed that agricultural wages in West Bengal had the upward trend during the period from early seventies to mid-eighties. On the other hand, Mallick (1992) observed that during the same period West Bengal had 'the lowest relative to agricultural real wages for male labourers of any Indian state'. But we get opposite view from Nayyar's study. Nayyar (1991) estimated that between 1960-84 both money wages and real wages had always been higher in West Bengal than in a majority of other States. However, we get confirmation of Nayyar's observation from a recent study in which Sengupta and Gazdar (1995) indicated clear increase in real wages in West Bengal during the period of rapid agricultural growth (i.e. since 1982-83).

There are also some studies which have directly considered the incidence of poverty and general well-beings of the rural inhabitants in West Bengal during LFG regime. From small sample village survey some scholars [Swaminathan (1990), Lieten (1990), Harriss (1992, 1993), etc.] observed that as a consequence of LFG's policies the rural poverty in West Bengal had declined with the improvement in general level of livings. This was, however, contested by Gazdar [1992]. Ninan (1994) observed that the poverty trend in West Bengal declined [$HCR_t = .9424^{*} - .0178^{*}t$] during 1969-70 and 1986-87 in contrast to rising trend [$HCR_t = .5124 + .0167^{**}t$] during the period from 1957-58 to 1968-69. From the comparison of the performance of broad range of economic, social and rural indicators [agricultural production, rural poverty, child mortality rate, literacy, etc.] between two years 1977-78 and 1987-88 as well as between 1980-81 and 1989-90 in West Bengal Dasgupta (1995) has established that " West Bengal, a very poor and very rural State, has vastly improv¬d it relative position" [p. 2691] in respect of these indicators. However, from an extensive study [covering 790 sample households of 67 villages] Bhattacharya *et al.* (1987, 1988) observed that the general levels of living [viz., private consumption, social consumption, housing conditions, stocks of consumer durables (along with clothing and footwear), etc.] of the rural poorers did not improve much in West Bengal between two years 1972-73 (NSS survey year) and 1985-86 (authors' reference year).

2.5 Weaknesses in the Earlier Researches

From an in-depth and extensive study of the existing literature on the measurement of growth, stability and equity in agriculture, it has been observed that almost all these works suffer from a combination or all of the following weaknesses :

(i) Sometimes trend equations for measuring the growth have been chosen either arbitrarily or on the basis of certain statistical criteria which, more often than not, lack economic justifications or which are not always sufficient from the view point for which trends are being fitted.

(ii) The arbitrary selection of periods and the comparison of growth rates among these periods very often lead to misleading results. Just a few significantly low or high values at the end or the beginning or even in the middle of a short period may be responsible for high growth or low growth during that period. If observations are not properly adjusted for the such low or high values, the estimated growth rate and its comparison to the growth rate estimated from similar type of a set of observations would give a wrong idea about the performance of agriculture during these periods.

(iii) In the selection of a particular growth curve out of a number of known growth curves, as well as in the interpretation of the estimated growth parameters, only some particular econometric criteria (like R^2 or $\bar{R}^2$, standard error of the estimated parameter and DW statistic) are considered, leaving aside other criteria (which have been spelt out in details in Chapter 3). They are left aside because consideration of them will involve problem for estimation and stand in the way of economic interpretation of the results. But fitting a trend line and using the estimated values of the parameters as measures of growth give misleading results if, for example, either the disturbance term is non-random or the variance of the disturbance term varies significantly (in the event of long term fluctuations in the data-set) or the disturbance term is serially correlated (In the existing studies DW statistics are presented in a casual manner, without taking any corrective measures for the presence of autocorrelation). Accordingly, the presence of these particular features in the disturbance term should be taken care of while estimating growth rates.

(iv) Similarly, in respect of measuring the influences of growth contributing factors, it has been observed that they have been examined in an isolated manner without giving due attention to their joint effect in agriculture and the related multicollinearity problem.

(v) Rainfall has been considered as a non-stochastic variable which is free from errors of measurement in the determination of its effect on agriculture. But it is practically a nature determined random variable (at least a part of this variable) that also suffers from the errors of measurement, mainly because it is used as a proxy of weather variable which has direct bearing upon agricultural production. Therefore, unlike in the existing studies, rainfall should be considered as a random regressor.

(vi) Pointing out the limitations of summary measures of instability (as were followed by the earlier studies), in the recent studies concentration has been given to the trend measures of instability. But these two measures are not substitutable for each other; rather from the summary measures we can know the level or magnitude of instability and from trend measure we get the nature or change of instability overtime. Therefore, in consideration of the usefulness of summary measure, this measurement technique has to be improved, specially in the presence of strong outliers. Further, in the existing studies linear trend is fitted to some forms of detrend statistics (i.e., residuals) to get trend measure of instability. But this technique often leads to statistically insignificant results as detrend statistic is more liable to volatile fluctuations than the original series. To overcome this, non-parametric tests should also be taken into account along with the parametric tests.

(vii) Besides these methodological issues, it has been observed, specially in West Bengal, that the agroecosystem functions (viz., growth, stability and equity) have been empirically used individually without giving any consideration to their interactions and interrelationships for proper assessment of the agricultural performance. More specifically, equity aspect of growth has not been properly evaluated in West Bengal (apart from the difference of opinion of the scholars in the fields of growth, stability and equity in West Bengal agriculture). For that an integrated approach is necessary.

In this study we have tried to cover up these research gaps apart from the consideration of different time period, different coverage and some new issues, like (a) determination of relative contributions of locational factors and individual crops towards overall agricultural output growth rate, (b) determination of growth contributing factors, through factor analysis, (c) statistical examination of the causes of instability, (d) application of sophisticated inequality measures using primary data, etc.

Inspite of these unique features, our study cannot avoid some inherent limitations. (1) Being an *expost time* series analysis this study solely depends on the reliability of official data. However, through data adjustment we have tried to minimize the dependability of official data, wherever necessary. (2) Though we have tried to carry on the analysis at disaggregated level taking districts and individual crops (even the villages for equity measurement) into consideration, this study is mainly aggregative in character and because of this, in some cases, the derived conclusions may be criticized as oversimplified and aggregative. Lastly, (3) due to non-availability of adequate secondary data we have to restrict ourselves in many counts (specially in the consideration of equity) and to take cross-section study (for factor analysis), though time series analysis is more reliable on economic justification.

3

Data Base and Methodology

The importance of any of the empirical studies is generally examined and valued by its data base and methodology. The analysis on data base, in general, includes (i) the choice and selection of the field, (ii) definition and recognition of the variables/parameters on which the data are to be collected, (iii) methods and sources of data collection and finally (iv) problems and limitations of the collected data. According to the objectives and hypotheses of our study, as mentioned earlier, we have chosen West Bengal as our field of analysis and the period of our analysis is from 1957-58 to 1994-95. Further, this study is mostly dependent on the data collected from secondary sources (i.e., from Government publications and also from earlier studies). Only in case of equity measurement we have additionally used primary data, collected from sample village survey. The types, sources and limitations of these data, to be used for the analyses of growth, stability and equity in agriculture, have been pointed out in the first three sections of this Chapter.

Methodology of a study, likewise, comprises proper choice and selection of the methods for the analysis and interpretation of the data set, consistent with the objective of the study and in conformity with the nature of the data base of the study. From the literature survey (Chapter 2) we have seen that in respect of the methodology used for the measurement of growth, stability and equity in agriculture the economists have differences of opinion in many cases and, further, in some cases the widely used existing methods are not free from limitations. All these necessitate devising some new methods and/or modification of the existing methods for the proper evaluation of the agricultural performance. The last four sections of this chapter discuss some new methods (along with the existing methods) that are used in this study for the proper evaluation of the agricultural performance in West Bengal.

3.1 Data Base of Growth and Stability Measures

The basic data that are required for the measurement of growth and stability in agriculture are area (i.e., gross cropped area), production and productivity (i.e., land productivity which implies production per unit of cultivated land) of the agricultural crops. We have collected officially published yearly data on area (in thousand hectares), production (in thousand tonnes) and productivity (in kg. per hectare) for altogether 20 major crops and crop-groups [aus rice, aman rice, boro rice, total rice, wheat, barley, other cereals, total cereals, gram, other pulses, total pulses, total foodgrains, mustard, linseed, sesamum, total oilseeds, jute, sugarcane, potatoes and tobacco] of West Bengal and two major crop-groups [total rice and total foodgrains] of all the constituent 15 districts of the State during the period 1957-58 to 1994-95. During the same period we have also collected data for total non-foodgrains and all crops combined of the State and its districts in the form of index numbers, as published officially, on area, production and productivity. For the State and its districts these indices have been expressed in terms of triennium ending crop year 1969-70 and 1971-72 as base years respectively. Besides these, we have collected officially published indices (with triennium ending crop year 1969-70 as base year) on net area sown and cropping intensity (as the components of gross cropped area) and on yield and cropping pattern (as the components of productivity) only for the all crops combined of the State.

The sources for these data are: (i) 'Economic Review', Statistical Appendix (various annual publications), Directorate of Agriculture, Government of West Bengal (GBWB), Alipore: WBGP., (ii) 'Statistical Abstract' (various issues), Bureau of Applied Economics and Statistics (BAES), West Bengal, Calcutta: GOWB and (iii)'Estimates of Area and Production of Principal Crops in West Bengal' (Annual issues), Socio-Economic Evaluation Branch, Calcutta, GOWB. In addition, we have used the price indices as constructed by Boyce [p. 64, Table 3.4 (1987)] for the measurement of crops' and districts' relative shares.

Unlike most of the other States in India where 'complete enumeration method' is followed by some unskilled government officials; the official estimates of the area, productivity and production of agricultural crops in West Bengal are based upon annual sample surveys and crop cuts, carried out by the BAES [following the methodology of Mahalanobis (1946)]. More specifically, from the sample grids (each grid consists of 2.25 acres of land) of each district the average proportion of area under each crop is estimated and then those proportions are multiplied by the total geographical area of the district (excluding municipal areas, reserve forests and large rivers) to arrive at estimates of crop areas for the district. Likewise, average yields are estimated from random crop cuts (the cut is a circular area of about 100 square feet) in the sample grids. The total production of each crop is then calculated by multiplying total area by average yield of each crop. The surveys and crop cuts are conducted three times per year: in the jute-aus rice season, in the aman rice season and in the rabi season. Different scholars [Sarma (1971), Maitra and Lahiri (1975), etc.] heavily criticised the 'complete enumeration method' and gave arguments in favour of sample surveys and crop cuts. Pointing out the independent estimates of National Sample Survey Organization (on cereals output from 1957 to 1969) Bansil (1974) argued that the West Bengal data were of relatively high quality, compared with other Indian States. Boyce (1987) concluded that, "there are strong grounds for believing that West Bengal's official crop production data are relatively accurate by national and international standards" [p. 89].

But after a close scrutiny within the State itself Boyce (1987, Pp. 56-57) found out that there was "systematic inflation in West Bengal's official figures from the mid-1960s until the late 1970s" [p. 80] due to often substitution of BAES's estimates by the estimates of Directorate of Agriculture (DOA). Citing an informal interview with a former director of the BAES, Boyce speculated that DOA's estimates had an upward bias [because government officials might inflate the figures to show their successful efforts in agricultural production]. Moreover, against DOA's estimates he gave the evidence that from late 1970s the BAES estimates were finally accepted in the official series. On these considerations Boyce revised the official series in West Bengal agriculture, taking BAES estimates as objective and correct ones and rejecting DOA. estimates as subjective and inflated ones.

Unlike most of the other States of India (where agricultural statistics are compiled by non-technical 'patwaris', local land revenue officials), DOA. in West Bengal has a separate statistical branch with district statistical offices), staffed with technical personnel who are entrusted to collect agricultural data from pilot surveys. So DOA. estimates should not be considered as subjective as alleged by Boyce. Further, the coincidence between official series and BAES estimates from late 1970s [more specifically, from 1980-81, as shown by Boyce himself in Table 3.1, p. 59] may not be due to the acceptance of BAES estimates but may be due to preparation of single official series on the basis of DOA. estimates of acreage and BAES estimates of yield. Even if we accept the accusation of upward bias in DOA. estimates (which is more apparent in the cases of boro rice, wheat and sugarcane), one cannot substitute them directly by BAES estimates which may also be charged with suffering from the problem of downward bias (though of lesser magnitude) specially because of the exclusion of hilly sub-divisions and aman acreage under HYVs (early harvested aman crops) from its estimation purview.

From 1986-87 and onwards the official series on West Bengal agriculture was again criticized on the ground of overestimation. Due to the working of 'multi-agency system' from 1986-87 onwards, Datta Roy (1994) pointed out that yield rates were collected both by BAES and DOA. and if there was any significant difference between these two estimates, the average of them was to be taken as official estimate; this led to misleading result. In support of his argument Datta Roy gave few instances (aman and foodgrains production for the year 1991-92 in comparison to the previous year) in which he showed that production figures could neither be explained in terms of rainfall nor in terms of the strategic input consumption. Like Boyce, he also suspected the biased attitude of the agricultural department that could inflate output figures. Though Datta Roy's arguments were strong and supportive, yet no valid conclusion could be derived only from few instances, specially in regard to the direction and extent of bias in the official series. For example, according to Datta Roy's own argument the data on foodgrains production had downward bias for the year 1990-91 and upward bias for the year 1991-92.

From the above discussion we can safely conclude that, in comparison to most of the other Indian States, agricultural data (on area, productivity and production) in West Bengal are relatively of high quality, specially for the period from (i) mid-1950s to mid-1960s [as established by NSSO] and (ii) late 1970s to mid-1980s [as recognized by Datta Roy]. For the periods (i) mid-1960s to late 1970s and (ii) 1986-87 onwards, however, there are allegations of upwards bias in the data sets. Unlike Boyce (who was fortunate enough to avail himself of the alternative source of data at that time) and like Saha and Swaminathan (1994) we have retained the official figures as the magnitude of bias in the recent data (1986-87 onwards) are not measurable from alternative available source and expecting that the discomforting periods will be spread more or less equally at the end of two sub-periods. Moreover, unlike Saha and Swaminathan, we have adjusted the data (discussed later) for the entire period in order to nullify, as much as possible, the effects of, and, therefore, the bias introduced by, a few extremely high or low observations (as observed by Datta Roy (1994)] present in the data set.

Apart from this general problem, some specific problems may also arise with the method of data collection for our study. First, all crops combined do not here imply total agricultural production because the excluded items are tea, mat-sticks, betel leaf, spices, vegetables, fruits and flowers. Tea has been omitted owing to the special circumstances of its cultivation (a plantation crop grown almost exclusively in Darjeeling district) and all other items represent very negligible proportion in the State's agricultural output. Though for some of the items data are available from the government offices (New Secretariat Building, Calcutta) we have not included them since prior to 1977-78 there were no data on these crops and for the later years the reliability of these unpublished data is questioned even by the Government officials. Second, we have taken only 15 districts without considering further division of 24-Parganas and West Dinajpur [as officially done recently, along with economic division of Midnapore district] to make them comparable over time. Third, the base periods of the State's indices and districts' indices are different, so they are not strictly comparable. But, in reality, this would not create much problem because the difference between these two bases is only two years and further between two base periods the average change in cropping pattern [from index value 100 to 102] and the average change in crop prices [for example, Rs 134.23 to Rs 134.58 per Quintal of aman rice] are very negligible [figures are calculated from Statistical Abstract, 1976]. Fourth, inspite of the availability of crop harvest prices at the district level we have used price weights for West Bengal to get the standardized value of agricultural production of the districts. We have done it deliberately because for the determination of district-shares in overall growth of State's agricultural output the growth weights (as explained in section 3.4) should be calculated on the basis of common measuring unit. Lastly, except total non-foodgrains and all crops combined, other crops and crop-groups have been taken in terms of physical quantity [production in thousand tonnes] as per official publications. Actually, this will create problem in case of comparison, specially to determine the effect of cropping pattern. But the problem has not been so acute as appears to be because the main objective of our study is to determine the effect of productivity instead of cropping pattern on the growth of output and further, theoretically the growth in real output can be estimated either from physical quantities or from index values, having common base year. Only the problem of aggregation arises in case of physical quantities; but index number is not altogether free from this limitation so far as the heterogeneity in quality within a single crop and the heterogeneity of prices of a crop and for a particular region are concerned (apart from the general problems of indexation). That's why, perhaps, in the earlier studies physical quantity has been given preference to index number (if not unavoidable, like the cases of non-foodgrains and all crops having different measurement units) for the measurement of agricultural growth rates.

3.2 Data Base for Determining Growth and Stability Contributing Factors

For the determination of factors contributing to productivity growth and output stability in agriculture one should collect data on those factors (technological, institutional, climatic, environmental, demographical, etc.) which directly affect and determine them. But here owing to non-availability of adequate and appropriate data we have to restrict and reorganize the analysis accordingly. As there are no official recorded data on institutional factors continuously even for a moderate period of time, we have considered only the influence of main technological factors from time series analysis on the productivity index of all crops combined in West Bengal during LFG regime (i.e., from 1977-78 to 1994-95). The technological factors on which data are available during this period are fertilizer consumption, area under HYV and area under irrigation from government canals (i.e., 'Water-Seed-Fertilizer' technology). As irrigation from govt. canals represents a part of total irrigation (whose time series data are not available) we have used a proxy variable of total irrigation. These data have been presented in Table 3.1. The sources and units of measurement of these data have been mentioned in the foot notes of Table 3.1, except the measurement procedure followed in case of estimated irrigation intensity. Irrigation intensity has been estimated from a proxy variable, defined as,

(Gross Cropped Area - Net Sown Area)/Gross Cropped Area

This proxy variable may also be defined as the modified cropping intensity [as the actual cropping intensity = (GCA-NSA)/NSA]. In West Bengal as the period of main rainfall (i.e., monsoon season) is mainly used for mono-crop culture, excess of GCA over NSA can be taken as proxy of irrigated area. But, strictly speaking, this modified cropping intensity may lead to underestimation of total irrigated area if single crop is also cultured under assured irrigation (i.e., govt. canals) or to overestimation if multiple crops are grown within the monsoon season (also taking winter rainfalls into consideration for rabi crops). We cannot rule out these possibilities in West Bengal; rather they happen so in West Bengal, though not on a very large scale and some parts of them may be off-setting in character. However, this proxy variable may be considered as better than the irrigation from only govt. canals (specially when the farmers are now extracting ground water on a large scale in West Bengal). We may point out another limitation that the use of total fertilizer (N+P+K) in estimating fertilizer intensity may not give accurate result specially when there is a common belief that nitrogen is more important for agriculture than other components of fertilizer. As we have no information on the crop-specific effects of different components of fertilizer at hand we could not use it in our analysis.

As an improvement over the time series analysis we have also made cross section studies (over districts, at three time points, namely, 1970-71, 1980-81 and 1990-91) for the determination of the relative effects of different factors on land productivity in West Bengal during and before LFG regime. For the cross section study we have collected data on altogether eight explanatory variables (four technological factors and four institutional factors) on and around the said three time points as the determinants of productivity of all crops combined of the districts. For three separate time points, data have been presented separately in three tables —Table 3.2, Table 3.3 and Table 3.4, along with the foot notes mentioning the sources and exact time periods of the collected data.

The following few points are to be noted on the explained and explanatory variables that have been used in these three tables [Tables 3.2—3.4].

(i) For Tables 3.2 and 3.3 we have collected data on the percentage of GCA irrigated [i.e., (total irrigated area/GCA) x 100] from the study of Boyce (1987). But as the data on total irrigated area at the district level are not available for the year 1990-91, we have estimated them in the following way:
(Percentage of GCA irrigated in 1990-91) = 100 x [irrigated area in 1980-81 + change in GCA between 1980-81 and 1990-91– change in NCA between 1980-81 and 1990-91] ÷ GCA in 1990-91. In this formula it has been implicitly assumed that the change in cropping intensity is synonymous with the change in irrigation intensity. But this is some sort of approximation as mentioned earlier.

(ii) Fertilizer intensity (kg. of N+P+K per hectare), presented in Tables 3.2 and 3.3, has been calculated from 3 years' (t-1, t and t+1) average data; but for Table 3.4 the data relate to the year 1990-91 only. This has been done as the later period's fertilizer data are more stable than those of early periods.

(iii) Owing to non-availability of data for the year 1980-81, we have estimated the percentage of GCA under HYV for that year using the district-wise data on the proportions of area under HYV for the year 1978-79 from Boyce [(1987), p. 178, Table 6.5] and the information that 100 percent of area of boro rice and wheat in the State were under HYV in the year 1980-81 (source: Economic Review, W.B), assuming that there were no significant changes in the proportions of HYV area of aus and aman rices between 1978-79 and 1980-81. It is to be noted in this connection that in West Bengal area under HYV implies those of rice and wheat [as per official publications for the years].

(iv) There are different types of agricultural machinery; but from the point of view of modern technological improvements we have taken only the tractors and power tillers. From the rough price estimates over the years, we have aggregated them by taking 3 power tillers = 1 tractor. Agricultural machinery intensity has been calculated by dividing the total agricultural machinery by GCA.

(v) Average operational holding of a district has been calculated by dividing NCA of the district by its total number of cultivators (collected from census reports). However, as the district-wise data on NCA are not available for the year 1990-91, we have replaced them by the figures of 1994-95, on the assumption that there was no significant change in NCA of the districts between these two years [which is evident from the very negligible growth of NCA in West Bengal, as shown later on]. In the same manner we have estimated the percentage of NCA under surplus land distributed for the year 1990-91 [taking NCA of 1994-95 and area under distributed surplus land as on 30-06-91].

(vi) From consecutive three census [1971,1981 and 1991] figures we have estimated district-wise percentage of agricultural labourers in total working force in agriculture as:

[Agricultural labourers ÷ (Agricultural labourers + Cultivators)] x 100.

(vii) For Table 3.4, percentage of NCA under tenancy in 1990-91 should have been calculated as:

(Area under tenancy in 1990-91÷NCA in 1990-91) x 100

But due to non-availability of district-wise data on area under tenancy in 1990-91, we have used data on area under recorded bargadars. Here also, instead of taking figures (on recorded bargadars) of 1991 we have taken the latest figures of 1995 (i.e., area under recorded bargadars as on 30-9-95) on the consideration that in the later years more and more tenants got themselves registered and, therefore, the area under recorded bargadars comes very close to area under tenancy. Further, we have used data on NCA of 1994-95, instead of 1990-91, because of the reasons, mentioned in v) above.

From the above analysis it is observed that owing to non-availability of appropriate data, in some cases we could not strictly adhere to the fixed time points and to the rigid definitions of the variables. However, in the process of data estimation in those cases we have tried to maintain objectivity as much as possible like the earlier studies in this field [Boyce (1987), Lieten (1990), etc.]. Further, to minimize the subjectivity, arising out of non-adherence to fixed time points in data collection, we have designed the explained variable in the following way:
For each district we have calculated the average gross cropped area and the average value of total agricultural production [using common price weights, as estimated by Boyce (1987)] from three consecutive years 1969-70, 1970-71 and 1971-72; symbolically, these are:

$$\Sigma_i \Sigma_j a_{ij} / 3 \text{ and } \Sigma_i \Sigma_j q_{ij} p_{io} / 3$$

[Where i denotes crop, j denotes year, a is cropped area, q is crop output and p_{io} is Boyce's price weight (calculated from average harvest prices in the triennium 1967-68 — 1969-70)].

After that, we have measured the value of agricultural output per hectare as:

$$[\Sigma_i \Sigma_j q_{ij} p_{io} / 3] \div [\Sigma_i \Sigma_j a_{ij} / 3]$$

On the basis of this value of agricultural output per hectare and the average productivity index (averaged over 1969-70, 1970-71 and 1971-72) we have estimated the adjustment factor for each district, taking Burdwan as standard one. For example, the value of agricultural output per hectare in Burdwan is 1622.3 and in Birbhum is 1376.3. And the average productivity indices in Burdwan and Birbhum are 99.15 and 100.43 respectively. Then the adjustment factor for Birbhum = [(99.15 x 1376.3) / (100.43 x 1622.3)] = 0.838.

The estimated adjustment factors for each district have been reported in the second column of Table 3.2. After estimating these adjustment factors, we have calculated the average productivity index (from official publications) for each district at three time points (t = 1970-71, 1980-81, 1990-91), taking average of 3 neighbouring years' (t-1, t and t+1) productivity indices. These average productivity indices of the districts have been finally multiplied by the respective adjustment factors for the districts to get district-wise productivity indices which are used as dependent variable [reported in col. 3 of Table 3.2 and col. 2 of Tables 3.3 and 3.4]. This makes the data on the dependent variable comparable over districts. Moreover, as a consequence of this, the dependent variable can be temporally related with the explanatory variable.

To determine the effect of rural population growth on agricultural production we have collected district-wise data on inter-census rates of growth of rural population from the study of Boyce [Table 5.1, p.139 (1987)] and census report, 1990-91, Govt of India. The rural population growth rates are presented in Table 3.5.

Like the analysis on growth contributing factors, we have made both time series and cross section (over districts) analyses to determine the causes of output instability in agriculture. In addition to the variables [average farm size and irrigation for cross section analysis and only irrigation for time series analysis], three more factors, viz., percentage of NCA under current fallow, crop-base and locational base have been considered as determinants of output instability, depending on the availability of data. More irrigation, lower farm size (i.e., higher intensive cultivation) and widening agricultural bases (crop-wise and location-wise) are commonly hypothesized as the factors that bring lower instability in agricultural output. We have also chosen percentage of area under current fallow as the proxy variable which signifies indirectly the nature of environmental degradation (by increasing soil salinity, water logging, etc.) and the extent of marginal lands brought under cultivation (which are vulnerable to water scarcity). These factors contribute to increase in current fallow, thus causing greater instability. The data on these three additional variables are presented in the tables 3.6 to 3.9. The sources and the definitions of the data are given in the foot notes of these tables. Owing to non-availability of data, however, the period for West Bengal has been reduced to 1957-58 to 1988-89 for time series analysis. In case of crop base (also locational base) for Table 3.6, we have, however, chosen those crops (districts) that have comparatively large shares in the State's agricultural production.

Finally, in order to measure the effect of rainfall on agricultural growth and instability, we have collected district-wise and monthly data on actual and normal rainfall (in millimeters) during the whole period from Directorate of Agriculture and Agricultural Meteorologist, West Bengal. From these data we have constructed district-wise rainfall indices by using the simple formula,

(actual annual rainfall ÷ annual normal rainfall) x 100

The annual rainfall is obtained by summing up the rainfall of April upto the rainfall of March next year, though crop calendar starts from July of one year to June of the next year. This has been done since the crops transplanted in the month of July normally require rainfall over a few previous months for their proper growth. From district-wise rainfall indices we have, lastly, calculated the State's rainfall index as the weighted average of the district level indices. Weights have been estimated from the share of each district in the net cropped area of the State. As these shares do often change over the years, we have estimated them from the simple average of 10 year-period, from 1977-78 to 1986-87. The estimated State level annual rainfall indices are presented in Table 3.10.

However, in estimating these rainfall indices (for the State as a whole) one cannot avoid some inherent approximations so far as the aggregation of rainfall over months/regions and choice of weights are concerned. These limitations are also encountered even at the disaggregated levels (such as sub-divisions, blocks, etc.). This measurement error has been taken care of while designing methodology for measuring the effect of rainfall (see methodology section, 3.4 below).

3.3 Data Base for the Measurement of Equity

3.3.1 *Secondary Data*

For the measurement of equity associated with agricultural development (specially during LFG regime) requisite data on wage rates, agricultural ownership land holding, rural consumption distributions and rural poverty have been collected from various secondary sources.

District-wise money wage rates of male and female agricultural labourers have been collected from Agricultural Wages in India (annual publications), Directorate of Economics and Statistics, Ministry of Agriculture, Govt. of India, over the period 1978-79 to 1991-92. Wage rate (per day per labourer) has been estimated in the form of average over the agricultural year (i.e., July to June) to smooth out the effect of seasonal fluctuations in wages. Further, to convert them into real wage rate, money wage rates have been deflated by the consumer price indices for agricultural labourers (CPIAL) for West Bengal (as district wise CPIALs are not obtained), collected from Labour Bureau, Chandigarh.

In addition, for the measurement of inequality in the distributions of ownership land holding and monthly per capita consumer expenditure of rural population of West Bengal uses have been made of different NSS reports, published for different NSS rounds. More specifically, land distribution data have been collected from 8th (1953-54), 16th (1961-62), 26th (1971-72), 37th (1982), 43rd (1988) and 50th (1993-94) NSS rounds. And for the distribution of consumer expenditure these rounds are 8th, 13th (1957-58), 16th, 28th (1973-74), 32nd (1977-78), 38th (1983), 40th (1985), 42nd (1987), 43rd, 45th (1989-90) and 50th. Considering the volume of our study we have not taken any separate measure on poverty. We have collected data on the incidence of rural poverty in West Bengal (as given in Table 3.11) from the study of Sen (1996).

3.3.2 *Primary Data*

To enrich our analysis on the equity measurement from more

classified data we have chosen purposely three villages of South Bengal, at different stages of agricultural development, for primary survey. Sample village I is from Hooghly District, named Balibela which may be recognized as so-called highly developed in the sense that, being situated on the bank of river Darakeswar and nearer to the sub-divisional town Arambag, three crop-rotations are widely cultured annually. Likewise, sample village II (named Ajodhya of Bankura District) and sample village III (Hoomgarh of Midnapore District) have been chosen as moderately developed and less developed villages having respectively two crop rotations and mono-crop cultivation. From these sample villages we have selected 153 households [53 from sample village II and 50 each from sample village I and III] as final units through proportionate random sampling. From primary survey we have collected data on annual income, consumption and land holding in the reference period 1994-95 by personal interviewing method through questionnaires.

The data on income pertain to farm income and off-farm income. Farm income includes agricultural income as entrepreneur and as labourers. Off-farm income is further split into three categories: (a) income from formal sector which includes income from Govt. and semi-govt. services, authorized institutions and registered businesses; (b) income from informal sector which includes income from working as maid-servant, cow-boy, small unregistered businesses, rickshaw pulling, etc. and (c) income from other sources that include income from poultry, dairy, cottage industries, community works, etc. All incomes have been calculated on net basis. However, due to conceptual and computational difficulties it was not possible for us to collect data on capital gains and imputed rents from residential structures. Further, the reported interest income from savings of the households are also excluded because of non-responsiveness of the households in most cases. The collected total income data for three sample villages have been classified in Table 3.12.

The consumption expenditure of the sample household is classified into two groups: (a) necessary consumption which includes food, shelter (running expenses), clothing, education and medical expenses and (b) other consumption which includes the consumption of luxury goods and intoxicants, expenditure to maintain social formalities, services of durable goods (i.e., rental value of the durables except residential structure), expenditure on higher education, etc. Size distributions of total annual consumption of the sample households are given in Table 3.13.

Lastly, data on the cultivable ownership land-holding of the rural households have also been collected. The size of ownership land holding in our study comprises owned land plus two-thirds of the registered leased-in land plus one-third of the registered leased-out land [on the basis of general crop sharing arrangement between the tenant and owner under 'Operation Barga' in West Bengal]. The veracity of the information and the data on land holding size furnished by the sample households have been checked with the information and data available from the neighbours, Land Revenue Department and Panchayat Records as far as practicable. The data on land holding distribution are presented in Table 3.14.

3.4 Methodology for Measurement of Agricultural Growth and Growth Decomposition

It is seen from the review of literature on the measurement of agricultural growth that the researchers have fitted a number of trend lines or growth curves to the agricultural production data. The trend lines or growth curves considered by them are linear, parabolic, exponential, log-quadratic and several S type curves such as gompertz, logistic, etc. The selection of a particular trend line is made on the basis of certain statistical criteria and some *apriori* economic reasonings. The statistical criterion that is widely used for selection of a particular trend is R^2 or $\bar{R}^2$ on the consideration that it is a measure of goodness of fit to the data. The t values or standard errors are also reported to indicate the probability levels at which the coefficients are statistically significant or not. Also are reported the F values and DW statistics to indicate the levels of significance of $\bar{R}^2$ and the presence of autocorrelation or not. But DW statistic has rarely been given due importance by making necessary adjustments in the equation, when it signifies the presence of autocorrelation in the disturbance term. It appears that reporting is done to caution that it is present and as such interpretation should be made cautiously. In the presence of autocorrelation the estimates are unbiased, but the sampling variances of these estimates may be unduly large compared with those achievable by a slightly different method of estimation. If the autocorrelation is positive, the calculated acceptance regions or confidence intervals will be often narrower than they should be for the specified level of significance or confidence. Therefore, the inferences drawn on the basis of the calculated results as regards the acceptance or rejection of the hypothesis that the trend coefficient and, therefore, the growth rate, which is derived from the trend coefficient and some other known valûes, is statistically significant or not will be obviously misleading. The comparison of the growth rates of, say, agricultural production of two periods, calculated from the estimated trend coefficients, with a view to drawing conclusions about the relative performances of agricultural production during these two periods may be acceptable, though with some doubt, if the disturbances in both the periods are either positively or negatively autocorrelated, thus rendering the biases to be of same nature in both the periods. But the comparison is highly improper if the disturbances in the two periods are autocorrelated in the opposite directions or if the disturbance term in one period is autocorrelated while that in the other period is not.

Therefore, in order to draw valid statistical inferences about significance or otherwise of the growth rates and to compare the agricultural performances during two or more than two periods on the basis of the trend coefficients one should make necessary corrections for the presence of autocorrelation, if there be any in the disturbance term, before fitting a trend line.

Only R^2 or $\bar{R}^2$ is an essential, but not a sufficient measure of goodness of fit in the presence of autocorrelation or heteroscedasticity in the disturbance term. The presence of autocorrelation in the disturbance term is indicative of the mis-specification of the equation. Thus, if the trend equations under consideration suffer from autocorrelation problem, choosing any one of them on the basis of the highest value of R^2 is in no way the best choice. The statistic R^2 is, of course, essential since its value shows whether the equation gives a statistically significant fit to the data or not. The R^2 is both an essential and sufficient, and as such the best, measure of goodness of fit to the data provided that the disturbance term in the equation satisfies all the classical normal regression assumptions.

The heteroscedasticity problem is rarely considered as a serious one in the time series data. Therefore, while fitting a trend line to the time series data, attention is rarely paid to this problem. But heteroscedasticity, the term used to denote the change in the variance of the disturbance term, may very well occur in the time series data, as in the cross section data. There are many factors that may contribute to the change in the variance of the disturbance term for the agricultural production. For example, technological change in the form of introduction of new HYV seeds, new inputs, etc. may lead to both comparatively high rise and fall in the agricultural production under favourable and adverse conditions respectively. In a number of studies it has been shown that the fluctuations in the agricultural production has either increased or decreased over time with the increase in the mean or the trend values of agricultural production. Thus, the time series agricultural production data may be quite often characterized by the problem of heteroscedasticity. As a result, the least square estimators of the time variable are, though unbiased and consistent, not BLUE, efficient or asymptotically efficient. Moreover, the conventional formula for the variance of the OLS estimate of the coefficient of the time variable is biased when the disturbance is heteroscedastic. Thus, in the presence of heteroscedasticity the problems regarding the test of hypothesis about the significance of growth rates remain the same as in the case of autocorrelation. But, unfortunately, no corrections are made in the data set for the presence of heteroscedasticity before fitting a trend line to these data.

Thus, it is contended here that the failure to take cognizance of the presence of autocorrelation and heteroscedasticity and to take necessary measures for correction in the data contributes to wrong statistical inferences about the statistical significance of growth rates as well as the statistical significance of the differences in the growth rates during two or more than two sub-periods.

Apart from these specific econometric problems, the time series agricultural production data may be characterized by significant volatile fluctuations over time. A trend line fitted to such data is unlikely to present the actual trend in the data. To capture the actual trend in the data appropriate smoothing is essential. But, this is also rarely done.

In this study we have dealt with these problems and estimated the trend coefficients after necessary corrections for the presence of autocorrelation, heteroscedasticity in the disturbance term and volatile fluctuations in the data set. We propose to proceed in the following way:

1. For selection of trend lines to be fitted to the data set the behavioural pattern of the data be first studied carefully.
2. The selection of a particular trend equation be guided by various statistical and econometric criteria.

The above procedures are elaborated below.

There is a large number of trend equations which may be fitted to the data to estimate the growth rate of a variable, say, agricultural production. One can try all the known equations, which is also normally done. But, it is proposed in this study that instead of trying with all known equations, one can consider only a few, leaving out other equations, on the basis of a careful study of the behavioural pattern of the data. For this purpose the point-to-point growth rates of production are estimated using the formula, $(Y_t - Y_{t-1}) \div Y_{t-1}$. Since the point-to-point growth rates exhibit wider fluctuations as compared to the original production series, an appropriate smoothing of the point-to-point growth rates is done. A linear trend equation of the form $gr_t = a + bt + u_t$ (t is the time variable normalized by shifting the origin to the centre of the series, gr_t—point-to-point growth rates at t=2,..,n) is then fitted to the smoothed data after necessary corrections for the presence of autocorrelation in the disturbance term, which is very likely to occur because of smoothing of the series. The estimated trend parameter, $\hat{b}$ indicates the behavioural pattern of the data set. If, for example, the hypothesis that $\hat{b}$ is not statistically significantly different from zero is accepted, the growth rate is constant. In such a situation the exponential trend is chosen as the most appropriate trend equation for estimating the growth rate. On the other hand, if $\hat{b}$ is positive and significantly different from zero, the behavioural pattern of the data indicates increasing growth rates. In such a situation the choice of trend equations has been restricted to (i) parabolic, (ii) log-quadratic and (iii) S-type growth curves. In the case of behavioural pattern indicating decreasing growth rate [$\hat{b}$ is negative and statistically significantly different from zero] the choice of trend equations has been restricted to the above three curves as well as the linear trend.

A particular equation is chosen out of these equations on the basis of its fulfilment of the criteria, namely, (i) non-autocorrelation (tested by DW statistic), (ii) randomness in the disturbance term (tested by non-parametric run test), and (iii) homoscedasticity in the disturbance term [tested by non parametric rank test (Johnston, 1960 p. 219) and Glejser test]. Apart from these the choice between parabolic and log-quadratic has also been made on the basis of the P. E. test [a variant of Box-Cox transformation (Kmenta,1971 pp. 521-522)].

In case none of these equations satisfy all these criteria, which, of course, rarely happens, the log-quadratic form has been chosen because of its certain special features, not possessed by others. For example, it takes care of multiplicative error term which is very likely to be present in most of the time series data. It accommodates varying nature of growth patterns and it smoothens the short term volatile fluctuations.

The following three points should be mentioned here:

(i) Autocorrelation in the disturbance term (arising out of smoothing of the point-to-point growth rates and also in other cases, if it arises) has been corrected by applying "Cochrane-Orcutt two-step Procedure" [Kmenta (1971) Pp. 314-317] through either first order or second order auto-regressive scheme. (ii) S-type growth curve has been taken in its simplified form [i.e., $Y_t=e^{a-b/t}$] to apply OLS method of estimation. (iii) $\bar{R}^2$ has been considered as the criterion of trend selection if other criteria (noted above) are fulfilled or if alternative trend equations give very poor fit.

Once an equation is selected following the above mentioned steps for the measurement of growth rates, appropriate smoothing of the original series, not the series of point-to-point growth rates, has been done if the original series is beset with short term fluctuations (determined by the magnitude of the standard deviation of the residuals) and long term volatile fluctuations (measured by the non-parametric rank test or Glejser test). Of the two smoothing techniques, namely, exponential smoothing and moving average smoothing, the later has been preferred to the former since the former is carried out after specifying an initial value and an arbitrarily chosen smoothing factor, α. A slight change in their values leads to a significant change in the smoothed series. In our case, the choice of the average of the first three values in the series as the initial value and the choice of α that minimizes the forecasting error did not improve the results. It was observed that the whole exponentially smoothed series lay either above or below the original series. The moving average smoothing did not suffer from this problem. Therefore, in our case the moving average smoothing has been preferred. The method is explained in the following way. [Makridakis *et al.* (1983), Pindyek and Rubinfeld (1981)].

The chosen equation has been fitted to the original series and the residuals are estimated. Then 3 year centered moving averages of the residuals are estimated. These smoothed values of the residuals are then added to the estimated values to form a new series. However, as described earlier, smoothing may introduce autocorrelation in the new séries. Necessary adjustments have been made to correct for the presence of autocorrelation in the disturbance term, in case autocorreletion is present.

The smoothing of the residuals may solve the problem of heteroscedasticity, i.e., varying variance of the residuals, to an appreciable extent. However, when a long series of observations appears to have come from two different populations characterized by significantly different means and residual variances, estimation of an unbiased (free from discontinuity bias) average growth rate for the whole period has been made by taking different periods of moving averages of the residuals depending on the magnitude of volatility of the fluctuations in the residuals. To illustrate, suppose a long period consists of two periods, 1 and 2. Let the fluctuations in period 2 be more volatile than those in period 1. Then depending on the magnitude of volatility of fluctuations, if a 3-period moving average or no smoothing is suggested for the period 1, a 5 or 7-period moving average may be taken for period 2.

After necessary corrections for the presence of auto-correlation, heteroscedasticity and volatile fluctuations (with constant variance) in the data set, the chosen equation has been fitted to the observations and the trend coefficients have been estimated and the growth rate has been calculated from the estimated trend coefficients. It has been observed that this procedure of fitting trend lines and estimating growth rates from the estimated trend lines has yielded better results in most of the cases and no worse result compared to those obtained from the unadjusted series. (For a comparison of the test results see Table 3.15).

The chosen trend equations have been fitted to the whole period to calculate the growth rate of the whole period. On the other hand, the kinked trend equations have been fitted to estimate the sub-period growth rates. More specifically, in the case of linear or parabolic trend (chosen for the whole period), the kinked linear trend and in the cases of other three trend equations the kinked exponential trend have been fitted to estimate sub-period growth rates on the realistic assumption that relatively long period's changing growth rate can be sub-grouped into relatively short period's constant growth rates. The generalized 'Kinked-Model' with m sub-periods and m-1 kinks is :

$$y_t \text{ (for linear kink) or } \log_e y_t \text{ (for exponential kink)} = a_1 + b_1(D_1 t + \sum_{j=2}^{m} D_j k_1) +$$

$$b_2 (D_2 t - \sum_{j=2}^{m} D_j k_1 + \sum_{j=3}^{m} D_j k_2) + \dots\dots\dots + b_i (D_i t - \sum_{j=i}^{m} D_j k_{i-1} + \sum_{j=i+1}^{m} D_j k_i)$$

$$+ \dots\dots\dots\dots + b_m (D_m t - D_m k_{m-1})$$

where a_1 and b_i are parameters, k_i s are kink points and D_j is the dummy variable for the sub-period j, such that $D_j=1$ for jth sub-period and $D_j=0$ for sub-periods other than j. Due to adjustment in the data set, kinks, arbitrarily placed on the break points (joining point of two adjacent sub-periods), will not lead to distortion in the estimated sub-period growth rates and empirically it has been observed that with this kinked model all the econometric criteria for estimation remain undisturbed. For example, in Table 3.16 are presented different estimated results on sub-period growth rates obtained by applying different approaches, taking all crops combined of West Bengal as a case. From Table 3.16 it is seen that our proposed method gives better results in comparison to other approaches of estimation of sub-period growth rates. Actually our kinked model is a generalized and improved version of Boyce's kinked model and this kinked model gets support from the general class of econometric models with "Join points" [Judge *et al.* (1980)].

Decomposition of Growth Rates

Customarily in agriculture, output growth is decomposed through multiplicative decomposition model by estimating components' growth rates separately and in doing so an unexplained error term crops up (which is termed as interaction term) in the decomposition model. This error or the interaction term arises because of (i) the use of different formulae in component index formations and (ii) biased estimation of components' growth rates (in the face of volatile fluctuations in the data sets). However, this term cannot be easily interpreted in economic sense. For example, applying extended decomposition model, as has been done in some earlier studies, to all crops combined of West Bengal we see from Table 3.17 that the magnitude of error term is quite high for all the periods. However, for our estimated adjusted output growth rates the magnitude of the error term is comparatively very low, but still quite high, in some cases, even higher than the contributions of one or the other or both the other components. Further, in West Bengal, as seen from Table 3.17, area growth implies mainly growth in cropping intensity and the share of growth of yield in the growth of productivity is increasing over time.

Consequently, we have not used the extended decomposition model and instead we have applied two-component simple multiplicative model where output growth has been decomposed into growth of area (i.e., of gross cropped area) and the growth of productivity for the whole period as well as for sub-periods. Out of these two components, one component has been chosen on the basis of relatively less volatile fluctuations [measured in term of standard deviation of point-to-point growth rates] to estimate trend growth rate and the other component's growth rate has been calculated by the residual method (i.e., output growth rate minus the chosen component's trend growth rate). For the estimation of the chosen component's trend growth rate we have followed the same procedure as outlined in the estimation of output growth rate but the choice of trend equation has been restricted to only exponential (when $\hat{b}$, as indicated earlier, is insignificant) and log-quadratic (when $\hat{b}$ is significant) for multiplicative model. With this technique of growth decomposition not only the calculation would be simplified but also the error term will not appear. To summarize the analysis on growth decomposition over diversified crops and districts the cross sectional (i.e., districtwise and cropwise) correlation coefficients among growth components have been calculated, as done by Boyce (1987).

Lastly, to get the ideas on locational and cropping pattern effects separately, the percentage shares of individual districts and individual crops in overall agricultural output growth in West Bengal for the whole period and for two sub-periods have been calculated by using the following formula (for each period):

$$\text{Percentage share of the crop/district} = \left[\frac{\text{output growth rate of the crop/district}}{\text{State's overall output growth rate}} \text{ X } \begin{array}{l}\text{growth weight}\\ \text{of the crop/}\\ \text{district}\end{array}\right] \text{X } 100$$

where growth weight of a crop for a particular period (say 1957-58 to 1976-77) has been calculated as :

$$\frac{\text{Annual average production of the crop during the period 1957-58 to 1976-77}}{\text{State's annual average total agricultural production during the period 1957-58 to 1976-77.}}$$

Here productions have been expressed in value term, using the price index as estimated by Boyce (1987, Table 3.4, col.5, P. 64).

Similarly, the growth weight of a district for a particular period (say 1957-58 to 1994-95) has been calculated in the following way:

We have first calculated the standardised production per unit of the index number of the district (using Boyce's price index) as:

$$\frac{\text{Total value of agricultural production of the district for the years 1969-70 to 1971-72}}{\text{(Total index number of the agricultural production of the district for the years 1969-70 to 1971-72).}}$$

After that we have calculated the annual average total agricultural production of the district during the period 1957-58 to 1994-95 as:

(yearly average index number of the total agricultural production of the district for the period 1957-58 to 1994-95) X (standardised production per unit of index number of the district, as calculated above).

In this way we have calculated annual average agricultural production for the period 1957-58 to 1994-95 of all districts and taking their summation (over districts) we get the State's annual average agricultural production. Now, the growth weight of the district for a particular period (say 1957-58 to 1994-95)

$$= \frac{\text{(Annual average total agricultural production of the district from 1957-58 to 1994-95)}}{\text{(State's annual average total agricultural production during the same period)}}$$

With these formulae, however, we do not get 100% accurate result as some minor crops have been excluded (because of non-availability of their price-weights), use of different base years for the districts and the State in the construction of index numbers, and the statistical discrepancies because of aggregation and averaging. However, these formulae can be safely used for calculation of the relative changes in the shares of the districts and crops over different sub-periods.

3.5 Methods for Measurement of Agricultural Stability and Related Issues

While assessing the state of stability in agricultural production (crop-wise and district-wise) in West Bengal we have considered the following four issues related to instability measurement. These are: (i) nature of instability, (ii) extent or magnitude of instability, (iii) decomposition of instability and (iv) association between growth and instability.

For the measurement of the nature of instability over time Boyce (1987) has applied trend method. He has used kinked exponential trend and calculated the detrend series, Z_t. Next $|Z_t|$ has been regressed upon time, t. The coefficient of t is used as a measure of the nature of instability. However, the arbitrary kink point on unadjusted data is likely to result in a high value of $|Z_t|$ at or near the 'break-point' which, in turn, may make the coefficient of t insignificant, thus forcing us to infer that there is no change in instability. Further, the linear trend that he has used may not be the proper specification for all types of data sets, as we have seen in our study. Moreover, the Glejser test, $[|Z_t|=a+b\ t+u_t]$, used by Boyce, is not a very powerful test of heteroscedasticity if the disturbance term is autocorrelated, which has occurred in quite a number of time series under study.

To overcome these problems we have estimated the detrend statistic, X_t, by fitting appropriate trend equation i.e., linear or parabolic or log-linear or log-quadratic (as selected for growth rate estimation) to the whole period adjusted data. X_t is defined as,

$$X_t = \frac{\text{(actual production at time t minus the production predicted at time t)}}{\text{(production predicted at time t)}}$$

Next, the parameters a and b have been estimated from the equation, $|X_t|=a+bt$. The statistically significant and positive (negative) b signifies increasing (decreasing) instability. A statistically insignificant b signifies constant or unchanged instability. Further, considering the possibility of existence of non-linear trend and the failure of Glejser test, the rank correlation test, a non-parametric test, has been applied, following Johnston's suggestion [Johnston (1972), Pp. 219-221], between $| X_t |$ and t as an additional measure of nature of instability.

For the measurement of the extent of instability in crop output over two sub-periods and the whole period, the period-wise averages of the observations with strong outliers replaced by the next higher value, denoted by $|Z_t|$, have been calculated. Here $|Z_t|$ is defined in the same way as $|X_t|$, stated above, excepting that it is expressed on the arithmetic scale, unlike X_t which is expressed on arithmetic as well as logarithmic scale, as and where applicable. The strong outliers have been detected by observing period-wise extreme values of $|Z_t|$. The existence of the strong outliers has been confirmed from the graphical exposition of the original series along with Z_t and the trend values.

This summary measure of instability is an improvement over the other measures, stated earlier, as: (i) it is unit free and comparable over crops and periods, (ii) it is not unduly affected by the strong outliers and (iii) it is based on the whole period's trend having been freed from discontinuity bias.

Following the model of Rao *et al.* (1988, Appendix II, Pp. 161-163) the output instability (measured as the variability in point-to-point output growth rates) has been decomposed into three sources, namely, variability in area growth rate, variability in productivity growth rate and their synchronous movement. That is,

$$Var(G_{ot})= Var(G_{at})+Var(G_{pt})+2Cov(G_{at},G_{pt})$$

where $Var(G_{ot})$, $Var(G_{at})$, $Var(G_{pt})$ are the variances of growth rates of output, area and productivity respectively and $Cov(G_{at},G_{pt})$ is the covariance of growth rates of area and productivity. $Var(G_{ot})$, $Var(G_{at})$ and $Var(G_{pt})$ have been calculated directly from the original series of production, area and productivity respectively and the other component [i.e., 2 $Cov(G_{at}, G_{pt})$] has been estimated indirectly through residual method [i.e., subtracting Var (G_{at}) and Var (G_{pt}) from Var (G_{ot})]. But the residual component does contain not only the effect of synchronous movements of area and productivity variabilities; but also the error term that arises from the instability decomposition model of Rao *et al.* However, to minimize the influence of error term we have taken the following steps :

(i) In the estimation of variances (for the whole period and two sub-periods), the trend growth rates (of the whole period and two sub-periods) have been used in place of the mean values to minimize the problem of underestimation or overestimation in output instability (as defined here alternatively) as well as its components. But the problem of underestimation/overestimation will not be totally avoided as each series does not precisely follow the log-linear trend.

(ii) For further improvement in these summary measures (i.e., variances) we have excluded those years from the calculations, that contain strong outlier in the point-to-point growth rates of the production series. Logically, strong outlier should be replaced by the next higher value (to avoid underestimation) and that should be detected from each series of area and productivity as well as production to get more accurate estimation of series-wise variabilities. But these could not be taken into consideration, because while doing so, other problems would crop up. For instance, in case of replacement of the outlier of a particular year, it is difficult to determine how much will be replaced for the other component's series that does not have any outlier for that year. Similarly, in case of detection of outlier from each series, quite a number of years would be deleted from the calculation and that leads to distortion in deriving true inference.

Therefore, even with these improvements (over the model of Rao *et al.*), some sort of inaccuracies in the instability decomposition model is unavoidable and these have been considered along with 2 Cov (G_{at}, G_{pt}) in the form of residual component. However, the actual magnitude of error term has also been estimated (Section 5.3, Chapter 5). Like growth decomposition, the correlation coefficients (both inter-and intra-temporal and cross-sectional over districts and crops) among the instability components have been calculated to summarize the instability decomposition results.

Lastly, to ascertain the nature of association between growth and instability (crop-wise and district-wise), we have used both tabular classifications and estimated correlation coefficients between:

(i) estimated growth rates and levels of instability (only for measuring correlation coefficients) and
(ii) estimated trend breaks in sub-period growth rates and the nature of instability.

3.6 Methods for Estimating Growth and Stability-Contributing Factors

For the determination of growth contributing factors, we have regressed productivity index both cross-sectionally (over districts at 3 time points) and over time (taking productivity index of all crops combined of West Bengal for the period 1977-78 to 1994-95) on some selected explanatory variables, as mentioned in the data base section. The principal component analysis, based on Kaiser's criterion, has been made (i) to club the factors belonging to the same group, thus avoiding the problem of multicollinearity among the explanatory variables [Koutsoyiannis (1977) Pp. 426-436] and (ii) to reduce the number of explanatory variables which are quite large in comparison to the number of sample observations. Moreover, because of non-availability of data on some of the explanatory variables used in the model under study, these variables have been approximated by some other closely related variables (see section 3.2 above), on which observations have been available. These variables, thus, suffer from measurement error and in this case, estimation of parameters of the approximated variables by the ordinary least squares method is not advisable. This problem can be solved, to a large extent, by clubbing the variables together through the principal component method. By this method the technological factors have been clubbed into a single factor and the institutional factors, excepting average farm size, into another factor. Average farm size has been used as a separate explanatory variable. To determine the causes behind instability in agricultural output, we have regressed $|Z_t|$ (as defined earlier) on some selected independent variables (explained in the data base section) using both time series and cross-section data. In these multiple regression analyses the multicollinearity problem has been taken care of by dropping the explanatory variable whose influence on the dependent variable is insignificant and which is collinear with the important explanatory variable(s). Importance of an explanatory variable has been determined apriori on the basis of measured partial correlation coefficient between it and the dependent variable.

The influence of rainfall on agricultural production in West Bengal has also been considered. Good rainfall has direct bearing upon agricultural production. Indirectly, it also helps agricultural production by raising the underground water level and thus ensuring the supply of water to the agricultural field during low rainfall season through various irrigation system. In a recent study, mentioned earlier, Saha and Swaminathan (1994) have noted that there was no statistically significant effect of rainfall on agricultural production in West Bengal during 1980s, despite the fact that there was good rainfall in most of the years during the period [Datta Roy (1994)]. This result of Saha and Swaminathan may be attributed to either (i) small sample size during which the variability in the rainfall was negligible or (ii) considering the variable as a non-stochastic one. Though in earlier works rainfall has been assumed to be non-stochastic, it should be instead considered as a stochastic variable since a large part of it is determined by the factors like humidity, temperature etc., which are in turn determined by afforestation/

deforestation, industrial and other pollutions, etc.

Moreover, the variable rainfall, as measured in millimeter, does not mean much as we want to measure its impact on agricultural production. It is the water in millimeter per hectare at different times that is important, not the rainfall as conventionally measured and used. Further, the approximation to total rainfall for the State or the district by summing up of rainfall of certain areas is also bound to suffer from measurement error.

These features of the variable render the estimates of coefficients (by ordinary least square method) both biased and inconsistent. To overcome this problem Durbin's ranking method [Koutsoyiannis (1977), p. 269] has been applied in regressing agricultural production (in absolute deviational form from trend values) on rainfall (in absolute deviational form from normal level) for all crops combined in West Bengal during 1957-58 to 1994-95 (along with sub-periods) for the determination of the later's effect on the former. The results have conformed to our expectations.

The effect of population growth on the growth of agricultural output (or per capita agricultural output) has been studied by researchers on the basis of three well known theories of population: Neo-Malthusian theory, Induced Innovation theory and Labour Surplus theory. According to Neo-Malthusian theory, population growth retards the growth of agricultural output owing to diversion of public and private resources from productive investment to consumption and also owing to degradation of the environment and agrarian structure. On the other hand, according to 'induced innovation theory', population growth boosts up the agricultural growth in the long-run by favourably inducing the technological and institutional changes. Under the operation of 'labour surplus theory', there will be no impact of the population growth on the growth of agricultural output. Agricultural growth may also affect the population growth in different ways. Here, however, to determine the effect of rural population growth on the growth of agricultural output and to discard the reverse causal relationship the growth of agricultural output has been regressed on the lagged values of the former (exact lag structure is determined by trial and error method) cross-sectionally (i.e., over districts) for the whole period and also for two sub-periods. Along with the growth of agricultural output, the per capita growth of agricultural output has been used as a dependent variable. For the analysis of this sub-section we are highly indebted to Boyce (1987, Pp. 138-159) in regard to the reasoning and adopted methodology of the analysis.

3.7 Methods of Equity Measurement and other Related Issues

In regard to the measurement of equity we have proceeded in the following way:

(i) The trend growth rates (by fitting exponential trend equation) of money and real wage rates of male and female agricultural labourers have been estimated over the districts during the period 1978-79 to 1991-92.

(ii) The inequalities in the distributions of ownership land holding and monthly per capita consumer expenditure among rural inhabitants of West Bengal over different NSS rounds have been measured by Gini coefficients, using the formula of Yntema and Morgan,

$$G = 1 - \sum_{t=1}^{T+1} f_t (q_t + q_{t-1})$$

[which has been explained in section 2.4, Chapter 2]. The linear trends have been estimated for (a) Gini coefficients of land distributions, (b) Gini coefficients of consumption distributions, (c) Shares of percentile distributions (lower 40%, middle 50% and upper 10%) of rural population in consumption distributions and (d) head count ratios of rural poverty to know their relative changes over time.

(iii) Actually, the formula, given by Yntema and Morgan, estimates Gini lower bound (GL) where the inequality within class is assumed to be zero. On the other hand, as derived by Gastwirth (1972), the Gini upper bound (GU) is obtained by maximizing the spread within each class and

$$GU = GL + \tilde{D} \text{ where } \tilde{D} = 1/\mu \sum_{t=1}^{T+1} f_t^{\,2} (\Delta x_t)\, \delta_t (1 - \delta_t).$$

Here $\Delta x_t = x_t - x_{t-1}$ and $\delta_t = (\mu_t - x_{t-1}) / \Delta x_t$

[μ =grand mean, δ_t = mean of the t th class and x_t = upper limit of the t th class].

These GL and GU act as test of goodness of fit for true Gini coefficient (G) which lies between GL and GU [Gastwirth and Smith (1972)]. The formula of true Gini coefficient, as derived by Kakwani (1980), is :

$$G = GL + (1/\mu) \sum_{t=1}^{T+1} f_t^{\,2} \mu_t G_t$$

where G_t is the Gini index for t th class. Using third degree polynomial Lorenz curve and the Pareto density function for the first and last open ended classes [which are also assumed by Kakwani (1980) as reasonable], one can derive

$G_1 = (x_1 - \mu_1)/(x_1 + \mu_1)$, $G_t = (1/6)\,[(\Delta x_t)/(\mu_t)]$ [for t = 2,3, ..., T]
and $G_{t+1} = (\mu_{T+1} - x_T) / (\mu_{T+1} + x_T)$

GL and G have been estimated from the classified primary data on the distributions of ownership land holding, consumption and income for the proper estimation of inequalities in these distributions.

(iv) Further to estimate the contributions and marginal effects of each income component on overall measured inequality in income distribution of the sample households in the sample villages taken together, a decomposition analysis of Gini coefficient, as derived by Lerman and Yitzhaki (1985) has been made. This can be explained briefly as follows:

Let Y_k be the kth income component, F be the cumulative distribution of income, G_k the Gini coefficient of income component k, S_k the kth source's share in total income, R_k be termed as "Gini correlation" [$R_k = cov(Y_k, F)/G_k$] and G be the true Gini coefficient in the distribution of total income. Then the contribution of kth income source towards overall income inequality is given by I_k

where $I_k = (R_k . G_k . S_k)/G$

And the relative marginal effect of kth income source to overall Gini is given by,

$$(dG/dl_k) / G = I_k - S_k$$

where l_k = percentage change in income from the kth source. Lastly, we have inquired into the nature of association among growth, stability and equity in the following ways:

(i) To know how far agricultural wage rates depend on the performance of agriculture, both money and real wage rates of agricultural male labourers have been regressed on one year's lagged values of agricultural production (all crops combined) for each district during the period 1977-78 to 1991-92. But from this time series regression result it is difficult to properly evaluate the dependence of agricultural wage rate on agricultural production when both the series increase overtime. For proper evaluation we have also considered the cross-sectional study in which districtwise growth rates in agricultural wages (money and real wages of agricultural male labourers) have been regressed on second sub-period's districtwise growth rates and levels of instability in agricultural output.

(ii) We have also performed listwise (i.e., omitting the missing data points) regression of logarithms (base e) of Gini coefficients for rural consumption distributions on the three year lagged moving averages (t period's value comes from the average of t-2, t-1 and t period's values) of both trend values and $|X_t|$ s (as defined earlier for the measurement of instability) of agricultural production in West Bengal during the period for which data are available for estimation of Gini coefficients . It is here expected that the moving averages of the explanatory variables would smooth out short term variations and also capture the lagged effect of agricultural performance on the Gini coefficients for consumption distribution.

(iii) Similarly, the head count rural poverty ratios have been regressed on the three year lagged moving averages of both trend values and $|X_t|$ s of agricultural production of West Bengal to empirically test the trickle down theory. However, unlike in the earlier studies, the price variable (CPIAL) has not been considered as an explanatory variable in our study, following the criticism of Ahluwalia (1986). The Gini coefficient for consumption has not also been considered as an explanatory variable in order to avoid the problem of multicollinearity. Further, the time variable is not used explicitly in the regression analysis, which was used by Ahluwalia, since the time factor has already been considered in the estimation of the trend values of agricultural production.

To summarize the above discussion it may be said that in the earlier studies various econometric problems relating to the estimation of trend coefficients have not been given due attention. As a result, the statistical inferences as regards the statistical significance of the growth rates implying significant rise or fall or constancy in the inter-temporal growth rates suffer from either upward or downward bias. Therefore, on the basis of these results no impartial judgment can be passed on the performance of agriculture during a period or on its performance relative to the same in some other period or a region or a State. Such statistical inference are to be taken with scepticisms.

Keeping these problems in mind various methods have been suggested in this study that will help us to draw inferences that will be free from biases—whether these inferences relate to the growth rates or the nature of instability or the association between growth rates and instability or the levels of equity.

In the following chapters results derived from the application of the proposed methods are analysed.

Table 3.1 : Productivity Index and Technological Factors in West Bengal, 1977-78 — 1994-95.

Year	Productivity Index a	Fertilizer Intensity b	HYV seed Intensity c	Irrigation Intensity d	Estimated Irrigation Intensity e
1977-78	114.7	21.786	0.2522	0.1293	0.2547
1978-79	111.5	35.059	0.2965	0.1533	0.2051
1979-80	101.8	34.586	0.2531	0.1224	0.2267
1980-81	114.9	39.414	0.2529	0.1418	0.2185
1981-82	98.7	37.135	0.2583	0.1300	0.2096
1982-83	94.3	39.703	0.3022	0.0874	0.1892
1983-84	121.7	51.189	0.3244	0.1356	0.2455
1984-85	126.2	56.992	0.3169	0.1390	0.2253
1985-86	127.7	57.011	0.3216	0.1391	0.2661
1986-87	123.6	67.650	0.3577	0.1363	0.2681
1987-88	129.7	74.591	0.4065	0.1381	0.2908
1988-89	145.6	87.302	0.4145	0.1365	0.2844
1989-90	148.63	89.208	0.4381	0.1497	0.2822
1990-91	141.17	97.429	0.4562	0.1419	0.3013
1991-92	159.18	97.851	0.4688	0.1372	0.3099
1992-93	153.52	96.431	0.4875	0.1294	0.2989
1993-94	158.27	95.574	0.5155	0.1404	0.3129
1994-95	164.66	97.454	0.5826	0.1357	0.2935

Notes :
a– Productivity index for all crops combined of W.B. adjusted to the common base year triennium ending 1969-70.
b– Fertilizer (in kilograms of N+P+K) per hectare of gross cropped area (GCA).
c– Area under HYV divided by GCA.
d– Area under irrigation as proportion of GCA
e– See text.

Source : Different issues of Economic Review & Statistical Abstract, G.O.W.B., Alipore; WBGP.

Table 3.2 : Productivity Index and its related Inputs over the Districts of West Bengal, 1970-71

District	Adjustment Factor for Productivity Index	Productivity Index	Technological Factors				Institutional Factors			
			% of GCA Irrigated	Fertilizer Intensity	% of GCA under HYV	Agri. Machinery Intensity	Avg.Size of Operational Holding	% of Agri. Labourers	% of NCA under Tenancy	% of NCA under Surplus Land Distributed
			a	b	c	d	e	f	g	h
Burdwan	1.000	99.15	66.7	21.2	34.69	0.2402	1.72	55.79	7.39	1.6
Birbhum	0.838	84.12	63.4	21.8	44.32	0.1384	1.94	53.30	10.68	1.3
Bankura	0.876	87.17	45.9	15.8	45.65	0.0296	1.58	48.85	6.04	2.2
Midnapore	0.793	76.32	16.7	8.1	8.33	0.0440	1.33	42.82	8.97	4.03
Howrah	0.994	92.08	24.4	23.9	49.31	0.3657	1.00	62.89	10.92	0.66
Hooghly	1.155	114.29	45.9	42.9	35.71	0.0671	1.11	54.83	6.98	0.79
24-Parganas	0.702	69.91	8.5	10.9	27.58	0.1598	1.21	53.19	13.20	1.7
Nadia	0.577	58.48	10.8	11.9	18.42	0.1252	1.43	42.21	7.10	0.8
Murshidabad	0.641	64.20	20.4	9.4	40.69	0.0597	1.32	47.79	6.86	1.2
W. Dinajpur	0.576	57.66	5.1	2.4	7.71	0.0322	1.66	33.11	12.64	3.7
Malda	0.482	47.99	13.1	8.3	24.24	0.0871	1.40	42.68	9.59	3.7

(Contd.)

Table 3.2 : *(Contd.)*

District	Adjustment	Productivity	Technological Factors				Institutional Factors			
	Factor for Productivity Index	Index	% of GCA Irrigated	Fertilizer Intensity	% of GCA under HYV	Agri. Machinery Intensity	Avg.Size of Operational Holding	% of Agri. Labourers	% of NCA under Tenancy	% of NCA under Surplus Land Distributed
			a	b	c	d	e	f	g	h
Jalpaiguri	0.688	68.82	6.3	3.7	0	0.1382	1.53	20.45	17.04	5.6
Darjeeling	0.685	68.39	38.9	15.5	0	0	1.20	23.05	5.29	4.2
Coochbehar	0.704	70.61	2.3	2.0	16.12	0.0474	1.04	18.67	11.30	3.6
Purulia	0.601	61.01	26.7	4.7	22.07	0.1013	1.25	42.88	0.80	2.7

Key : GCA = Gross Cropped Area, HYV = (area under) High Yielding Varities, NCA = Net Cropped Area.

Sources : a– Boyce, 1987, Table 6.1, page 168, data for the year 1971-72;

b– Boyce, 1987, Table 6.5 page 178, data for the year 1970-72.

c– Agricultural census, 1970-71 for estimates of Area under HYV and Statistical Abstract (1976) for GCA in the year 1970-71;

d– Statistical Abstract (1976) for agricultural machinery, data relate to the year 1972;

e– Statistical Abstract (1976) for NCA in the year 1970-71 & Census, 1971 (GOI) for total number of cultivators in the year 1970-71.

f– Census, 1971, GOI.

g– Agricultural census 1970-71, Govt. of W.B.

h– Calculated from Ratan Ghosh and K Nagraj (1978), P. 58 and Board of Revenue, yearly publications. Data relate to the year 1972.

Table 3.3 : Productivity Index and its Determinants Over the Districts of West Bengal, 1980-81

District	Productivity Index	Technological Factors				Institutional Factors			
		% of GCA Irrigated	Fertilizer Intensity	% of GCA under HYV	Agri. Machinery Intensity	Avg.Size of Operational Holding	% of Agri. Labourers	% of NCA under Tenancy	% of NCA under Surplus Land Distributed
		a	b	c	d	e	f	g	h
Burdwan	119.62	81.3	59.6	51.26	1.4519	1.46	56.79	16.66	3.141
Birbhum	85.26	58.1	42.9	42.80	0.1664	1.50	50.35	13.67	2.507
Bankura	76.62	40.8	33.7	26.14	0.0608	1.20	45.82	8.91	4.733
Midnapore	78.19	28.6	24.1	24.10	0.1328	1.09	41.73	10.19	8.301
Howrah	90.33	42.1	176.8	27.70	1.1664	1.19	58.50	18.44	1.048
Hooghly	148.28	62.1	101.5	45.52	2.3340	0.95	53.74	15.85	1.419
24-Parganas	76.51	21.5	33.4	24.47	0.3642	1.12	34.36	15.15	3.204
Nadia	72.93	23.4	49.3	25.25	0.1954	1.30	47.78	7.68	1.480
Murshidabad	70.92	52.0	33.8	32.56	0.0457	1.21	48.65	8.51	2.237
W. Dinajpur	53.61	18.0	18.7	20.25	0.0428	1.37	42.31	10.82	7.614
Malda	64.06	21.2	31.9	26.68	0.0905	1.27	46.20	15.68	7.182

(Contd.)

Table 3.3 : *(Contd.)*

District	Productivity	Technological Factors				Institutional Factors			
	Index	% of GCA Irrigated	Fertilizer Intensity	% of GCA under HYV	Agri. Machinery Intensity	Avg.Size of Operational Holding	% of Agri. Labourers	% of NCA under Tenancy	% of NCA under Surplus Land Distributed
		a	b	c	d	e	f	g	h
Jalpaiguri	70.66	7.7	11.2	20.45	0	1.38	32.84	35.02	11.155
Darjeeling	78.10	57.5	47.6	19.70	0	1.53	25.32	12.69	5.781
Coochbehar	62.48	11.1	14.0	23.22	0.0394	0.97	34.37	23.17	7.421
Purulia	57.92	17.1	18.7	23.48	0.0490	0.91	32.94	1.01	5.607

Key : GCA = Gross Cropped Area, HYV = (Area under) High Yielding Varities, NCA = Net Cropped Area.

Sources : a & b– Boyce, 1987, Tables 6.1(page 168) & 6.5 (page 178); data for a & b relate to the years 1980-81 & 1978-80 respectively.

c & d– Calculated from Boyce, 1987 (Table 6.5) & Economic Review, 1981-82 and Statistical Abstract, 1985-86. Data relate to the year 1980-81 for c & 1982 for d.

e & f– Census, 1981 Govt. of India & Statistical Abstract, Govt. of W.B., 1985-86. Data relate to the years 1980-81 for e & 1981 for f.

g– Datta, 1981, pages 21-22 & D.O.A., W.B., 1980-81; data for 1981.

h– Economic Review & Board of Revenue, yearly, data for 1981.

Table 3.4 : Productivity Index and its Determinants Over the Districts of West Bengal, 1990-91

District	Productivity Index	Technological Factors				Institutional Factors			
		% of GCA Irrigated	Fertilizer Intensity	% of GCA under HYV	Agri. Machinery Intensity	Avg.Size of Operational Holding	% of Agri. Labourers	% of NCA under Tenancy	% of NCA under Surplus Land Distributed
		a	b	c	d	e	f	g	h
Burdwan	178.79	80.31	125.24	74.51	4.1356	1.21	58.46	9.42	4.53
Birbhum	127.88	65.46	117.75	54.13	0.7641	1.15	53.60	14.33	4.31
Bankura	140.69	41.31	67.98	40.56	0.5278	1.01	45.21	7.13	5.39
Midnapore	136.86	37.49	75.47	42.58	0.8203	0.78	37.61	5.73	10.34
Howrah	148.12	64.08	279.51	54.87	1.7895	0.69	57.91	12.80	1.56
Hooghly	223.99	64.05	206.79	67.91	4.0232	0.81	55.76	10.50	1.94
24-Parganas	118.52	26.11	96.62	42.74	0.7752	0.88	34.80	7.98	4.43
Nadia	124.44	36.60	89.33	46.52	0.7799	0.92	48.36	6.18	2.15
Murshidabad	116.01	56.52	81.69	48.66	0.2654	0.91	48.20	6.42	3.42
W.Dinajpur	90.27	13.81	61.00	39.90	0.2282	1.02	45.12	6.77	9.61
Malda	95.38	33.01	101.12	36.61	0.1075	0.99	50.04	11.06	9.48

(Contd.)

Table 3.4 : *(Contd.)*

District	Productivity	Technological Factors				Institutional Factors			
	Index	% of GCA Irrigated	Fertilizer Intensity	% of GCA under HYV	Agri. Machinery Intensity	Avg Size of Operational Holding	% of Agri. Labourers	% of NCA under Tenancy	% of NCA under Surplus Land Distributed
		a	b	c	d	e	f	g	h
Jalpaiguri	84.14	11.68	53.63	25.36	0.3251	1.24	36.60	12.09	13.62
Darjeeling	95.76	61.01	110.62	39.26	0	1.32	32.32	4.89	7.68
Coochbehar	84.06	17.21	78.10	27.09	0.1507	0.78	35.09	13.40	8.67
Purulia	81.38	9.76	82.26	31.55	0.1609	0.85	33.97	0.96	8.03

Key : Same as those in table 3.3

Sources : Statistical Abstract 1995, Economic Review, annual issues, Census(GOI) 1991 and Statistical Bulletins (yearly), District Statistical Offices, D.O.A., W.B.

Notes : Data relate to the year (s):
a– 1990-91, b– 1990-91, c– 1990-91, d– 1989, e– 1990-91, f– 1991, g– 1994-95 and h– 1991.

Table 3.5 : Inter-Census Rates of Growth of Rural Population, West Bengal by Districts, 1921-91.

(percent per annum)

District	1921-31	1931-41	1941-51	1951-61	1961-71	1971-81	1981-91
(1)	(2)	(3)	(4)	(5)	(6)	(7)	(8)
Burdwan	0.94	1.46	1.13	3.00	1.81	1.12	1.50
Birbhum	1.07	0.91	0.10	2.98	2.05	1.51	2.10
Bankura	0.86	1.38	0.22	2.30	1.98	1.55	1.73
Midnapore	0.48	1.21	0.34	2.55	2.39	1.89	2.17
Howrah	0.97	2.29	0.26	1.08	1.46	1.43	1.55
Hooghly	0.32	1.84	0.99	3.12	2.46	1.69	1.96
24-Parganas	0.82	2.01	1.48	2.78	2.46	1.79	2.95
Nadia	0.13	1.30	2.58	4.00	2.59	2.55	2.82
Murshidabad	1.13	1.73	0.39	2.81	2.51	2.21	2.67
West Dinajpur	0.70	1.02	1.19	2.52	3.19	2.36	2.69
Malda	0.48	1.53	0.99	2.61	2.77	2.27	2.67
Jalpaiguri	0.63	1.28	0.36	3.76	2.48	1.81	2.30
Darjeeling	1.23	1.42	0.99	3.13	2.27	1.87	2.18
Coochbehar	-0.02	0.69	0.11	4.23	3.30	2.24	2.13
Purulia	1.57	0.97	0.61	1.50	1.48	1.38	1.94

Sources : Boyce (1987), Table 5.1, Page139, for columns 2 to 7 and for column 8, Census reports, 1980-81 & 1990-91, Govt. of India.

Table 3.6 : Factors determining Output Instability, West Bengal, 1957-58 — 1988-89

Year	% of NCA under current fallow	Crop base a	Locational base b	Year	% of NCA under current fallow	Crop base a	Locational base b
1957-58	10.37	76.25	40.93	1973-74	1.6	70.56	40.93
1958-59	10.68	73.91	37.74	1974-75	1.69	68.14	41.93
1959-60	8.63	72.52	38	1975-76	1.46	67.66	43
1960-61	7.03	72.52	42.06	1976-77	2.34	69.57	41.43
1961-62	6.24	73.74	41.21	1977-78	1.23	71.46	41.34
1962-63	6.96	74.39	40.61	1978-79	1.16	70.27	45.13
1963-64	5.94	73.68	42.13	1979-80	3	71.9	43.32
1964-65	5.01	73.9	41.69	1980-81	1.47	74.87	45.32
1965-66	6.03	74.28	39.47	1981-82	1.54	76.1	41.09
1966-67	6.71	72.91	41.31	1982-83	3.1	74.97	44.06
1967-68	5.96	72.85	40.53	1983-84	1.86	75.57	44.65
1968-69	7.9	69.32	39.98	1984-85	1.14	77.19	44.36
1969-70	3.91	71.16	40.04	1985-86	1.21	79.45	42.39
1970-71	6.63	69.53	40.01	1986-87	6.45	77.5	44.05
1971-72	3.21	69.88	40.43	1987-88	6.74	77.64	47.6
1972-73	2.82	70.89	42.55	1988-89	7.39	78.94	46.61

Sources: Statistical Abstract (various issues) and Economic Review (annual publications).

Notes : a– percentage of GCA under 5 principal crops, *viz.*, aman rice, boro rice, jute, potatoes and mustard.

b– percentage share of main 4 districts (*viz.*, Burdwan, Midnapore, Hooghly and 24-Parganas) in State's total agricultural production.

Table 3.7 : Factors Determining Output Instability, West Bengal by Districts, 1970-71.

District	% of NCA under current fallow	Crop base a	Locational base b	District	% of NCA under current fallow	Crop base a	Locational base b
Burdwan	9.56	83.92	11.87	Murshidabad	2.87	43.23	9.23
Birbhum	3.11	60.83	9.70	West Dinajpur	0.45	65.60	7.68
Bankura	14.39	81.23	6.82	Malda	7.98	41.59	4.51
Midnapore	6.71	84.10	13.65	Jalpaiguri	0.48	64.14	4.38
Howrah	29.81	82.59	1.73	Darjeeling	1.84	37.28	0.71
Hooghly	12.50	82.15	6.23	Coochbehar	0.72	66.43	4.84
24-Parganas	7.74	84.28	8.26	Purulia	13.19	84.18	3.63
Nadia	5.17	35.38	6.76				

Sources : Statistical Abstract and Economic Review (different publications), G.O.W.B., Calcutta.

Notes : a– Same as those in Table 3.6.

b– Percentage share of the district in State's total agricultural production.

Table 3.8 : Factors Determining Output Instability, West Bengal by Districts, 1980-81.

District	% of NCA under current fallow	Crop base a	Locational base b	District	% of NCA under current fallow	Crop base a	Locational base b
Burdwan	2.75	88.00	12.16	Murshidabad	0.63	50.15	8.89
Birbhum	1.01	82.77	6.52	West Dinajpur	0.43	67.31	6.68
Bankura	2.88	83.03	6.86	Malda	0.33	55.58	3.63
Midnapore	2.04	85.70	14.48	Jalpaiguri	0.67	61.93	4.02
Howrah	0.74	79.68	2.04	Darjeeling	0.26	35.80	0.75
Hooghly	0.49	87.30	9.07	Coochbehar	0.07	56.28	4.52
24-Parganas	0.15	87.97	9.61	Purulia	10.20	87.42	3.18
Nadia	0.38	51.73	7.59				

Sources and Notes : Same as those in Table 3.7

Table 3.9 : Factors Determining Output Instability, West Bengal by Districts, 1990-91

District	% of NCA under current fallow k	Crop base a	Locational base b	District	% of NCA under current fallow k	Crop base a	Locational base b
Burdwan	3.76	92.30	13.17	Murshidabad	1.53	64.97	8.05
Birbhum	3.69	89.62	6.26	West Dinajpur	3.60	82.61	6.29
Bankura	7.73	84.86	7.56	Malda	10.58	60.71	4.08
Midnapore	2.93	88.59	15.15	Jalpaiguri	0.88	59.53	3.28
Howrah	10.48	94.16	2.12	Darjeeling	6.84	34.56	0.60
Hooghly	1.63	98.14	9.14	Coochbehar	1.05	68.33	4 01
24-Parganas	4.86	89.74	9.37	Purulia	25.92	84.43	3.90
Nadia	1.88	65.63	7.02				

Sources and Notes : Same as those in Table 3.7
k– data relate to the year 1994-95.

Table 3.10 : Estimated Annual Rainfall Index, West Bengal, 1957-58 to 1994-95.

Year	Rainfall Index	Year	Rainfall Index	Year	Rainfall Index	Year	Rainfall Index
1957-58	76.6	1967-68	87.0	1977-78	90.6	1987-88	113.4
1958-59	89.1	1968-69	111.0	1978-79	91.6	1988-89	108.7
1959-60	117.9	1969-70	98.7	1979-80	80.5	1989-90	109.9
1960-61	90.2	1970-71	106.3	1980-81	109.9	1990-91	115.1
1961-62	83.4	1971-72	136.3	1981-82	105.9	1991-92	103.6
1962-63	85.9	1972-73	77.9	1982-83	78.2	1992-93	86.9
1963-64	90.8	1973-74	95.8	1983-84	95.3	1993-94	114.6
1964-65	98.2	1974-75	74.7	1984-85	120.9	1994-95	101.7
1965-66	85.8	1975-76	62.0	1985-86	89.3		
1966-67	73.2	1976-77	64.9	1986-87	112.5		

Source and Estimation Procedure : See text.

Table3.11: Estimates of Rural Head Count Poverty Ratio in West Bengal

Year	Head Count Ratio	Year	Head Count Ratio
1977-78	68.3	1989-90	37.2
1983	63.1	1990-91	49.5
1986-87	47.3	1992	44.0
1987-88	48.3	1993-94	40.3

Source : A. Sen (1996), EPW, Spl. No., Table 7, Page 2466.

Table 3.12 : Size Distribution of Total Annual Income of the Rural Households

Income Class (Rs.)	Sample Village I		Sample Village II		Sample Village III	
	Mean Income (Rs.)	No. of Households	Mean Income (Rs.)	No. of Households	Mean Income (Rs.)	No. of Households
Below 8,500	7,122.00	03	6,780.00	02	5,921.10	13
8,500-15,000	11,918.00	12	11,300.70	10	9,672.52	11
15,000-25,000	21,431.15	08	21,776.00	07	16,984.00	07
25,000-35,000	31,668.21	05	32,212.13	08	28,563.14	06
35,000-50,000	42,123.00	03	40,838.13	08	38,312.00	03
50,000-75,000	66,200.16	06	62,152.85	07	58,834.50	05
75,000-1,00,000	86,452.11	09	80,210.00	06	82,421.20	03
1,00,000 & above	1,22,342.00	04	1,27,057.00	05	1,11,322.00	02
Grand mean / Total	45,703.58	50	45,566.34	53	27,053.10	50

Table 3.13 : Size Distribution of Total Annual Consumption of the Rural Households

Total Consumption Class (Rs.)	Sample Village I		Sample Village II		Sample Village III	
	Mean Consumption (Rs.)	No. of Households	Mean Consumption (Rs.)	No. of Households	Mean Consumption (Rs.)	No. of Households
Below 7,000	6,392.50	01	6,040.50	02	6,012.00	11
7,000-10,000	9,482.00	08	8,900.00	02	8,321.50	03
10,000-13,000	11,633.50	03	11,187.64	05	10,510.50	07
13,000-23,000	19,652.00	09	16,985.29	09	16,612.00	10
23,000-33,000	31,876.50	04	29,981.14	07	28,657.50	06
33,000-43,000	39,837.50	07	34,500.94	11	33,752.00	03
43,000-53,000	50,432.00	14	49,394.00	12	50,138.00	08
53,000 and above	67,374.50	04	65,532.70	05	62,668.50	02
Grand mean / Total	33,518.63	50	32,989.75	53	22,608.64	50

Table 3.14:Size Distribution of Cultivable Land Holdings of the Rural Households

Land Holding Class (in acre)*	Sample Village I		Sample Village II		Sample Village III	
	Avg. Land Holding (in acre)	No. of Households	Avg. Land Holding (in acre)	No. of Households	Avg. Land Holding (in acre)	No. of Households
Below 0.5	0.16	15	0.22	15	0.31	10
0.5–1.5	0.79	08	0.99	04	0.93	15
1.5–3	2.58	13	2.44	09	2.58	06
3–5	4.05	04	3.85	12	4.55	04
5–7	6.35	06	5.72	02	6.40	07
7–11	10.50	03	10.08	09	9.75	06
11 & above	12.00	01	11.50	02	14.50	02
Grand mean / Total	2.796	50	3.785	53	3.661	50

Note : * One acre is equivalent to 0.4048 hectare.

Table 3.15 : Different Approaches of Trend[1] Estimation and their Test Results, West Bengal, All Crops Combined, 1957-78 — 1994-95

Approaches of trend estimation ↓	Estimated results →	$\bar{R}^2$ [F Value]	DW	Rank	∂_e (S.D. of residuals)	Runs	Estimated growth rate[c] [percent per annum]
Unadjusted series		.9195* [212.299]	1.381	-.3710**	.0908	17	2.821* (20.418)
Inclusion of rainfall in trend		.9233* [149.539]	1.270	-.3017***	.0874	18	2.746* (19.296)
Omission of strong outliers[a]		.9483* [330.964]	1.747+	-.3992*	.0735	17	2.870* (25.531)
3-year moving average on the original series		.9580* [400.139]	0.437	.0530	.0603	09*	2.787* (27.988)
Proposed technique [b]		.9894* [634.238]	1.843+	.1552	.0296	16	2.684* (53.682)

Notes : [1] Using Log- quadratic trend equation in all cases,
[a] outside the 5% band limit, [b] adjustment through 3-year centred moving average of the residuals followed by second order auto-regressive scheme (for the correction of autocorrelation), [c] t values within parentheses, * significant at 1% level, ** significant at 5% level, *** significant at 10% level, + non-autocorrelated at 5% level, significant rank and runs imply respectively heteroscedasticity and non-random disturbance term.

Table 3.16 : Estimated Sub-period Growth Rates and Test Results of Different Approaches, All Crops Combined of West Bengal, 1957-58—1976-77 & 1977-78—1994-95

Approaches of estimating sub-period growth rates ↓	Estimated results →	$\bar{R}^2$ (F Value)	DW	Rank	∂_e	Runs	Growth rates (t value): First Sub-period (1957-58-1976-77)	Growth rates (t value): Second Sub-period (1977-78-1994-95)
Separate exponential trend for each sub-period [a]	1st Sub-period	.8111* (85.562)	1.666+	.1699	.0688	09	2.491* (9.086)	4.097* (8.234)
	2nd Sub-period	.7971* (67.794)	1.306	-.5789*	.1063	10		
Boyce's kinked exponential trend [a]		.9189* (210.766)	1.374	-.3098**	.0911	20	2.122* (7.271)	3.641* (10.995)
Proposed method [b]		.9883* (1474.109)	1.769+	-.0227	.0312	20	2.081* (18.824)	3.649* (29.441)

Notes: [a] Fitted to the unadjusted data, [b] kinked exponential estimated on adjusted data, + implies non-autocorrelated disturbance term, *&*** imply significant at 1% and 10% level respectively.

Table 3.17 : Extended Decomposition Analysis of Growth Rates in Total Agricultural Output in West Bengal, 1957-58 to 1994-95

Period	Component Growth Rates [a]						Interaction Term [b]	
	Net Sown Area (1)	Cropping Intensity (2)	Gross Cropped Area (3)=(1)+(2)	Yield (4)	Cropping Pattern (5)	Productivity (6)=(4)+(5)	For unadjusted output growth rates	For adjusted output growth rates
Whole period i.e., 1957-58 to 1994-95	-.013 (.0004)	.758* (.0004)	.745* (.0005)	.967* (.0012)	.587* (.0003)	1.554* (.0013)	.522	.385
First Sub-period i.e., 1957-58 to 1976-77	.278* (.0005)	.790* (.0009)	1.068* (.0009)	.314 (.0022)	.339* (.0005)	.653* (.0021)	.770	.360
Second Sub-period i.e., 1977-78 to 1994-95	-.355* (.0006)	.720* (.0011)	.365* (.0011)	1.731* (.0025)	.879* (.0006)	2.610* (.0024)	1.122	.674

Notes : Standard errors within parentheses,
* implies significant at 1% level, [a]Component growth rates have been calculated from exponential trends, [b]Interaction term is

4

Measurement of Agricultural Growth

This chapter is devoted to the measurement of growth of agricultural production in West Bengal during the period 1957-58 to 1994-95 and for the sub-periods 1957-58 to 1976-77 and 1977-78 to 1994-95. The growth of agricultural production has then been decomposed into the growth of area and productivity during the whole period as well as the sub-periods. The contributions of individual crops as well as the districts' contributions to the State's overall growth in crop output have also been determined.

4.1 Choice of Trend Equation

From the estimated linear trends on the smoothed values of point-to-point (i.e., annual) growth rates (calculated from the original production series), presented in Table 4.1, it is seen that in 28 cases (out of total 82 cases : 22 crops/crop-groups of West Bengal and 4 major crop-groups in each of 15 districts) the trend coefficient (i.e., $\hat{b}$) remains statistically insignificant, which implies constancy of the annual growth rates over time. Hence, in these 28 cases, exponential trend equation may be chosen. But in some cases exponential trend equation gives poor fit in terms of the low values of either DW statistic or runs or $\bar{R}^2$ (the results reported in Table 4.1). Therefore, other alternative forms of trend equations have been fitted to all these 28 crop production series (results not reported here). It is seen that the situations either remain unaltered or deteriorate, except in two cases, *viz.*, other pulses of West Bengal and total non-foodgrains of West Dinajpur. Therefore, the exponential trend equation has been retained as the best fitting trend equation to 26 out of 28 crops. In case of the production of other pulses in West Bengal, which is affected by the 'white-noise', no trend is obtained as both the coefficients of linear trend on point-to-point growth rates remain statistically insignificant

$$[gr_t = \underset{(.0045)}{-.0053} \quad \underset{(.00055)}{-.00085t} \; ;$$

terms within parentheses imply standard errors]. In this case, however, the log-quadratic equation is selected as it gives a better fit (DW = 1.256, $\bar{R}^2$= 0.5968 and runs =17) than others. In the case of non-foodgrains in West Dinajpur parabolic trend has been chosen on the basis of its relatively better results of various test criteria (DW =1.483, $\bar{R}^2$= 0.6616 and runs = 15). The logic behind fitting the parabolic trend in place of exponential trend in this case is that if the trend line, $gr_t = a+bt+u_t$, is fitted to the eleven point centered moving average, instead of nine point centered moving average, both a and b are found to be positive and statistically significant at 1% and 5% probability levels respectively.

Table 4.2 presents a list of those crops/crop-groups (altogether 36 cases) in which $\hat{b}$ is statistically significant and positive, implying that the point-to-point growth rates are increasing over time. So in these cases the trend equations are chosen from amongst parabola, log-quadratic and S-type growth curve on the basis of the estimated test criteria. The test results are presented in Table 4.3. It is observed from Table 4.3 that S-type growth curve gives poor fit to all cases. Now, as between parabola and log-quadratic, the former has been chosen for nine crops [*viz.*, W.B.–aman rice, Coochbehar–TNFG, Purulia–rice, Purulia–TFG and Purulia–all crops on the basis of higher $\bar{R}^2$; West Dinajpur–all crops and Midnapore–TFG on the basis of Runs and W.B.–TNFG and Hooghly–TNFG on the basis of DW statistics] and the log-quadratic for sixteen crops [viz., W.B.–rice, Birbhum–rice, Bankura–rice, Bankura–all crops, Nadia–rice, West Dinajpur–rice, Burdwan–rice, Burdwan–TFG, Burdwan–all crops and Midnapore–TNFG on the basis of DW statistic; Murshidabad–rice on the basis of both runs and DW statistic; Darjeeling–TFG on the basis of $\bar{R}^2$ and DW statistic; Birbhum–TNFG on the basis of $\bar{R}^2$ and runs and three crops of W.B. (total cereals,

TFG and all crops) on the basis of runs]. For other eleven crops, it was difficult to choose any particular equation on the basis of the results of these three criteria, namely, DW, $\bar{R}^2$ and runs. Therefore, other criteria have been considered to deal with these cases (to be discussed later on).

On the other hand, linear, parabola, log-quadratic and S-type growth curves have been fitted to the production data of those 18 crops that have registered decreasing growth rates over time (Table 4.4). Comparing the results of various test statistics for alternative trend equations, reported in Table 4.5, log-quadratic trend equation has been finally selected for eight crops [*viz.*, W.B.–barley, W.B.–total pulses, Howrah—TNFG, Howrah—all crops and Hooghly—rice on the basis of DW statistic; W.B.—wheat and Purulia–TNFG on the basis of clearly relatively higher value of $\bar{R}^2$ and W.B.–sugarcane on the basis of elimination process*] and parabola for three crops [*viz.*, Darjeeling—TNFG and Jalpaiguri—TNFG on the basis of DW statistic and W.B.–linseed on the basis of elimination process]. For other seven crops no equation could be chosen on the basis of DW statistic, $\bar{R}^2$ and runs. Out of the seven crops (for which no trend equation could be selected), in only two cases (namely, W.B.—Jute and Jalpaiguri—all crops) linear trend equation competes with parabola and log-quadratic and for other five crops the linear trend equation is discarded on the consideration of one criterion or other. The S-type curve is also discarded as it gives poor results of the statistics for all these seven crops.

From the above analysis it is observed that no trend equation could be selected for eighteen crops, (eleven crops have registered increasing growth rate and seven decreasing growth rate as measured by the sign of $\hat{b}$ coefficient in the equation $gr_t = a+bt+u_t$) on the basis of $\bar{R}^2$, DW or runs. These eighteen crops have been classified according to (i) the absence or presence of autocorrelation and (ii) alternative forms of trend equations (Table 4.6). For these eighteen crops an additional criterion of heteroscedasticity test has been employed to detect that particular form of trend equation (over others) in which the long term variations in the data set remain insignificant.

*By the elimination process the trend equations having statistically significant autocorrelation in the disturbance term are discarded in favour of the equation with DW lying in the inconclusive range.

For non-autocorrelated crops both non-parametric (i.e., rank) and parametric (i.e., Glejser) tests of heteroscedasticity have been done (See Table 4.7).

On the basis of the presence of homoscedasticity in the data set, parabolic trend equation for four crops [*viz.*, 24-Parganas–all crops, Malda–all crops, Murshidabad–TNFG and Nadia–TNFG] and log–quadratic trend equation for only one crop, i.e., TNFG of Bankura, have been chosen. Further, for two other crops, namely, W.B.–Jute and Jalpaiguri–all crops, as straight line trend equation is also one of the options, the rank and Glejser test have also been carried out on the residuals of linear trend. It is seen that in both the cases Glejser tests remain insignificant along with rank test [rank = -.1058 for W.B.—jute and rank = .0682 for Jalpaiguri–all crops]. Consequently, we have chosen linear trend equation for W.B.—jute, in which homoscedasticity is now restored. And in case of Jalpaiguri—all crops choice of trend equation remains yet still indecisive as homoscedasticity is restored for all competing trend equations.

In the case of other six non-autocorrelated crops in which heteroscedasticity test also fails to specify the trend equation, P.E. test has been carried out to select the trend equation between parabola and log-quadratic (Table 4.8). On the basis of P.E. test results, parabolic trend for total foodgrains of Bankura and log-quadratic trend for all crops of Jalpaiguri* have been chosen.

However, P.E. test can be applied only when there is no problem of autocorrelation. For other crops, namely, total rice and total foodgrains in 24-Parganas, and total rice and all crops in Midnapore, as P.E. test fails to yield any conclusive result, log-quadratic trend equation is chosen because of its certain advantages over other equations (as mentioned in the methodology section) and also because of its fulfillment of the requisite econometric criteria, like non-autocorrelation, random disturbance term, homoscedasticity, etc. (this type of choice has been termed here as 'other criterion').

For the crops which are not decisively free from autocorrelation,

---

*P.E. test is used to choose between equations measured on natural scale and logarithmic scale. For all crops combined in Jalpaiguri, P.E. test prefers logarithmic scale to natural scale. Therefore, linear trend as well as parabolic trend which are measured on the natural scale are discarded for all crops in Jalpaiguri.

only non-parametric rank test has been employed for detection of heteroscedasticity and the results are presented in Table 4.9. From these estimated results we have chosen parabolic trend equation for TFG–Hooghly and log-quadratic trend equation for gram–W.B. and TFG–Malda, since in all these cases the residuals of the respective trend equations are free from heteroscedasticity problem. For the other three crops (*viz.*, mustard, total oilseeds and boro rice of West Bengal) log-quadratic trend equation has been selected on the basis of 'other criteria', which implies that, while all the above mentioned tests fail to select a particular form of trend equation, the log-quadratic trend may be chosen owing to its special characteristics and because the other test results are not worse than those of other competing equations.

The various statistical and econometric results of the equations fitted to the State's and the districts' crop production data are presented as a whole in Table 4.10.

4.2 Necessary Adjustments in the Data Set

Inspite of the careful selection of the trend equation (as was done in the previous section) the problems of autocorrelation and long-term volatile fluctuations in the data set could not be solved in many cases. After measurement of autocorrelation it is observed that only 36 crops out of total 82 crops considered in this study are free from autocorrelation problem and for other 46 crops no decisive statistical inference could be drawn on the absence of this problem. More specifically, it is seen from Table 4.11 that out of total 46 crops, 25 crops suffer from positive autocorrelation problem (12 of them are autocorrelated event at 1% level) and for other 21 crops, the DW statistic lies within the indecisive range. Further, it is interesting to note that all crops and crop-groups, except only three crops, namely, aman rice, barley and jute, in the State are suffering from autocorrelation problem. This phenomenon indirectly supports the earlier observations of Boyce (1987) and Datta Roy (1995) about the unreliability of agricultural data in West Bengal, since one of the reasons for the presence of autocorrelation is the interpolations in the statistical observations.

Long-term variations in the data sets have been measured by

the conventional heteroscedasticity tests on the residuals of the fitted chosen trend equations. The crops suffering from heteroscedasticity problem have been enlisted in Tables 4.12 to 4.14. From these tables it is observed that, out of total 82 crops, 20 crops are suffering from significant long-term volatile fluctuations in the data sets. Of these 20 crops, 7 crops (Table 4.12) are non-autocorrelated (where heteroscedasticity has been established by considering both rank test and Glejser test) and other 7 crops (Table 4.14) suffer from autocorrelation (where only rank test has been considered for the detection of heteroscedasticity). In the cases of remaining 6 crops (Table 4.13), heteroscedasticity has been established either by only Glejser test (in case of 4 crops) or by only rank test (in case of two crops). But as in these crops DW statistics lie in the indecisive range, the possibility of autocorrelation problem could not be discarded and hence the Glejser test can not be relied upon for arriving at a conclusion about the presence of heteroscedasticity. To resolve this matter, the period-wise variations (in term of standard deviation) have been measured in the point-to-point growth rates for those four crops where only Glejser test is significant (see Table 4.15). From the measured period-wise standard deviations in point-to-point growth rates, we can safely presume the existence of heteroscedasticity for three crops (except W.B.–total pulses) where the differences in these period-wise standard deviations are significant. On the other hand for total pulses in West Bengal, the assumption of homoscedasticity is confirmed by other estimated results, like (a) the sub-period-wise difference in the variances of the original production data (*e.g.*, var_1 =.022 and var_2 = .030), (b) the autoregressive conditional heteroscedasticity test result, i.e., $\bar{R}^2$ = .0787 (which is insignificant) for the estimated equation $e^2_t = a + b\, e^2_{t-1} + c\, e^2_{t-2}$ and (c) the Goldfeld-Quandt test result, i.e., RSS_1 / RSS_2 = .1957 / .1402 = 1.396 which is insignificant. Therefore, the presence of long-term variations in the data set remains significant for altogether 19 crops (out of total 82 crops).

Apart from these two problems of autocorrelation and heteroscedasticity, time series data on agricultural production, in general, suffer from short-term volatile fluctuations (measured by the standard deviation of the residuals, 6_e which has not been presented here separately). To take care of these problems necessary adjustments in the data need to be made. In Tables 4.16 and 4.17 are enlisted total 46 crops whose time-series data on production have been adjusted by taking 3-period centered moving averages of the residuals and subsequently corrected for auto-correlation (for the smoothed production series), assuming first order auto-regressive scheme for 9 crops (Table 4.16) and second order auto-regressive scheme for 37 crops (Table 4.17). After these adjustments all the desirable criteria (such as, non-autocorrelation, random and homoscedastic disturbance terms, minimisation of short term volatile fluctuations and sufficient improvement in the values of $\bar{R}^2$) are fulfilled for all the crops, except only one-crop, i.e., TFG in Malda, where only runs remain significant as before. It is important to note here that many of these criteria were not fulfilled when the trend lines were fitted to the unadjusted data. Further, after adjustment Glejser test also becomes insignificant (results have not been reported here) for those four crops (namely, Darjeeling–TFG, Malda—rice, West Dinajpur–TNFG and West Dinajpur–rice) where it was significant before adjustment. However, this adjustment in the form of three period centered moving average of the original series followed by other types of adjustment such as correction for autocorrelation may sometimes lead to worse results. This problem is encountered in as many as 36 crops in our case. To cope with this problem only corrections for autocorrelation, or heteroscedasticity, or both autocorrelation and heteroscedasticity, wherever they are presented in the data, have been made. For example, in 9 (nine) out of these 36 cases necessary corrections for the presence of autocorrelation have been made (Table 4.18). In the case of seven other crops where the residuals pertaining to the production data for either the first sub-period or the second sub-period are characterised by significant fluctuations, three year moving averages have been calculated for that period only before making necessary adjustments for the presence of autocorrelation if there were any (Table 4.19). It is seen from Tables 4.18 and 4.19 that in all the 16 cases there has been significant improvement in the test results after necessary adjustments in the observations. The disturbances do not any more suffer from autocorrelation, heteroscedasticity and non-randomness. Moreover, there has been improvement in the values of $\bar{R}^2$ and decline in ∂_es.

Lastly, for the remaining twenty crops as the original production data did not suffer from any of the above mentioned problems, no adjustments were made to these data.

To sum up this section it may be observed that while fitting a trend line for estimation of growth rates and determining nature of growth (declining, increasing or constant), due attention should be paid to the nature of the data. There may be wide fluctuations in the data. The unsystematic component of the time series may be autocorrelated. If no proper adjustments in the data are made through smoothing or such other process and no proper corrections are made for the presence of autocorrelation in the disturbance term, fitting trends to such data do not always give us consistent and efficient estimates and, therefore, no valid statistical inferences can be made on the basis of these estimates. Different crops may exhibit different characteristics and, therefore, corrections are to be made accordingly. Sometimes, visual inspection of the data presented in graphical forms may help us in identifying these characteristics and choosing appropriate adjustment and correction procedure.

4.3 Estimated Growth Rates in Agricultural Production: Crop-wise and District-wise Disaggregation (1957-58—1994-95)

After making necessary adjustments in the data sets the growth rates have been estimated for the whole period (i.e., 1957-58 to 1994-95) from the estimated coefficients of the chosen trend equations. More specifically, the growth rates have been directly measured from the estimated coefficients of t (i.e., time) in case of exponential and log-quadratic (with normalisation of time, i.e., shifting origin to the mid-point of the time period) trend equations. In case of linear and parabolic (with normalisation of time) trend equations the estimated coefficient of t has been divided by the harmonic mean of the series to get trend growth rate. The growth rates have been expressed in percent per annum. The Tables 4.21 to 4.23 present the estimated growth rates (calculated on both adjusted and unadjusted series) with acceleration and deceleration (i.e., $\hat{c}$, the estimated coefficient of t^2), as the case may be, for all crops/crop-groups of West Bengal during the period 1957-58 to 1994-95. From these three tables, it is observed that unadjusted growth rates (along with $\hat{c}$s) differ from adjusted growth rates (along with $\hat{c}$s); but the pattern of these differences is not uniform. In as many as ten cases, the unadjusted growth rates suffer from upward bias and in eleven cases they suffer from downward bias. On the other hand, the estimated values of c from unadjusted series suffer from downward bias in 11 out of 15 cases. But the coefficients estimated from adjusted series do not suffer from any such bias, since they satisfy all the basic criteria of estimation. Hence, they are more reliable than those estimated from unadjusted series (as was established in the previous section). Hence forth, only adjusted growth rates will be considered as the estimated growth rates for the period under study.

From Table 4.21 it is seen that the growth rates in the productions of all three major crop-groups (i.e., TFG, TNFG and all crops) in West Bengal for the whole period are significantly positive with statistically significant acceleration. Further, the growth rate in TNFG is much higher (4.351%) than the growth rate in TFG (2.515%), thus, making the growth rate in the production of all crops (2.684%) lie in between above two growth rates. However, the growth rate in all crop is close to the growth rate of TFG as the weightage of TFG in total agricultural production is much higher than that of TNFG in West Bengal [for example, in the year 1990-91, the percentage share of TFG in total acreage is 82.65%]. The growth rate in the production of TFG has been disaggregated crop-wise in Table 4.22 and it is observed that the growth in TFG has been due to growth in the production of total cereals; that, in turn, has been contributed mainly by the growth in the productions of boro rice and wheat, followed by that in the productions of other cereals, aman rice and aus rice. Actually, the growth rates in the productions of pulses and barley are negative with significant deceleration. The other crops in which the growth rates (though very high and significantly positive) have decelerated over time are boro rice and wheat. There is only one crop (i.e., aman rice) whose growth in production has significantly accelerated over time and because of its large share in the cropping pattern (*e.g.*, 66.3% of TFG acreage was shared by aman rice in 1990-91), this has led to the significant accelerations of the growth rates in the productions of crop-groups, like, total rice, total cereals and total foodgrains (TFG). (In this connection it is to be noted that $\hat{c}$ coefficient of parabolic trend equation is not comparable with that of log-quadratic trend equation, except the sign, i.e., the nature).

The growth rate in production of TNFG has been disaggregated crop-wise in Table 4.23. Here we see that the main crops behind the growth in TNFG are sesamum, mustard and potatoes, followed by jute. The growth rate in the production of tobacco is very low and actually the growth rates in the productions of linseed and sugarcane are negative with further decelerations. Here also there is only one crop (i.e., mustard) whose growth rate in the production has accelerated significantly over time and this acceleration leads to accelerations in the growth rates of the productions of crop-groups, like total oilseeds and TNFG.

To know the spatial pattern of agricultural performance in West Bengal we have also estimated the growth rates of four major crop-groups, namely, total rice, TFG, TNFG and all crops, over the districts for the whole period. The estimated results have been presented in Table 4.24. From the estimated whole period's growth rates in the productions of all crops of the districts we see that Hooghly, Howrah, Nadia, Burdwan, Murshidabad, Midnapore and Malda (arranged rank-wise) are experiencing high growth rates (growth rate $\geq$ 3%) and the other district whose growth rate is higher than State average is Darjeeling. In Jalpaiguri, this growth rate is lowest and for other six districts the growth rates are moderate [i.e., 2% $\leq$ growth rate $<$ 3%] but lower than State average. Further, there are only seven districts, namely, Burdwan, Bankura, Midnapore, 24-Parganas, West Dinajpur, Malda and Purulia, where output growth rates in all crops have significantly accelerated over time; for other districts either there is no acceleration or acceleration is insignificant and in case of Jalpaiguri, actually, $\hat{c}$ is negative. In comparison with production growth rates in TFG and TNFG it is seen that growth rate in the production of all crops of the districts is dominated mainly by that of TFG and further, the growth in TNFG is sufficiently higher than the growth in TFG except for only four districts, namely, West Dinajpur, Malda, Darjeeling and Coochbehar. The pattern of growth rates in rice production over the districts remains more or less same as that in the production of TFG except for Darjeeling where the growth rate in rice production is sufficiently low compared to that in TFG. From Table 4.24, in general, it can be said that there are six districts, namely, Burdwan, Midnapore, Howrah, Hooghly, Nadia and

Murshidabad, whose agricultural performance (in terms of growth rates in all four major crop-groups) is better than that of the State, and further, there are only two districts, namely, Midnapore and Bankura where growth rates have been increasing at an increasing rate ($\hat{c}>0$ and statistically significant) for all the crop-groups and total agricultural production.

4.4 Sub-period Growth Rates in Crop Output: Crop-wise and District-wise Disaggregation

In consideration of the main objective of our study we have estimated the sub-periods' growth rates in crop-outputs in West Bengal to evaluate the performance of agriculture before (i.e., 1957-58 to 1976-77) and during (i.e., 1977-78 to 1994-95) Left Front Government regime. The sub-period growth rates have been estimated from the fitted kinked trend equations on the adjusted series, dividing the whole period (i.e., 1957-58 to 1994-95) into two sub-periods: (i) 1957-58 to 1976-77 and (ii) 1977-78 to 1994-95. In some cases for the adjusted series each sub-period has been reduced by one year (deleting the first and the last years). Along with the sub-periods' growth rates the trend breaks have also been estimated to know the statistical difference between two sub-periods' growth rates. The sub-periods' growth rates have been expressed in percent per annum (for kinked linear trend the estimated coefficients have been divided by the respective harmonic means, reason of which has been discussed earlier).

In Tables 4.25 to 4.29 the estimated growth rates of agricultural production of West Bengal for the sub-periods and the whole period, trend breaks and the criteria ($\bar{R}^2$, DW and runs) of the fitted kinked trend equations are presented. From these five tables it is seen that, in general, the kinked trend equations, fitted to the adjusted series, give good fits on the following considerations :

(i) The basic criteria of estimation have been satisfied, i.e., the values of $\bar{R}^2$ are sufficiently high and there is no autocorrelation in the disturbance term. However, in few exceptional cases runs are significant (in only 5 cases, that too at 10% level) and DW statistics lie in the indecisive range only in eight cases.

(ii) In all cases the whole period's growth rates are lying between respective two sub-periods' growth rates; this implies that there is no discontinuity bias in the estimation of sub-period growth rates.

(iii) It is further seen, from the values of the estimated trend coefficients of point-to-point growth rates (i.e., $\hat{b}$s of Tables 4.1, 4.2 & 4.4), coefficients of acceleration/declaration in whole period's growth rates (i.e., $\hat{c}$s of Tables 4.21 to 4.24) and the estimated trend breaks, that kinked trend equations restore the consistency with respect to the nature, movement and statistical significance of these estimated parameters. However, only in three cases, namely, W.B.–til, Nadia–all crops and Howrah–total rice, there is inconsistency with respect to the levels of significance (10%). In other six cases [3 unadjusted series, namely, Jalpaiguri-all crops, Nadia—TNFG and Murshidabad–TNFG and 3 adjusted series, namely, Howrah—TNFG, Howrah–all crops and Hooghly–TFG] also, inconsistency (regarding sign in some cases and statistical significance in other cases) arises between $\hat{b}$s and trend breaks. But greater confidence can be placed on trend breaks as point-to-point growth series (by which $\hat{b}$s have been calculated) are more volatile than the original series.

From the estimated sub-period growth rates in crop-outputs in West Bengal (Table 4.25) it is observed that during first sub-period the high (≥3%) growth experiencing crops were boro rice, wheat, sesamum, potatoes, linseed and barley (arranged in descending order). In the second sub-period three of these 6 crops (namely, wheat, linseed and barley) are facing negative growth rates in their productions and two crops (*viz.*, aman rice and mustard) have improved their positions from very low growth category (during first sub-period) to very high growth category (during second sub-period). During second sub-period high (≥3%) growth experiencing crops can be arranged as mustard, sesamum, potatoes, boro rice, aman rice and other cereals. In both the sub-periods, the growth rates in the productions of pulses are negative and the growth rates in aus rice and jute productions are low (<2%) and moderate (i.e., 2%≤ gr.<3%) respectively.

As regards the changing pattern of production growth rates between two sub-periods (i.e., the trend breaks), it is seen that for most of the crops the trend breaks are significantly negative. There are only two crops, namely, aman rice and mustard, significant improvements in which have led to higher growth rates in the productions of total foodgrains and total oil seeds and as such total non-foodgrains respectively during the second sub-period (i.e., the period of LFG regime) compared to the previous sub-period (1957-58 to 1976-77). Thus, these two crops have significantly contributed to the higher growth rate of total agricultural production during the second sub-period. The annual growth rate in the total agricultural production in West Bengal during the second sub-period is estimated to be 3.649% with the trend break of 1.568 per cent. This is much below the annual growth rate of 6.50 per cent, as estimated by Saha and Swaminathan (1994).

To enquire into the regional differences in the agricultural production of West Bengal over two sub-periods, sub-period growth rates in the productions of all crops of the districts have been estimated for these sub-periods. These estimated growth rates are presented in Table 4.26. From these district-wise estimated sub-period growth rates in agricultural productions, it is seen that there are only four districts, namely, Hooghly, Howrah, Nadia and Murshidabad whose growth rates were high ($\geq$3%) in the first sub-period and the districts of Burdwan, Malda, Darjeeling, Birbhum, Jalpaiguri and 24-Parganas experienced moderate growth rates between 2% and 3%. In all other districts, growth rates were below 2%. During second sub-period, all the districts belonging to the class of high growth rates in the first sub-period have also remained in the high growth class. The districts which have improved their position from moderate growth category of first sub-period to high growth category in second sub-period are Burdwan, Malda and Darjeeling; and from low to high growth category are Midnapore, Bankura, Purulia and West Dinajpur. Over these two sub-periods, the position of Jalpaiguri has deteriorated (from moderate to low category) and the position of Coochbehar has slightly improved (from low to moderate category). The positions of the other two districts, namely, Birbhum and 24-Parganas have remained unchanged. These are also evident from the estimated trend breaks.

In general, the districts' growth rates in agricultural production have improved during second sub-period (in comparison to first sub-period) and the remarkable improvement has taken place in three

Classification of Crops/Crop-groups of West Bengal on the Basis of Sub-period Growth Rates in Production

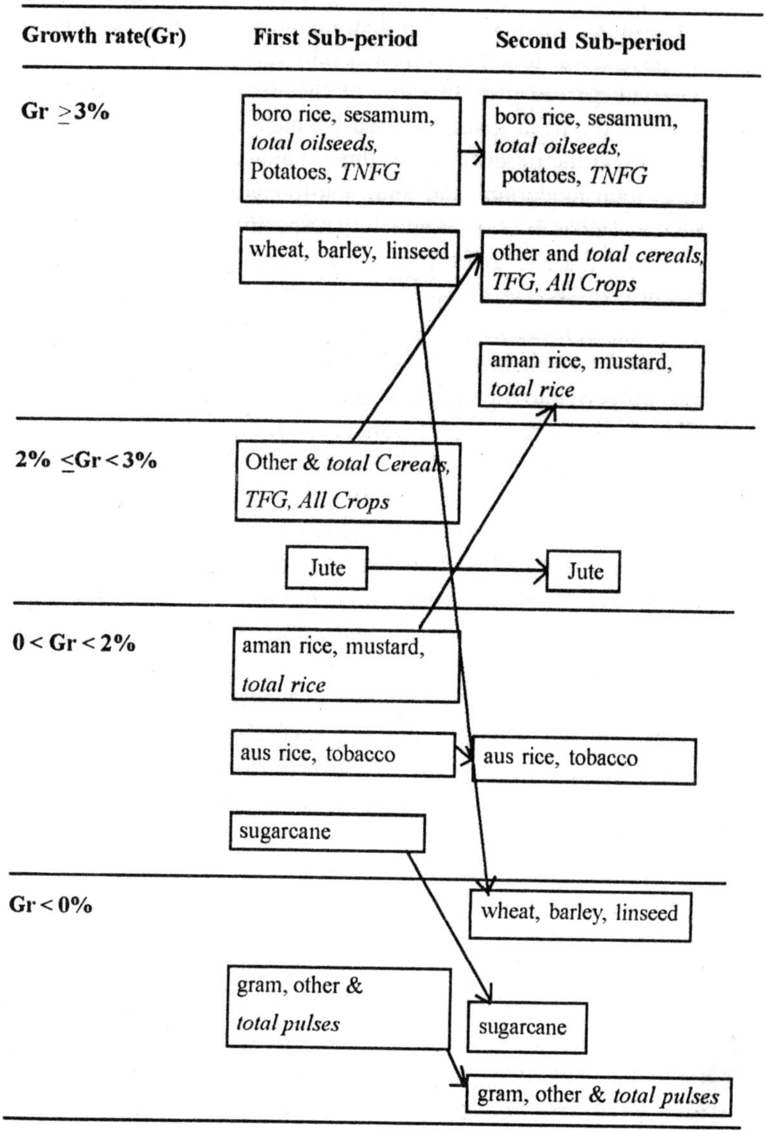

districts, namely, Bankura, Midnapore and Purulia that may lead to high growth in agricultural production in West Bengal during LFG regime. The following table (page 117) gives a diagrammatic exposition of the positions of the districts in respect of growth rates in total agricultural production in two sub-periods.

The sub-period growth rates of the districts in agricultural production have been disaggregated by measuring the sub-period growth rates in the productions of three major crop-groups, such as total rice, total foodgrains and total non-foodgrains. The results are presented in Tables 4.27, 4.28 and 4.29 respectively. From Table 4.27 it is observed that the growth rate in rice production has improved significantly in second sub-period over the first sub-period in 10 districts. Among these 10 districts, the growth rate in rice production has increased significantly (i) from low growth category (<2%) in the first sub-period to high growth category (≥3%) in the second sub-period in 6 districts, namely, Bankura, Midnapore, 24-Parganas, Murshidabad, West Dinajpur and Purulia, (ii) from moderate growth category (2% ≤ gr.rate <3%) of first sub-period to high growth category of second sub-period in 2 districts, namely, Burdwan and Nadia and (iii) from high to high and from low to moderate categories in Howrah and Birbhum respectively. In other five districts either growth rates have not increased significantly or they have declined, thus, retaining same positions in two sub-periods (Hooghly and Malda in High growth category, Coochbehar in moderate growth category and Darjeeling and Jalpaiguri in low growth category).

From the estimated sub-period growth rates in the productions of TFG of the districts, as reported in Table 4.28, it is seen that the movement of the growth rates and the positions of the districts in respect of their annual growth rates remain more or less the same as those in case of rice production.

But the situation is altogether different in case of the productions of TNFG over the districts, as is evident from their estimated sub-period growth rates, presented in Table 4.29. Most of the districts, namely, Birbhum, Bankura, Midnapore, Hooghly, Nadia and Murshidabad belonging to high growth category in both the sub-periods have experienced increase in their growth rates in the productions of TNFG from first sub-period to second sub-period. Two districts, namely, West Dinajpur and Coochbehar have also

Annual Growth Rates of Total Agricultural Production in the Districts of West Bengal in Two Sub-periods

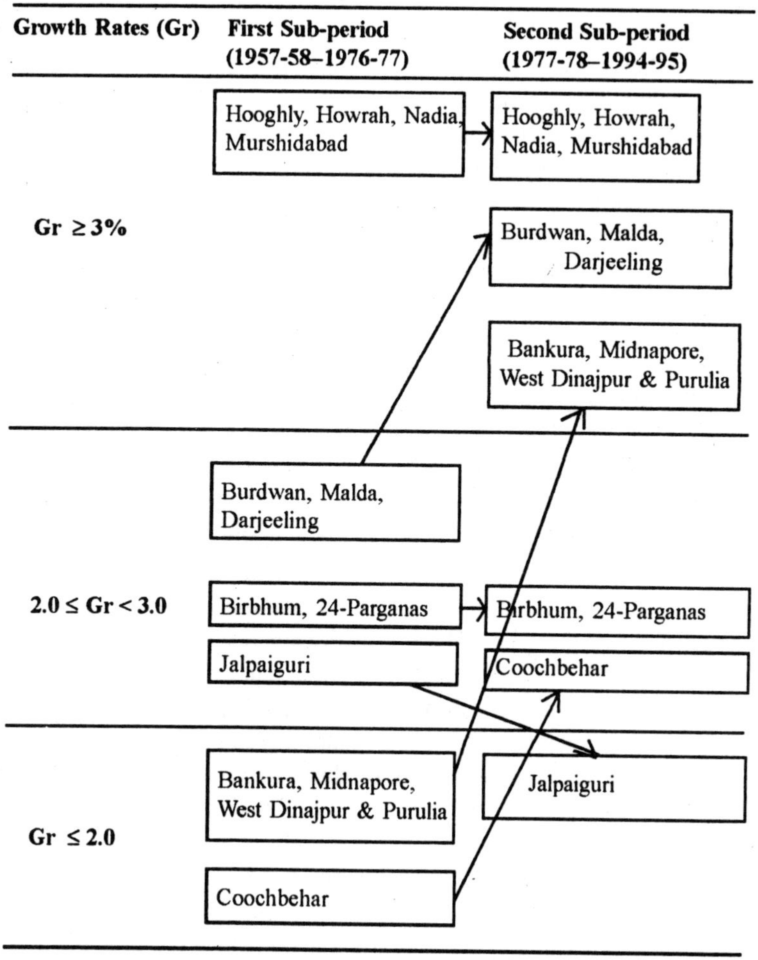

experienced significant increase in growth rates in TNFG, thus, either moving from low growth category during first sub-period to high growth category during second sub-period (in case of West Dinajpur) or retaining the same low growth category in both the sub-periods with higher growth rate in the second sub-period, as in case of Coochbehar. The growth rate in the production of TNFG has declined significantly from first sub-period to second sub-period in three districts, namely, Jalpaiguri, Darjeeling and Purulia. In the remaining four districts [3 of them in high growth category for both the sub-periods: Burdwan, Howrah and 24-Parganas] growth situation remains more or less stagnant.

From the above discussion (by comparing Tables 4.26 to 4.29) we can conclude that there are only two districts, *viz.*, Bankura and Midnapore, whose positions during LFG regime have improved very significantly and consistently over all crop-groups; likewise, the position of Jalpaiguri district has deteriorated on all fronts. The growth rates in agricultural productions during LFG regime have gone up in Burdwan, 24-Parganas, Malda and Purulia mainly due to growth in TFG and in Nadia and West Dinajpur mainly due to growth in TNFG. There are two districts, namely, Howrah and Hooghly whose growth rates in agricultural production remain more or less stagnant, though both of them lie in high growth category for all major crop-groups in both the sub-periods. For the remaining four districts, namely, Birbhum, Murshidabad, Darjeeling and Coochbehar, agricultural growth rates remain stagnant due to mixed results on the growth rates in the productions of three major crop-groups (*viz.*, rice, TFG and TNFG) between two sub-periods.

4.5 Decomposition of Output Growth Rates

For the decomposition of output growth rate (into the growth rates of the components, area and productivity), first, one of the two components has been chosen for each crop/crop-group to estimate its trend growth rate on the basis of its relatively lower value of the standard deviation of point-to-point growth rates (calculated from original series) in relation to the other. Comparing the values of measured standard deviations, not reported here, we have selected productivity in 16 cases (boro rice, wheat, linseed, sesamum, total oilseeds, jute, sugarcane, tobacco and total non-foodgrains of the State and total foodgrains of Birbhum and total non-foodgrains of Hooghly, 24-Parganas, Nadia, Malda, Darjeeling and Purulia) and area in all other 66 cases for estimating their trend growth rates. Then to estimate their trend growth rates for the whole period as well as for two sub-periods, we have followed the same techniques (related to the choice of trend equation, necessary adjustments in the data sets etc.) as were applied in the cases of measuring output growth rates.

After estimation of the trend growth rates of area under individual crops/crop-groups (productivity of individual crops/crop-groups), the growth rates of productivity (area) have been determined by subtracting the former from the output growth rates of these crops. Thus, the output growth rate is sum of area growth rate and productivity growth rate. The results are given in Tables 4.30 to 4.34.

From Table 4.30 it is observed that productivity has made greater contribution than the area to the output growth of all crops in West Bengal during the second sub-period as well as the whole period. This high contribution of productivity growth in case of all crops is mainly due to that of the TFG (also of total rice and total cereals). In case of TNFG, the share of productivity growth in output growth is much less than that of area for the first sub-period which dominates the whole period's shares. As regard relative contributions of these two components to the growth rates of individual crops the following observations may be noted. (i) The share of productivity growth is much higher than the share of area growth for all the periods [i.e., whole period and two sub-periods] for aus rice, aman rice, other cereals and tobacco (as well as jute, expect for first sub-period). (ii) The negative output growth of the individual crops, wherever applicable, may be attributed to the negative growth in area. (iii) For other crops, such as boro rice, wheat, mustard, sesamum and potatoes, area growth is more prominent than productivity growth for the whole period and for at least one of the sub-periods. In general, we can conclude that the productivity made greater contribution than the area to the growth of output in the second sub-period than in the first sub-period.

The decomposition of the growth rates of output of the crop-groups, such as total rice, total foodgrains, total non-foodgrains and all crops combined, into the area growth rate and the productivity growth rate has been made for each district in the State. The results are presented in Tables 4.31 to 4.34. This helps us to get an insight into the inter-district and inter-crop-groups variations in the growth components' contributions to output growth rates.

It is seen from Table 4.31 that in six districts–Howrah, 24-Parganas, Nadia, Murshidabad, Malda and Darjeeling–the share of productivity growth rate of all crops combined in the output growth rate has been higher than that for the State during the whole period as well as the two sub-periods. On the other hand, in the districts of Burdwan, Hooghly, West Dinajpur and Coochbehar the share of area growth rate in the output growth rate has been higher than that for the State during the whole period as well as the two sub-periods. Also it is observed that the share of productivity in the output growth rate of all crops combined has always been higher than that of the area growth rate in all the districts in the State, except (i) Jalpaiguri, during the period from 1957-58 to 1994-95, (ii) Bankura, Jalpaiguri, Coochbehar and Purulia during the period 1957-58 – 1976-77. The share of productivity growth rate has been increasing over the two sub-periods in all the districts in the State with the exception of Howrah, Hooghly and 24-Parganas where, of course, the decreases were marginal.

As far as the annual growth rates of the area and productivity components are concerned, it is seen from Table 4.31 that in three northern districts (West Dinajpur, Jalpaiguri and Coochbehar) and in two Western districts (Birbhum and Purulia) the productivity growth rates of all crops combined have been lower than that for the State during the whole period as well as during two sub-periods. In Bankura, the productivity growth rate of all crops combined has been lower than that for the State during the whole period and the first sub-period. But it was significantly higher in the second sub-period.

As regards the major crop groups, such as total rice, total foodgrains and total non-foodgrains, it is seen in the case of total rice that the share of productivity in the output growth rate has been higher than the area growth rate in all the districts other than (i) Jalpaiguri and Darjeeling during 1957-58—1994-95, (ii) Bankura, West Dinajpur, Jalpaiguri, Darjeeling, Coochbehar and Purulia during the first sub-period, 1957-58—1976-77 and (iii) Only Darjeeling during the second sub-period, 1977-78—1994-95. For the State, the contribution of productivity has been higher than that of area during the whole period and the second sub-period. In the first sub-period, these two components had almost equal shares. It is further observed that the productivity growth rate of total rice in eight districts, namely, Burdwan, Midnapore, Howrah, Hooghly, 24-Parganas, Nadia, Murshidabad and Malda, was higher and in the remaining seven districts it was lower than the State productivity growth rate during the period 1957-58 – 1994-95. During the first sub-period eight districts, namely, Malda, Murshidabad, Nadia, 24-Parganas, Hooghly, Howrah, Birbhum and Burdwan had productivity growth rates higher than that for the State as a whole, while in the second sub-period seven districts, namely, Malda, West Dinajpur, Murshidabad, Nadia, 24-Parganas, Midnapore and Bankura experienced productivity growth rates higher than that for the State.

In the case of total foodgrains (TFG) the productivity growth rate is found to be higher than that of the State for the same districts as in the case of total rice, during the first sub-period. However, in the second sub-period eight districts, namely, Bankura, Burdwan, Midnapore, Howrah, Nadia, Murshidabad, Malda and Darjeeling registered higher productivity growth rate than the State productivity growth rate during second sub-period (1977-78 – 1994-95). Of these eight districts, five districts–Bankura, Midnapore, Nadia, Murshidabad and Malda–experienced higher productivity growth rate of total rice during this second sub-period. Thus, in these districts, the productivity growth of total rice makes a significant contribution to the productivity growth of the total foodgrains. Further, it is observed that for the foodgrains the share of the productivity component in the output growth has been larger than that of the area component for (i) all districts in the second sub-period (1977-78 – 1994-95), (ii) in all the districts, except Jalpaiguri, during the period 1957-58 – 1994-95 and (iii) except Bankura, Jalpaiguri, Coochbehar and Purulia in the first sub-period (1957-58 – 1976-77). In the case of the State, the contribution of productivity has been higher in all the periods.

In the case of the total non-foodgrains (TNFG) in five districts, namely, Burdwan, Birbhum, Bankura, Midnapore and West Dinajpur the contribution of area to the output growth rate has been higher than that of the productivity during 1957/8 – 1994/5. During the first sub-period and the second sub-period, the contribution of area to the output growth rate has been higher than that of productivity in the above mentioned districts. In addition, in Purulia in the first sub-period and in Jalpaiguri in the second sub-period, the contribution of area was higher than that of productivity to the output growth rate. For the State, however, the contribution of productivity has been higher than that of the area, in the case of total non-foodgrains, in the second sub-period and the contribution of area is higher than that of the productivity in the first sub-period as well as in the whole period.

To summarize, the productivity made significant contribution in relation to the area to the growth of output of total rice, total foodgrains and total non-foodgrains in most of the districts during the period 1957-58 – 1994-95 and the sub-periods, 1957-58 – 1976-77 and 1977-78 – 1994-95. It is further observed that the productivity plays a significant role in the growth of output in the eastern and south-eastern districts of the State, namely, Murshidabad, Malda, Nadia and 24-Parganas while the area is the principal contributing factor to the output growth in western district of Purulia and northern districts of Jalpaiguri, Coochbehar and West Dinajpur. For the other districts, the relative contributions of area and productivity vary among crop-groups and over time periods.

The result on growth decomposition is summarized by measuring correlation coefficients cross-sectionally (over districts) for each of four major crop-groups (total rice, TFG, TNFG and all crops) between output growth rate and the component's (area/productivity) growth rate during each period. The results are presented in Table 4.35. From Table 4.35 it is observed that all estimated correlation coefficients are positive, implying that higher output growth occurs with higher growth in each of the components for all periods and for all crop-groups, though at different levels of statistical significance. For the whole period, output growth is strongly correlated with productivity growth and weakly with area growth in the cases of total rice, TFG and all crops; but the reverse is true in case of TNFG. This relationship holds for the first sub-period also. But in the second sub-period the relative strength of the correlations undergoes a striking change. The area correlations as well as the productivity correlations are very high, but the latter is higher for all crops combined and crop-groups except total non-foodgrains. Therefore, on the whole, it can be said that productivity has been the main contributor to the growth in the total agricultural production and in the output of all crop-groups, excepting total non-foodgrains. In the case of total non-foodgrains area is found to have made significant contribution to the output growth in many of the districts in the State.

4.6 Relative Contributions of the Crops and Districts in the State's Agricultural Output Growth

In this section an attempt is made to measure the shares of individual crops as well as of the districts in the growth of total agricultural output in the State, following the methodology mentioned in the section 3.4 of Chapter 3. The estimated growth weights of individual crops in the State and of the districts are presented in Tables 4.36 and 4.37 respectively. The crop-wise price weights, as calculated by Boyce (1987), are used here (col. 2 of Table 4.36) with slight modifications, such as (i) price weight for jute has been expressed in terms of bale (1 bale = 0.18 tonne) and (ii) price weights have been considered not in percentage forms but in rates, taking price weight for aman rice equal to unity. Using these price weights, period-wise average productions of the crops, expressed in terms of standard unit of aman crop production in '000 tonnes (columns 3, 5 and 7 of Table 4.36), have been calculated. It is observed that for West Bengal the total agricultural production calculated on the basis of crop aggregation differs from that calculated on the basis of State's index number for each period. However, the magnitudes of these differences [2.3% for the whole period, 1.7% for the first sub-period and 1.4% for the second sub-period] are very negligible. These differences may be attributed to (i) the possible statistical errors in transforming data from averages and aggregates to indices and (ii) the exclusion of 'other cereals' and 'other oilseeds' from the crop aggregate because of non-availability of their actual price weights. On the basis of the agricultural production from crop aggregate, the growth weight of each crop has been estimated by dividing each crop's production by total production. The results are given in cols. 4, 6 and 8 of Table 4.36 for the whole period, first sub-period and second sub-period respectively.

In col. 2 of Table 4.37 are presented the district-wise measured agricultural productions (in the unit of '000 tonnes of aman rice, using price weights of Table 4.36, col. 2) per unit of index number [i.e., total agricultural production (in terms of aman rice) during the period 1969-70 to 1971-72 divided by total index numbers of agricultural production during the same period]. Here there is wide difference between State's figure and district's total, which is due to the difference in the base period of the State's index and the base period of the districts' indices. When the base period of the State's index is adjusted to the base period of the districts indices, this difference becomes negligible. Multiplying standardized production per unit of index number by the average index number for a particular period the standardized average production of each district for the whole period (col. 3), first sub-period (col. 5) and second sub-period (col. 7) in Table 4.37 have been calculated. Here also, there is difference between the State-figure and the district-totals. However, this difference is very negligible [0.05 per cent for the whole period, 0.87 per cent for the first sub-period and -1.75 per cent for the second sub-period] and is due to the rounding up the decimal figures upto two points, in additions to the reasons mentioned earlier (in case of Table 4.36). Now expressing the standardized average production of each district as ratio of the district-total, the district-wise growth weights for all three periods (cols. 4,6 and 8 of Table 4.37) have been calculated.

The shares of the individual crops and the districts in the growth rate of agricultural production are presented in Tables 4.38 and 4.39. In cols. 3, 6 and 9 of Table 4.38 the period-wise weighted output growth rates (output growth rate x respective growth weight) for each of the crops have been estimated and it is observed that there is wide difference between the State's trend growth rate and the State's weighted average growth rate (obtained by adding weighted growth rates of the crops). A part of this difference can be minimized if the weighted output growth rate of boro rice is calculated by the residual method (i.e., weighted output growth rate of total rice minus weighted output growth rates of aus rice and aman rice). This residual method should be accepted reasonably as sudden spurt in the boro production during mid-sixties, reflected in the form of its very high trend growth rate, leads to overestimation in its measured weighted output growth rate. This has happened in the case of wheat also. But because of non-availability of price index of 'other cereals', the residual method could not be applied to estimate the weighted output growth rate of wheat. The other part of the difference (between trend and weighted average growth rates) may be ascribed to the reasons stated above and also to the aggregation of the crops to get total agricultural output by which the uneven growth pattern of the crops is somewhat smoothed. Likewise, the difference between the State's trend growth rate and the total of the districts' weighted output growth rates in the production of all crops (cols. 3, 6 and 9 of Table 4.39) can also be explained by the factors mentioned above. From the weighted output growth rates the shares of the crops and districts for the whole period and also for two sub-periods have been estimated. The results are given in cols. 4, 7 and 10 of Tables 4.38 and 4.39 respectively.

From Table 4.38 it is seen that for the whole period aman rice has been the main contributor to the output growth in the State, followed by boro rice, potatoes and wheat. Among other crops, jute, aus rice, mustard and sesamum have also made substantial contributions to the output growth in the State. On the other hand, barley, pulses, linseed and sugarcane exert negative pressure on the agricultural output growth rate in the State. Considering the changing pattern of the crops' contributions between two sub-periods, it is observed that the crops which have gained in importance over time are aman rice, mustard, sesamum and potatoes and the crops which are losing their importance are wheat, aus rice, barley, linseed, sugarcane, jute and boro rice. The positions of pulses and tobacco remain more or less constant. Moreover, the share of the aman rice has gone up so much so that its contribution to the output growth in the second sub-period is more than 50 per cent, as compared to only 17.57% during the first sub-period and nearly equal to the output growth rates of boro rice and wheat during the first sub-period.

From the estimated shares of the districts (Table 4.39) for the whole period it is seen that 70 per cent of the State's agricultural production growth comes from only six districts, *viz.*, Midnapore, Burdwan, Hooghly, Nadia, 24-Parganas and Murshidabad (arranged in descending order of the districts' shares); other nine districts' contribution is only 30 per cent. Among these nine districts, again, nearly 20 per cent of the State's agricultural growth comes from 4 districts, namely, Bankura, West Dinajpur, Birbhum and Malda; the remaining 5 districts' share is only 10%. If the shares of these districts are compared between two sub-periods it is seen that the shares of only three districts (*viz.*, Midnapore, Bankura and Purulia) have increased remarkably from the first sub-period to the second sub-period; the shares of five districts–Birbhum, 24-Parganas, Murshidabad, Jalpaiguri and Coochbehar–have actually declined while the shares of as many as seven districts have remained more or less constant over these two sub-periods. If the districts are grouped region-wise it is seen that the respective shares of the southern (Midnapore, Howrah and 24-Parganas), central (Burdwan, Hooghly, Nadia and Murshidabad), Western (Purulia, Bankura and Birbhum) and northern (Malda, West Dinajpur, Darjeeling, Jalpaiguri and Coochbehar) districts in the State's agricultural output growth were 23.27 per cent, 46.38 per cent, 11.18 per cent and 19.17 per cent during the first sub-period and 31.39 per cent, 40.26 per cent, 14.79 per cent and 13.56 per cent during the second sub-period. Therefore, though the share of the central districts as a whole remains the highest among all regions, it is found to have declined from the first sub-period to the second sub-period, while the shares of the southern and western districts have increased over these periods.

Table 4.1 : Exponential Trend Equation and its Estimated Results for the Crops, West Bengal and the Districts, 1957-58 - 1994-95 (On the basis of constancy of point-to-point growth rates)

State/ District Crop	$\hat{b}^{++}$	Results of Test Statistics		
		DW	$\bar{R}^2$	runs
West Bengal				
Aus rice[k]	0.00041 (0.00072)	0.866	0.4730	9
Other cereals[k]	0.00083 (0.00104)	0.953	0.5450	11
Other pulses[k]	-0.00085 (0.00055)	0.933	0.4463	19⁺
Til[k]	-0.00146 (0.00118)	0.489	0.9397	12
Potatoes[k]	0.00014 (0.00062)	1.221	0.9521	11
Tobacco[k]	-0.00051 (0.00078)	0.937	0.0616	11
24-Parganas				
Total non-food-grains	0.00035 (0.00065)	1.624[a]	0.8127	16⁺
Birbhum				
Total foodgrains	0.00039 (0.00067)	1.452[c]	0.6155	16⁺
All crops	0.00085 (0.00062)	1.452[c]	0.7084	18⁺
Howrah				
Total rice [k]	0.00204 (0.00135)	1.910[a]	0.6767	16⁺
Total food-grains [k]	0.001698 (0.001152)	1.831[a]	0.6716	16⁺
Nadia				
Total food-grains [k]	0.00036 (0.00054)	1.3591[c]	0.9095	17⁺
All crops[k]	0.00077 (0.00050)	1.232α	0.9277	15⁺
Murshidabad				
Total foodgrains	0.00076 (0.00053)	1.417[c]	0.8361	17⁺
All crops[k]	0.00024 (.00048)	1.634[a]	0.8996	16⁺

(Contd.)

Table 4.1 : *(Contd.)*

State/ District Crop	$\hat{b}$ ++	Results of Test Statistics		
		DW	$\bar{R}^2$	runs
West Dinajpur				
Total food-grains[k]	0.00053 (0.00053)	1.541[a]	0.8243	18[+]
Total non-food-grains[k]	0.000477 (0.000701)	0.977	0.5645	15[+]
Darjeeling				
Total rice	0.00078 (0.00065)	1.799[a]	0.5813	19[+]
All crops[k]	0.00007 (0.00038)	1.946[a]	0.8924	18[+]
Coochbehar				
Total rice[k]	0.000103 (0.000602)	1.237α	0.7476	15[+]
Total food-grains[k]	-0.00076 (0.00049)	1.175	0.8131	16[+]
All crops	0.00017 (0.00041)	1.810[a]	0.8215	21[+]
Hooghly				
All crops[k]	0.00048 (0.00044)	1.619[a]	0.9422	19[+]
Malda				
Total rice[k]	0.00028 (0.00043)	1.715[a]	0.9090	18[+]
Total non-foodgrains	0.00047 (0.00088)	1.622[a]	0.4468	17[+]
Jalpaiguri				
Total rice[k]	-0.000234 (0.000586)	1.592[a]	0.3957	21[+]
Total food-grains[k]	-0.00024 (0.00055)	1.704[a]	0.5340	21[+]
Burdwan				
Total non-food-grains[k]	-0.00089 (0.00078)	0.957	0.9123	11

Notes : ++ $\hat{b}$ is the estimated coefficient in the equation : $gr_t = a + bt$ [here gr_t implies smoothed value of point-to-point growth rate at period t]

[k] $\hat{b}$ is estimated after autocorrelation correction,

[a] free from autocorrelation at 5% level, [c] free from autocorrelation at 1% level, α inconclusive range at 1% level, [+]implies runs are statistically insignificant (i.e, random disturbance term), standard errors are in the parentheses.

Table 4.2 : List of Crops Having Increasing Growth Rates over time, West Bengal and the Districts, 1957-58 — 1994-95

(On the basis of point-to-point growth rates)

State/District Crop	$\hat{b}^{++}$	State/District Crop	$\hat{b}^{++}$
West Bengal		*24-Parganas*	
Aman rice[k]	0.00228*	Total foodgrains[k]	0.00124*
	(0.00055)		(0.00057)
Total rice	0.00185*	All crops	0.0009214***
	(0.00052)		(0.000482)
		Birbhum	
Total cereals[k]	0.00085***	Total rice	0.001776*
	(0.00049)		(0.00067)
Mustard[k]	0.00594*	Total non-foodgrains[k]	0.01173*
	(0.00104)		(0.00254)
		Bankura	
Total foodgrains[k]	0.000863***	Total rice[k]	0.00232**
	(0.000477)		(0.00097)
Total oil seeds[k]	0.004504*	Total foodgrains[k]	0.001873**
	(0.000824)		(0.00091)
Total non-food-grains[k]	0.000964*	Total non-foodgrains[k]	0.002843*
	(0.000301)		(0.00120)
All crops[k]	0.00115*	All crops[k]	0.00238*
	(0.000389)		(0.00088)
24-Parganas		*Nadia*	
Total rice[k]	0.00161*		
	(0.00059)	Total rice	0.001973**
			(0.000547)
Murshidabad		*Purulia*	
Total rice	0.0020*	All crops[k]	0.002989*
	(0.00063)		(0.001352)
West Dinajpur		*Malda*	
Total rice[k]	0.001664*	Total foodgrains[k]	0.001471**
	(0.00054)		(0.00063)

(Contd.)

Table 4.2 : *(Contd.)*

State/District Crop	$\hat{b}^{++}$	State/District Crop	$\hat{b}^{++}$
West Dinajpur		*Malda*	
All crops^k	.00101**	All crops^k	0.00085**
	(0.000414)		(0.000328)
Darjeeling		*Burdwan*	
Total foodgrains^k	0.006601*		
	(0.000883)	Total rice	0.001689*
			(0.000408)
Coochbehar			
Total non-foodgrains^k	0.001456***	Total foodgrains^k	0.001206*
	(0.000787)		(0.00045)
Purulia			
Total rice^k	0.00292**	All crops^k	0.001074*
	(0.00144)		(0.000344)
		Hooghly	
Total foodgrains^k	0.003013**	Total non-foodgrains	0.0088*
	(0.00141)		(0.00232)
Midnapore		*Midnapore*	
Total non-foodgrains	0.003072*	Total rice^k	0.00218*
	(0.00072)		(0.000624)
All crops^k	0.002738*	Total foodgrains^k	0.00175*
	(0.0006005)		(0.000536)

Notes : * Significant at 1% level. ** Significant at 5% level. *** Significant at 10% level.

++ $\hat{b}$ is the estimated coefficient in the equation : $gr_t = a + bt$ [here gr_t implies smoothed value of point-to-point growth rate at period t]

^{k} $\hat{b}$ is estimated after autocorrelation correction,

^{a} free from autocorrelation at 5% level. ^{c} free from autocorrelation at 1% level , α inconclusive range at 1% level. $^+$implies runs are statistically insignificant (i.e, random disturbance term), standard errors are in the parentheses.

Table 4.3 : Alternative Trend Equations and the Values of Test Statistics for the Crops Having Increasing Growth Rates, West Bengal and the Districts, 1957-58 - 1994-95

State/District crop	Parabola			Log-quadratic			S-type growth curve		
	DW	$\bar{R}^2$	runs	DW	$\bar{R}^2$	runs	DW	$\bar{R}^2$	runs
West Bengal									
Aman rice	2.138^{a}	0.6375	19^{++}	2.238^{a}	0.3884	19^{++}	1.412^{c}	.0404	15^{++}
Total rice	1.261^{k}	0.8865	17^{++}	1.453^{c}	0.8629	18^{++}	1.035	0.2834	12
Total cereals	1.158	0.8843	12	1.263^{k}	0.8681	16^{++}	0.363	0.2984	11
TFG	1.129	0.8745	12	1.230^{k}	0.8574	16^{++}	0.371	0.3039	11
Mustard	0.554	0.8740	7	0.730	0.8901	13	0.118	0.1364	5
Total oil seeds	0.985	0.9006	7	1.027	0.9353	13	0.121	0.2101	5
TNFG	1.523^{c}	0.9741	14	1.271^{k}	0.9567	17^{++}	0.144	0.3165	3
All crops	1.245^{k}	0.9281	14	1.381^{k}	0.9195	17^{++}	0.254	0.2958	11
24-Parganas									
Total rice	1.658^{a}	0.8174	21^{++}	1.808^{a}	0.7887	21^{++}	0.592	0.2401	13

(Contd.)

Table 4.3 : (Contd.)

State/District crop	Parabola			Log-quadratic			S-type growth curve		
	DW	$\bar{R}^2$	runs	DW	$\bar{R}^2$	runs	DW	$\bar{R}^2$	runs
24-Parganas									
TFG	1.701[a]	0.8021	21[++]	1.839[a]	0.7718	21[++]	0.639	0.2436	13
All crops	1.819[a]	0.8423	21[++]	1.939[a]	0.8147	21[++]	0.597	0.2844	13
Birbhum									
Total rice	1.543[c]	0.7336	18[++]	1.694[a]	0.6866	22[++]	0.771	0.1853	9
TNFG	1.873[a]	0.4096	6	1.445[c]	0.7493	15[++]	0.987	0.2377	7
Bankura									
Total rice	1.387[k]	0.7725	18[++]	1.557[c]	0.6684	17[++]	0.672	0.1360	9
Bankura									
TFG	1.401[c]	0.7542	18[++]	1.535[c]	0.6525	17[++]	0.707	0.1519	13
TNFG	1.830[a]	0.9403	21[++]	1.654[a]	0.9396	20[++]	0.182	0.1862	5
All crops	1.337[k]	0.8396	16[++]	1.571[c]	0.7637	17[++]	0.519	0.1551	13
Nadia									
Total rice	1.318[k]	0.9370	17[++]	1.927[a]	0.9189	19[++]	0.286	0.2327	11

(Contd.)

Table 4.3 : (Contd.)

State/District crop	Parabola			Log-quadratic			S-type growth curve		
	DW	$\bar{R}^2$	runs	DW	$\bar{R}^2$	runs	DW	$\bar{R}^2$	runs
Murshidabad									
Total rice	1.525^{c}	0.8512	13	2.032^{a}	0.8138	20^{++}	0.539	0.1796	13
West Dinajpur									
Total rice	1.324^{k}	0.8029	16^{++}	1.491^{c}	0.7576	18^{++}	0.473	0.3361	19^{++}
All crops	1.229^{k}	0.8509	16^{++}	1.306^{k}	0.8146	12	0.389	0.4754	9
Darjeeling									
TFG	2.625^{k}	0.7069	18^{++}	2.447^{c}	0.8225	16^{++}	0.651	0.2776	9
Coochbehar									
TNFG	1.883^{a}	0.5148	22^{++}	1.803^{a}	0.4355	23^{++}	0.967	-0.0278	7
Purulia									
Total rice	1.833^{a}	0.5591	21^{++}	1.938^{a}	0.4077	21^{++}	1.241^{k}	0.0443	15^{++}
TFG	1.773^{a}	0.5506	21^{++}	1.884^{a}	0.3986	21^{++}	1.232^{k}	0.0514	13
All crops	1.736^{a}	0.5601	19^{++}	1.852^{a}	0.4079	19^{++}	1.205^{k}	0.0575	21^{++}
Burdwan									
Total rice	1.528^{c}	0.9054	16^{++}	1.693^{a}	0.8887	22^{++}	0.347	0.2038	9
TFG	1.332^{k}	0.8806	16^{++}	1.455^{c}	0.8615	19^{++}	0.364	0.2258	9
All crops	1.389^{k}	0.9472	17^{++}	1.448^{c}	0.9446	16^{++}	0.216	0.2822	7

(Contd.)

Table 4.3 : *(Contd.)*

State/District crop	Parabola			Log-quadratic			S-type growth curve		
	DW	$\bar{R}^2$	runs	DW	$\bar{R}^2$	runs	DW	$\bar{R}^2$	runs
Midnapore									
Total rice	1.480[c]	0.8332	16[++]	1.493[c]	0.7972	20[++]	0.472	0.2143	15[++]
TFG	1.471[c]	0.8284	16[++]	1.482[c]	0.7980	14	0.475	0.2293	15[++]
TNFG	1.087	0.9396	11	1.575[c]	0.9664	17[++]	0.183	0.2733	7
All crops	1.392[c]	0.8816	16[++]	1.437[c]	0.8657	18[++]	0.338	0.2264	11
Hooghly									
TNFG	1.871[a]	0.9644	15[++]	1.324[k]	0.9252	17[++]	0.208	0.2535	3
Malda									
TFG	1.313[k]	0.7489	11	1.358[k]	0.8507	13	0.324	0.4188	9
All crops	1.485[c]	0.9354	19[++]	1.479[c]	0.9155	19[++]	0.183	0.4436	5

Notes : [a]implies no autocorrelation at 5% level.
[c]implies no autocorrelation at 1% level.
[k]implies inconclusive range at 1% level.
[++]implies that the runs are insignificant.

Key : TFG = Total Foodgrains
TNFG = Total Non-foodgrains

Table 4.4 : List of Crops Having Decreasing Growth Rates over time, West Bengal and the Districts, 1957-58—1994-95

State/District Crop	$\hat{b}^{++}$	State/District Crop	$\hat{b}^{++}$
West Bengal		***Darjeeling***	
Boro rice ^{k}	-0.014997*	TNFG ^{k}	-0.002476*
Wheat ^{k}	(0.001539)		(0.000524)
	-0.023984*	***Purulia***	
	(0.001965)	TNFG ^{k}	-0.005326*
Barley ^{k}	-0.009404*		(0.001642)
	(0.0009385)	***Howrah***	
		TNFG ^{k}	-0.006524*
			(0.001054)
Gram ^{k}	-0.004668*	All crops ^{k}	-0.002765*
	(0.000564		(0.0009815)
Total Pulses ^{k}	-0.00153*	***Nadia***	
	(0.000426)	TNFG ^{k}	-0.004712*
Linseed ^{k}	-0.007743*		(0.001147)
	(0.001188)	***Hooghly***	
Jute ^{k}	-0.003188*	Total rice ^{k}	-0.001476**
	(0.00099)		(0.000678)
Sugarcane ^{k}	-0.00206**	TFG ^{k}	-0.001988
	(0.00077)		(0.000695)
Murshidabad		***Jalpaiguri***	
TNFG	-0.00228**	TNFG ^{k}	-0.002891*
	(0.001031)		(0.000458)
		All crops	-0.000342***
			(0.000194)

Notes : Same as those in Table 4.2

Table 4.5 : Alternative Trend Equations and the Results of Related Test Statistics for the Crops Having Decreasing Growth rates. West Bengal and the District 1957-58—1994-95

State/District crop	Straight Line			Parabola			Log-quadratic			S-type growth curve		
	DW	$\bar{R}^2$	runs	DW	$\bar{R}^2$	runs	DW	$\bar{R}^2$	runs	DW	$\bar{R}^2$	runs
West Bengal												
Boro Rice	0.529	0.8576	9	1.178	0.9363	13	0.403	0.9175	11	0.209	0.4156	9
Wheat	0.259	0.3775	6	0.464	0.6341	13	0.302	0.8344	7	0.237	0.4069	7
Barley	0.693	0.2755	5	1.466^{c}	0.6572	12	1.849^{a}	0.8543	18^{++}	0.282	-0.0261	9
Gram	0.949	0.7319	8	1.144	0.7722	16^{++}	0.929	0.8632	15^{++}	0.194	0.1049	5
Total pulses	0.851	0.6369	17^{++}	1.220^{k}	0.7499	16^{++}	1.426^{c}	0.8205	22^{++}	0.313	0.0034	5
Linseed	0.499	0.0393	7	1.225^{k}	0.6091	13	0.963	0.6857	14	0.286	-0.2238	14
Jute	1.914^{a}	0.6451	20^{++}	2.097^{a}	0.6666	24^{++}	2.281^{a}	0.6181	21^{++}	1.129	0.2461	15^{++}
Sugarcane	0.731	0.3165	11	1.128	0.5587	14	1.212	0.6186	14	0.452	-0.2754	9
Murshidabad												
TNFG	1.471^{c}	0.8036	16^{++}	1.977^{a}	0.8495	21^{++}	2.206^{a}	0.8079	19^{++}	0.625	0.3629	7

(Contd.)

Table 4.5 : *(Contd.)*

State/District crop	Straight Line			Parabola			Log-quadratic			S-type growth curve		
	DW	$\bar{R}^2$	runs	DW	$\bar{R}^2$	runs	DW	$\bar{R}^2$	runs	DW	$\bar{R}^2$	runs
Darjeeling												
TNFG	1.163	08277	16^{++}	1.712^{a}	0.8241	16^{++}	1.123	0.7253	17^{++}	0.392	0.2499	5
Purulia												
TNFG	0.724	0.3982	9	0.828	0.4661	8	0.959	0.6016	11	0.623	0.3355	8
Howrah												
TNFG	1.539^{c}	0.7489	16^{++}	1.570^{c}	0.7464	20^{++}	1.973^{a}	0.7572	19^{++}	0.769	0.3119	7
All crops	1.045	0.8221	14	1.454^{c}	0.8679	11	1.787^{a}	0.7916	15^{++}	0.593	0.2428	11
Nadia												
TNFG	1.142	0.8056	11	1.755^{a}	0.8698	21^{++}	2.003^{a}	0.8133	21^{++}	0.592	0.2880	5
Hooghly												
Total rice	1.379^{c}	0.8396	19^{++}	1.570^{c}	0.8566	19^{++}	1.625^{a}	0.8350	19^{++}	0.541	0.2812	11
TFG	1.229	0.8143	15^{++}	1.249^{k}	0.8128	15^{++}	1.291^{k}	0.7970	15^{++}	0.537	0.3015	11
Jalpaiguri												
TNFG	1.255^{k}	0.2694	11	1.417^{c}	0.3427	13	1.357^{k}	0.4236	13	1.005	0.2296	6
All crops	2.082^{a}	0.9236	20^{++}	2.084^{a}	0.9215	20^{++}	2.113^{a}	0.9190	20^{++}	0.355	0.4309	7

Notes: Same as those in Table 4.3

Table 4.6: Classification of Crops[a] on the basis of DW Statistic and Specified Trend Equation, West Bengal by District, 1957-58—1994-95.

Alternative form of the Trend Equation [b]	Crop with region		
	Non-autocorrelated	Autocorrelated	Inconclusive range
Parabola or log-quadratic	*24-Parganas*-Total rice, TFG, All crops. *Bankura*-TFG, TNFG. *Midnapur*-Total rice, All Crops. *Malda*-All crops. *Murshidabad*-TNFG. *Nadia*-TNFG.	*West Bengal*-Mustard, Total oil seeds, Boro rice and Gram.	*Malda*-TFG. *Hooghly*-TFG.
Straight Line or Parabola or Log-quadratic	*West Bengal*-Jute *Jalpaiguri*-All crops		

Notes: a Trend equation could not be selected for these crops on the basis of either DW statistic or $\bar{R}^2$ or runs.

b Choice is restricted on the basis of point-to-point growth rates.

Table 4.7: Measurements of Heteroscedasticity for Non-autocorrelated Crops[a], West Bengal and the Districts, 1957-58—1994-95

District/State	Parabola		Log-quadratic	
Crop	Rank test	Glejser test	Rank test	Glejser test
24 Parganas				
Total rice	0.1220	Insignificant[+]	-0.1432	Insignificant[+]
TFG	0.1369	Insignificant[+]	-0.1542	Insignificant[+]
All crops	0.0389	Insignificant[+]	-0.2650***	Insignificant[+]
Bankura				
TFG	-0.0441	Insignificant[+]	-0.1874	Insignificant[+]
TNFG	0.5730*	$\mid e_t \mid^{++}$ = -0.406+1.622*t (9.35) (0.418) DW=2.26	0.0130	Insignificant[+]
Midnapore				
Total rice	-0.0100	Insignificant[+]	-0.1747	Insignificant[+]
All crops	0.0712	Insignificant[+]	-0.1682	Insignificant[+]
Malda				
All crops	0.1710	Insignificant[+]	-0.2636***	$\mid e_t \mid^{+++}$ = 0.049*+0.268*1/t (0.011) (.051) DW=1.69
Murshidabad				
TNFG	-0.0605	Insignificant[+]	-0.3198**	$\mid e_t \mid^{++}$ = 0.221*-0.0037**t (0.041) (.0018) DW=2.07
Nadia				
TNFG	-0.2003	Insignificant[+]	-0.3632**	$\mid e_t \mid^{\wedge}$ = 0.285*-0.006**t (0.053) (.002) DW=2.17

(Contd.)

Table 4.7: *(Contd.)*

District/State	Parabola		Log-quadratic	
Crop	Rank test	Glejser test	Rank test	Glejser test
West Bengal				
Jute	-0.2658***	Insignificant+	-0.4951*	$\lvert e_t \rvert^{\lambda}$ = 0.246-0.0046**t (0.047) (.002) DW=1.98
Jalpaiguri				
All crops	-0.0406	Insignificant+	-0.1102	Insignificant+

Notes : * Significant at 1% level; ** Significant at 5% level; *** Significant at 10% level.

\+ Trend coefficient is insignificant in all forms of Glejser test (i.e, 1/t, t^2, $\sqrt{t}$ & t).

++ Significant trend coefficients also in the forms of 1/t, t^2, $\sqrt{t}$;

+++ Significant trend coefficient in all cases except t^2,

λ Significant trend coefficient in all cases except 1/t; Terms within parentheses imply standard errors;

ª Trend equation is to be chosen from parabola and log-quadratic form for all cases except the last two and also from linear equation for last two cases.

Table 4.8 : Results of P.E. Test for Non-autocorrelated and Homoscedastic Crops, West Bengal by Districts, 1957-58—1994-95

District/State	Parabola[a]			Logquadratic[a]		
Crop	$\bar{R}^2$	DW	ϕ_0 [b]	$\bar{R}^2$	DW	ϕ_1 [c]
24-Parganas						
Total rice	0.8545*	1.97+	2490.306** (987.115)	0.8178*	1.99+	-0.00288*** (0.001585)
TFG	0.8375*	1.96+	2889.043** (1270.318)	0.7834*	2.01+	-0.003122*** (0.00184)
Bankura						
TFG	0.7911*	1.69+	1956.677* (729.890)	0.6680*	1.66+	-0.00460 (.00284)
Midnapore						
Total rice	0.8711*	1.96+	3521.962* (1055.422)	0.8159*	1.69+	-0.001821** (0.00085)
All crops	0.9162*	2.00+	243.645* (62.015)	0.8786*	1.64+	-0.01159** (0.00533)
Jalpaiguri						
All crops	0.9193*	2.08+	-125.690 (641.943)	0.9169*	2.12+	-0.00015*** (0.000093)

Notes : [a] artificial regressions,

[b] ϕ_0 be the coefficient of l_{1t} in the artificial regression equation $y_t = a + bt + ct^2 + \phi_0 l_{1t}$
[Where l_{1t} = predicted value of log-quadratic minus logarithm of predicted value of parabolic].

[c] ϕ_1 be the coefficient of l_{2t} in the artificial regression equation $\ln y_t = a + bt + ct^2 + \phi_1 l_{2t}$;
[Where l_{2t} = predicted value of parabolic minus the exponential of the predicted value of log-quadratic]. Here predicted values have been calculated from the original series, fitting the respective trend equation.

* Significant at 1% level; ** Significant at 5% level;

*** Significant at 10% level; + Non-autocorrelated at 5% level;

\# trend equation for these crops is to be chosen from parabola and log-quadratic.

Table 4.9 : Measurement of Heteroscedasticity for Autocorrelated[++] Crops#, West Bengal by Districts, 1957-58—1994-95

Crop with region	Estimated Equation		Equation selected
	Parabola	Log-quadratic	
	Rank Test	Rank Test	
West Bengal			
Mustard	0.5691*	0.03308**	Log-quadratic[k]
Total Oil Seeds	0.6632*	0.3104***	Log-quadratic[k]
Boro rice	0.5481*	-0.5433*	Log-quadratic[k]
Gram	-0.3468*	0.2458	Log-quadratic
Malda			
TFG	0.6188*	0.1047	Log-quadratic
Hooghly			
TFG	0.1102	-0.3345**	Parabola

Notes : * Significant at 1% level.

** Significant at 5% level.

*** Significant at 10% level.

Trend equation is to be selected from parabola and log-quadratic.

++ Also the crops for which DW statistic lies within the indecisive range.

k Heteroscedasticity test remains inconclusive regarding choice of trend equation, yet log-quadratic has been chosen for other reason (see text).

Table 4.10 : Cropwise List of the Chosen Trend Equation and Results of Test Statistics, West Bengal and the Districts, 1957-58 — 1994-95

Trend Equation	Criterion	Name of the crop (with region)
Exponential	Constancy of point-to-point growth rates	*West Bengal*-aus rice, other cereals, til, potatoes and tobacco; *Burwan*-TNFG; *Birbhum* - TFG and all crops; *Howrah*-rice and TFG: *Hooghly*-all crops; *24-Parganas*-TNFG; *Nadia*-TFG and all crops; *Murshidabad*-TFG and all crops; *West Dinajpur*-TFG; *Malda* -rice and TNFG; *Jalpaiguri*-rice and TFG; *Coochbehar*-rice, TFG and all crops; *Darjeeling*-rice and all crops.
Parabola	DW Statistic	*W.B.*-TNFG and linseed; *Hooghly*-TNFG; *West Dinajpur*-TNFG; *Darjeeling*-TNFG.
	Run Test	*Midnapore*-TFG; *West Dinajpur*-all crops.
	$\bar{R}^2$	*W.B.*-aman rice; *Coochbehar* - TNFG; *Purulia* - rice, TFG and all crops.
	Heteroscedasticity (rank/Glejser) Test	*Hooghly*-TFG; *24-Parganas*-all crops; *Murshidabad*-TNFG; *Nadia*-TNFG; *Malda*-all crops.
	PE Test	*Jalpaiguri*-all crops.
Log-quadratic	DW Statistic	*W.B.*-rice, barley, other pulses, total pulses, and sugarcane; *Burdwan* -rice, TFG and all crops; *Birbhum*-rice; *Bankura*-rice and all crops; *Midnapore*-TNFG; *Howrah*-TNFG and all crops; *Hooghly*-rice; *Nadia*-rice; *West Dinajpur*-rice, *Darjeeling*-TFG;
	Run Test	*W.B.*-total cereals, TFG and all crops; *Bhirbhum*-TNFG; *Murshidabad* - rice.
	$\bar{R}^2$	*WB.*-wheat; *Purulia*-TNFG.
	Heteroscedasticity (rank/Glejser) Test	*W.B.*-gram; *Bankura*-TNFG; *Malda*-TFG
	PE Test	*Bankura*-TFG.
	Others[1]	*W.B.*-boro rice, mustard and total oilseeds; *Midnapore*-rice and all crops; *24-Parganas*-rice and TFG.
Straight Line	Heteroscedasticity (rank/Glejser) Test	*W.B.*-Jute

Key : rice = total rice, TFG = total foodgrains, TNFG = total non-foodgrains, W.B. = West Bengal.

Note: [1] See text.

Table 4.11: List of Crops with Disturbance Term Decisively not Free from Autocorrelation[k], West Bengal and Districts, 1957-58—1994-95

Crops (with region) having autocorrelation in the disturbance term at 5% level	Crops (with region) having DW Statistic in indecisive range at 5% level
West Bengal Aus-rice+, Boro-rice+, Wheat+, Other Cereals+, Total Cereals, Gram+, Other Pulses, Total Foodgrains, Mustard+, Linseed, Til+, Total Oilseeds+, Sugarcane, Potatoes+, Tobacco+.	*West Bengal* Total Rice*, Total Pulses*, All Crops**, Total Non-Foodgrains*.
Burdwan Total Non-Foodgrains+.	*Bankura* Total Rice*, Total Foodgrains*, All Crops*.
Hooghly Total Foodgrains.	*Birbhum* Total Foodgrains*, All Crops*, Total Non-Foodgrains*.
Malda Total Foodgrains.	*West Dinajpur* Total Rice*, Total Non-Foodgrains*.
Nadia Total Foodgrains*, All Crops.	*Darjeeling* Total Foodgrains***.
Murshidabad Total Foodgrains*.	*Burdwan* Total Foodgrains*, All Crops*.
Coochbehar Total Rice, Total Foodgrains*.	*Midnapore* Total Rice*, Total Foodgrains*, Total Non-Foodgrains*, All Crops*.
West Dinajpur All Crops.	*Malda* All Crops*.
Purulia Total Non-Foodgrains+.	*Jalpaiguri* Total Non-Foodgrains*.

Notes: k autocorrelation should be detected at 5% level.
+ autocorrelation also at 1% level.
*Non-autocorrelated at 1% level, ** indecisive at 1% level.
***possibility of negative autocorrelation.

Table 4.12 : Heteroscedasticity Tests for Crops[a] with Non-autocorrelated Disturbances, West Bengal by Districts, 1957-58—1994-95

District/ Crop	Parametric (Glejser) Test	Non-Parametric (Rank) Test	Trend Equation selected
Howrah			
Rice	$\lvert e_t \rvert^c = 0.308^{*} - 0.0054^{**}t$ (0.054) (0.0024)	-0.314***	Exponential
TFG	$\lvert e_t \rvert^c = 0.295^{*} - 0.0052^{**}t$ (0.049) (0.0022)	-0.3196**	Exponential
All Crops	$\lvert e_t \rvert^c = 0.286^{*} - 0.0059^{*}t$ (0.045) (0.0020)	-0.377**	Log-quadratic
Hooghly			
Rice	$\lvert e_t \rvert^b = 0.179^{*} - 0.0025^{***}t$ (0.032) (0.0014)	-0.339**	Log-quadratic
All Crops	$\lvert e \rvert_t^c = 0.144^{*} - 0.0029^{**}t$ (0.025) (0.0011)	-0.4704*	Exponential
24-Parganas			
TNFG	$\lvert e_t \rvert^d = 0.258^{*} - 0.0052^{*}t$ (0.035) (0.0015)	-0.431	Exponential
Malda			
Rice	$\lvert e_t \rvert^d = 0.152^{*} - 0.0024^{**}t$ (0.026) (0.0011)	-0.216+	Exponential

Notes: + insignificant; * significant at 1% level; ** significant at 5% level;
*** significant at 10% level.
[a] Crops suffering from Heteroscedasticity
[b] Heteroscedasticity also at t^2
[c] Heteroscedasticity also at t^2 & $\sqrt{t}$
[d] Heteroscedasticity also at $1/t$, t^2 & $\sqrt{t}$

Table 4.13 : Heteroscedasticity Tests for Crops [K] Having DW Statistic in Inconclusive Region. West Bengal by Districts, 1957-58—1994-95

State/District Crop	Parametric(Glejser) Test	Non-Parametric (Rank) Test	Trend Equation selected
West Bengal			
Total Pulses	$\lvert e_t \rvert^a$ = 0.081*+ 0.1643**1/t (0.014) (0.068) Heteroscedasticity	-0.043+ Homoscedastic	Log-quadratic
All Crops	Insignificant (in all forms of t) Homoscedastic	-0.371** Heteroscedasticity	Log-quadratic
West Dinajpur			
Rice	$\lvert e_t \rvert^a$ =0.089* + 0.208**1/t (0.017) (0.084) Heteroscedasticity	-0.0896+ Homoscedastic	Log-quadratic
TNFG	$\lvert e_t \rvert^a$ = 13.107*+22.429**1/t (2.079) (10.075) Heteroscedasticity	-0.232+ Homoscedastic	Parabola
Jalpaiguri			
TNFG	Insignificant (in all forms of t) Homoscedastic	0.267*** Heteroscedasticity	Parabola
Darjeeling			
TFG	$\lvert e_t \rvert^b$ =0.072***+ 0.0029***t (0.038) (0.0017) Heteroscedasticity	0.156+ Homoscedastic	Log-quadratic

Notes : + Insignificant; * significant at 1% level; ** significant at 5 % level; *** significant at 10 % level.

K Crops suspected to suffer from Heteroscedasticity.

a Homoscedastic in other forms of t.

b Heteroscedasticity also at t^2.

Table 4.14 : Heteroscedasticity Test for Autocorrelated Crops[a], West Bengal and the Districts, 1957-58 to 1994-95

State/District Crop	Chosen Trend Equation	Non-Parametric(rank) test
West Bengal		
Boro	Log-quadratic	-0.5433*
Til	Exponential	0.5218*
Wheat	Log-quadratic	-0.446*
Mustard	Log-quadratic	0.331**
Total Oil-seeds	Log-quadratic	0.3104***
Potatoes	Exponential	-0.547*
Murshidabad		
TFG	Exponential	-0.276***

Notes : * Significant at 1 % level ; ** significant at 5 % level;
*** significant at 10 % level.
[a] crops suffering from Heteroscedasticity.

Table 4.15 : Period-wise Measured Standard Deviation of Point-to-Point Growth Rates in Production of some Selected Crops, Presented in Table 4.14.

State/District Crop	Standard Deviation of Point-to-Point Growth Rates		
	Whole Period (1957-58—1994-95)	Ist Sub-period (1957-58—1976-77)	2nd Sub-Period (1977-78—1994-95)
West Bengal			
Total Pulses	0.1655	0.1780	0.1465
West Dinajpur			
Rice	0.1910	0.2239	0.1551
TNFG	0.3058	0.3744	0.2172
Darjeeling			
TFG	0.3199	0.1468	0.4381

Table 4.16 : Regression Results of the Crop Data with and without Adjustment through 3-Period Centred Moving Average and First Order Autoregressive Scheme, West Bengal and the Districts, 1957-58—1994-95

State/District	'hosen trend	Criteria before Adjustment $^{\grave{A}}$					Criteria after Adjustment $^{\grave{A}\grave{A}}$					
Crop	equation	DW	Runs	σ_e^{+}	$\bar{R}^2$	rank^{a}	$\hat{b}^{++}$	DW	Runs	$\bar{R}^2$	σ_e^{+}	rank^{a}
West Bengal												
Aman rice	Parabola	2.138^{c}	19^{x}	948.812	.6375*	-.1432^{d}	.6576* (.1231)	1.627^{c}	17^{x}	.9180*	352.049	.0692^{d}
Total Pulses	Log-quadratic	1.426	22^{x}	.128	.8205*	-.0432^{d}	.6818* (.1129)	1.666^{c}	19^{x}	.9592*	.054	-.2239^{d}
Jute	Straight line	1.914^{c}	20^{x}	773.738	.6451*	-.1058^{d}	.6629* (.1288)	1.603^{c}	16^{x}	.8934*	342.823	.1488^{d}
Howrah												
Rice	Exponential	1.910^{c}	16^{x}	.268	.6767*	-.3135***		.7814*	1.591^{c}		15^{x}	.9366*
.099	-.1133^{d}						(.1063)					
TNFG	Log-quadratic	1.973^{c}	19^{x}	.273	.7572*	-.0879^{d}	.6256* (.1409)	1.605^{c}	18^{x}	.9246*	.1291	-.1449^{d}
24-Parganas												
TNFG	Exponential	1.624^{c}	16^{x}	.203	.8127*	-.4314*	.7645* (.1062)	1.789^{c}	14^{x}	.9518*	.085	-.1506^{d}
Murshidabad												
Rice	Log-quadratic	2.032^{c}	20^{x}	.168	.8138*	.0513^{d}	.6386* (.1283)	1.857^{c}	21^{x}	.9521*	.0745	.1701^{d}

(Contd.)

Table 4.16: *(Contd.)*

State/District	Chosen trend	Criteria before Adjustment À					Criteria after Adjustment À À					
Crop	equation	DW	Runs	σ_e^+	$\bar{R}^2$	rank[a]	$\hat{b}^{++}$	DW	Runs	$\bar{R}^2$	σ_e^+	rank[a]
Murshidabad												
TFG	Exponential	1.416	17[x]	.145	.8361*	-.2758***	.7893* (.0998)	1.873[c]	20[x]	.9590*	.063	.1017[d]
Darjeeling												
TFG	Log-quadratic	2.447[e]	16[x]	.176	.8225*	.1555[d]	.5350* (.1446)	1.744[c]	15[x]	.9606*	.0712	.1838[d]

Notes : + ∂_e denotes the standard deviation of the residuals around trend equation, which is a measure of short-term variation,
[a] rank denotes rank correlation between time and absolute value of residual, ++$\hat{b}$ is the coefficient of first order autoregressive scheme $e_t = b\, e_{t-1}$; À number of observations in the original series before adjustment is 38, À À number of observations in the series after adjustment is 36 (deleting first and last one); Terms within the parentheses denote S.E.,
[c] implies the absence of autocorrelation; [d] implies presence of homoscedasticity.
[e] implies presence of negative autocorrelation; [x] implies insignificant runs.
* significant at 1% level.
** significant at 5% level.
*** significant at 10% level.

Table 4.17 : Regression Results of the Crop Data with and without Adjustment through 3-Period Centred Moving Average and Second Order Autoregressive Scheme, West Bengal and the Districts, 1957-58—1994-95

Name of the crop with region	Chosen trend equation	Criteria before Adjustment ^{À}					Coefficients of second order autoregressive scheme^{++}		Criteria after Adjustment $^{\text{À À}}$				
		DW	Runs	$\bar{R}^2$	σ_e^+	rank[a]	$\hat{b}^{++}$	$\hat{c}$	DW	Runs	$\bar{R}^2$	σ_e^+	rank[a]
Aus-W.B.	Exponential	.866	9*	.4730*	.2053	-.1494[d]	1.247* (.162)	-.3883** (.1591)	2.190[c]	18[x]	.8731*	.0653	.0461[d]
Other cereals-W.B.	Do	.953	11**	.5450*	.3152	-.1789[d]	1.352* (.123)	-.7159* (.124)	2.275[c]	20[x]	.9025*	.1079	.2453[d]
Potatoes-W.B.	Do	1.221	11**	.9521*	.1741	-.5470*	1.321* (.148)	-.5169* (.147)	2.194[c]	22[x]	.9951*	.0519	-.1094[d]
TNFG-Burdwan	Do	.957	11**	.9123*	.2082	-.1686[d]	1.343* (.146)	-.5262* (.149)	2.181[c]	15[x]	.9898*	.0647	.0337[d]
TFG-Birbhum	Do	1.452	16[x]	.6155*	.1673	-.0498[d]	1.105* (.161)	-.4063** (.159)	2.161[c]	20[x]	.9035*	.0641	.2085[d]

(Contd.)

Table 4.17 : (Contd.)

Name of the crop with region	Chosen trend equation	Criteria before Adjustment À					Coefficients of second order autoregressive scheme[++]		Criteria after Adjustment À À				
		DW	Runs	$\bar{R}^2$	σ_e^+	rank[a]	$\hat{b}^{++}$	$\hat{c}$	DW	Runs	$\bar{R}^2$	σ_e^+	rank[a]
AC-Birbhum	Exponential	1.452	18[x]	.7084*	.1566	.0456[d]	1.144* (.156)	-.4652* (.155)	2.227[c]	22[x]	.9368*	.0594	.2250[d]
TFG-Howrah	Do	1.831[c]	16[x]	.6716*	.2513	-.3196**	1.085* (.162)	-.4063** (.165)	2.283[c]	20[x]	.9408*	.0889	-.0010[d]
TFG-Nadia	Do	1.359	17[x]	.9095*	.1345	-.1266[d]	1.035* (.151)	-.4034** (.151)	2.023[c]	20[x]	.9816*	.0551	-.1351[d]
AC-Nadia	Do	1.232	15[x]	.9277*	.1284	-.2474[d]	1.173* (.153)	-.4223* (.153)	2.284[c]	20[x]	.9876*	.0483	-.0746[d]
Rice-Malda	Do	1.715[c]	18[x]	.9090*	.1343	-.2163[d]	.939* (.144)	-.3125** (.144)	1.853[c]	17[x]	.9897*	.0399	-.0196[d]
TNFG-Malda	Do	1.622	17[x]	.4468*	.2168	.0533[d]	1.029* (.176)	-.3694** (.176)	2.182[c]	21[x]	.8160*	.0896	.0914[d]

(Contd.)

Table 4.17 : (Contd.)

Name of the crop with region	Chosen trend equation	Criteria before Adjustment À					Coefficients of second order autoregressive scheme++		Criteria after Adjustment À À				
		DW	Runs	$\bar{R}^2$	σ_e^+	rank[a]	$\hat{b}^{++}$	$\hat{c}$	DW	Runs	$\bar{R}^2$	σ_e^+	rank[a]
Rice-Coochbehar	Exponential	1.237	15[x]	.7476*	.1582	-.1870[d]	1.213* (.159)	-.4279* (.154)	1.819[c]	19[x]	.9586*	.0518	-.1436[d]
TFG-Coochbehar	Do	1.175	16[x]	.8131*	.1165	-.0933[d]	1.221* (.152)	-.5212* (.146)	1.854[c]	17[x]	.9690*	.0482	-.0646[d]
Rice-Darjeeling	Do	1.799[c]	19[x]	.5813*	.1564	-.2351[d]	.964* (.150)	-.3706* (.152)	2.095[c]	20[x]	.9110*	.0541	.0167[d]
Linseed-WB	Parabola	1.225	13***	.6091*	3.2028	.1701[d]	1.261* (.159)	-.4863* (.155)	2.129[c]	17[x]	.9386*	1.012	.0178[d]
TFG-Hooghly	Do	1.249	15[x]	.8128*	64.632	.1102[d]	1.301* (.156)	-.4757* (.158)	2.069[c]	18[x]	.9758*	19.879	.1691[d]
TNFG-W.Dinajpur	Do	1.483	15[x]	.6616*	19.473	-.2321[d]	1.044* (.139)	-.5421* (.124)	2.343[c]	17[x]	.9364*	6.095	-.1547[d]

(Contd.)

Table 4.17 : *(Contd.)*

Name of the crop with region	Chosen trend equation	Criteria before Adjustment [λ]					Coefficients of second order autoregressive scheme[++]		Criteria after Adjustment [λλ]				
		DW	Runs	$\bar{R}^2$	σ_e^+	rank[a]	$\hat{b}^{++}$	$\hat{c}$	DW	Runs	$\bar{R}^2$	σ_e^+	rank[a]
AC-West Dinajpur	Parabola	1.229	16[x]	.8509*	11.868	.0487[d]	1.184* (.149)	-.4418* (.140)	1.779[c]	14[x]	.9791*	3.768	.0682[d]
TNFG-Jalpaiguri	Do	1.417	13***	.3422*	21.065	.2665***	1.079* (.174)	-.3755** (.177)	2.078[c]	14[x]	.7844*	8.177	.2005[d]
Rice-W.B.	Log-quadratic	1.453	18[x]	.8629*	.1003	-.1610[d]	1.249* (.149)	-.5119* (.144)	1.782[c]	15[x]	.9824*	.0361	.1552[d]
Total cereals-W.B.	Do	1.263	16[x]	.8681*	.1179	-.2590[d]	1.339* (.139)	-.6012* (.138)	1.702[c]	18[x]	.9834*	.0368	.0520[d]
Other pulses-W.B.	Do	1.256	17[x]	.5968*	.1647	-.0257[d]	1.007* (.155)	-.2661** (.145)	2.116[c]	19[x]	.9101*	.0599	-.0664[d]

(Contd.)

Table 4.17 : *(Contd.)*

Name of the crop with region	Chosen trend equation	Criteria before Adjustment [À]					Coefficients of second order autoregressive scheme[++]		Criteria after Adjustment [À À]				
		DW	Runs	$\bar{R}^2$	σ_e^+	rank[a]	$\hat{b}^{++}$	$\hat{c}$	DW	Runs	$\bar{R}^2$	σ_e^+	rank[a]
TFG-W.B.	Log-quadratic	1.230	16[x]	.8574*	.1153	-.2518[d]	1.336* (.139)	-.5899* (.138)	1.698[c]	18[x]	.9815*	.0360	.0757[d]
AC-W.B.	Do	1.381	17[x]	.9195*	.0908	-.3710**	1.252* (.139)	-.5791* (.136)	1.843[c]	16[x]	.9894*	.0296	.1552[d]
Rice-Burdwan	Do	1.693[c]	22[x]	.8887*	.1098	-.2238[d]	1.191* (.150)	-.5229* (.149)	1.906[c]	19[x]	.9847*	.0366	.0278[d]
TFG-Burdwan	Do	1.455	19[x]	.8615*	.1151	-.0566[d]	1.284* (.145)	-.5682* (.145)	1.869[c]	19[x]	.9812*	.0374	-.0299[d]
AC-Burdwan	Do	1.448	16[x]	.9446*	.0878	.1406[d]	1.220* (.132)	-.6463* (.133)	1.710[c]	16[x]	.9929*	.0293	-.0203[d]
Rice-Bankura	Do	1.557	17[x]	.6684*	.1933	-.2043[d]	1.192* (.149)	-.5353* (.147)	2.016[c]	19[x]	.9374*	.0659	-.0324[d]

(Contd.)

Table 4.17 : *(Contd.)*

Name of the crop with region	Chosen trend equation	Criteria before Adjustment ^À^					Coefficients of second order autoregressive scheme++		Criteria after Adjustment ^ÀÀ^				
		DW	Runs	$\bar{R}^2$	σ_e^+	rank[a]	$\hat{b}^{++}$	$\hat{c}$	DW	Runs	$\bar{R}^2$	σ_e^+	rank[a]
TFG-Bankura	Log-quadratic	1.535	17[x]	.6525*	.1943	-.1874[d]	1.214* (.148)	-.5457* (.147)	2.031[c]	19[x]	.9333*	.0659	-.0471[d]
TNFG-Bankura	Do	1.654[c]	20[x]	.9396*	.1963	.0130[d]	1.018* (.163)	-.4807* (.161)	2.217[c]	20[x]	.9913*	.0707	-.0512[d]
AC-Bankura	Do	1.571	17[x]	.7637*	.1842	-.1303[d]	1.223* (.145)	-.5733* (.144)	2.039[c]	23[x]	.9618*	.0617	-.0376[d]
TNFG-Midnapore	Do	1.575	17[x]	.9664*	.1586	-.0369[d]	1.206* (.144)	-.6144* (.143)	2.050[c]	17[x]	.9959*	.0514	.0770[d]
Rice-Hooghly	Do	1.625[c]	19[x]	.8350*	.1652	-.3394**	1.107* (.164)	-.3690** (.165)	2.045[c]	17[x]	.9735*	.0576	.0592[d]
Rice-Nadia	Do	1.927[c]	19[x]	.9189*	.1521	-.1412[d]	.7506* (.168)	-.3129*** (.161)	2.092[c]	17[x]	.9839*	.0623	.0543[d]

(Contd.)

Table 4.17 : *(Contd.)*

Name of the crop with region	Chosen trend equation	Criteria before Adjustment À					Coefficients of second order autoregressive scheme++		Criteria after Adjustment À À				
		DW	Runs	$\bar{R}^2$	σ_e^+	rank[a]	$\hat{b}^{++}$	$\hat{c}$	DW	Runs	$\bar{R}^2$	σ_e^+	rank[a]
TFG-Malda	Log-quadratic	1.358	13***	.8507*	.1659	.1047[d]	1.233* (.149)	-.5936* (.147)	1.728[c]	12**	.9785*	.0544	.1514[d]
Rice-West Dinajpur	Do	1.491	18[x]	.7576*	.1487	-.0896[d]	1.006* (.154)	-.2408*** (.141)	1.762[c]	15[x]	.9649*	.0458	-.2441[d]
TNFG-Purulia	Do	.959	11**	.6016*	.3749	.1257[d]	1.482* (.134)	-.6931* (.134)	2.025[c]	18[x]	.9458*	.1082	.0126[d]

Notes : 1. Notations have the same meanings as those in the Notes of Table 4.16.

2. ++ $\hat{b}$ & $\hat{c}$ are the coefficients of second order autoregressive scheme, $e_t = b\, e_{t-1} + c\, e_{t-2}$.

Key : W.B = West Bengal ; AC = All crops, Other as before.

Table 4.18 : Regression Results of the Crop Data with and without Adjustment through Autoregressive Scheme, West Bengal and the Districts, 1957-58—1994-95

Name of the Crop with Region	Chosen trend equation	Criteria before Adjustment $^{\lambda}$					Criteria after Adjustment $^{\lambda\lambda}$					
		DW	Runs	$\bar{R}^2$	σ_e^+	rank^a	$\hat{b}^{++}$	DW	Runs	$\bar{R}^2$	σ_e^+	rank^a
Wheat - W.B.^{S} .2560^d	Log-quadratic	0.302	7*	.8344*	.5680	-.4457*	1.341* (.142)	2.216^c	19^x	.9666*	.2427	-
Gram-W.B.	Do	0.929	15^x	.8632*	.2277	.2485^d	.5384* (.1485)	1.714^c	22^x	.8922*	.1941	.2173^d
Sugarcane-W.B.	Do	1.212	14***	.6186*	.2268	.1542^d	.3548** (.1493)	1.739^c	16^x	.6736*	.2004	.2302^d
TNFG-Birbhum	Do	1.445	15^x	.7493*	.3288	.1467^d	.2599*** (.160)	2.023^c	19^x	.7587*	.2121	.1846^d
Rice-Midnapore	Do	1.493	20^x	.7972*	.1462	-.1747^d	.2486*** (.160)	2.006^c	22^x	.8072*	.1408	-.1362^d
All Crops-Midnapore	Do	1.437	18^x	.8657*	.1396	-.1682^d	.2734*** (.159)	2.018^c	22^x	.8753*	.1335	-.1253^d

(Contd.)

Table 4.18 : (Contd.)

Name of the Crop with Region	Chosen trend equation	Criteria before Adjustment [À]					Criteria after Adjustment [À À]					
		DW	Runs	$\bar{R}^2$	σ_e^+	rank[a]	$\hat{b}^{++}$	DW	Runs	$\bar{R}^2$	σ_e^+	rank[a]
TNFG-W.B.	Parabola	1.523	14**	.9741*	10.9032	.0150[d] (.1565)	.1982[I]	2.104[c]	20[x]	.9772*	10.134	.1248[d]
TFG-Midnapore	Do	1.471	16[x]	.8284*	175.813	-.0005[d]	.2518*** (.143)	2.075[c]	22[x]	.8403*	168.14	-.0496[d]
AC- Malda	Do	1.485	19[x]	.9354*	11.616	.1710[d]	.1884[I] (.150)	1.963[c]	21[x]	.9468*	10.310	.1056[d]

Notes : Same as those in Table 4.16, except,

À À the number of observations after adjustment remain as it is, i.e., 38.

S Corrected by second order autoregressive scheme and coefficient $\hat{c}$ = -.5808* (.140)

I implies significant at 20% level.

Table 4.19 : Regression Results of the Crop Data with and without Adjustment through Varying Periods of Moving Average, West Bengal and Districts, 1957-58—1994-95

Name of the Crop with region	Trend Equation	Chosen Criteria before adjustment					Estimated coefficient of auto-regressive scheme;	Nature of varying period of moving average	Criteria after adjustment				
		DW	Runs	$\bar{R}^2$	σ_e^+	rank[a]	$\hat{b}$++		DW	Runs	$\bar{R}^2$	σ_e^+	rank[a]
Til-WB	Exponential	0.489	12**	0.9397*	0.3662	0.5218*	0.7906* (0.1018)	Second Period Smoothed by MA(3)[i.e., 1977-78 to 1994-95]	1.737[c]	13***	.9810*	.1975	.2344[d]
Tobacco-W.B.	Do	0.937	11**	.0616***	0.2088	0.2511[d]	0.5842* (0.1354)	Do	1.793[c]	12***	.1963*	.1502	.1238[d]
AC-Hooghly	Do	1.619[c]	19[x],	0.9422*	0.1206	-0.4704*	0.4487* (0.1499)	First period smoothed by MA(3)[i.e., 1957-58 to 1976-77]	1.956[c]	22[x]	.9746*	.0792	.0544[d]
Boro-rice W.B.[s]	Log-quadratic	0.430	11**	0. 9175*	0.4887	-0.5433*	1.124* (0.154)	Do	2.272[c]	20[x]	.9828*	.2208	.0618[d]

(*Contd.*)

Table 4.19 : *(Contd.)*

Name of the Crop with region	Chosen Trend Equation	Criteria before adjustment DW	Runs	$\bar{R}^2$	σ_e^+	rank[a]	Estimated Nature of coefficient varying of auto-regressive moving average scheme; $\hat{b}^{++}$	period of	Criteria after adjustment DW	Runs	$\bar{R}^2$	σ_e^+	rank[a]
Mustard-W.B.	Log-quadratic	0.730	13***	o.8901*	0.2897	0.3308**	0.5925* (0.1321)	Second period smoothed by MA(3)[i.e., 1977-78 to 1994-95	2.103[c]	20[x]	.9448*	.2046	.0032[d]
Total Oil Seeds- W.B.	Do	1.027	13***	0.9353*	0.2131	0.3104***	0.4371* (0.1476)	Do	2.101[c]	16[x]	.9645*	.1584	-.0938[d]
AC- Howrah	Do	1.787[c]	15[x]	0.7916*	0.2293	-0.3774**	0.5313* (0.1445)	First period smoothed by MA(3)[i.e., 1957-58 to 1976-77]	1.788[c]	17[x]	.9149*	.1370	-.0504[d]

Note: s corrected by second order autoregressive scheme and coefficient $\hat{c}$ = -0.3522**; others are same as those in Table 4.16
(0.154)

Table 4.20 : Regression Results of the Crop Data without Adjustment, West Bengal and the Districts, 1957-58—1994-95

Name of the Crop with Region	Chosen Trend Equation	Estimated Regression Results				
		DW	Runs	$\bar{R}^2$	σ_e^+	Rank[a]
AC-Murshidabad	Exponential	1.634[c]	16[x]	.8996*	.1183	-.0445[d]
TFG-West-Dinajpur	Do	1.542[c]	18[x]	.8243*	.1285	-0.0026[d]
Rice-Jalpaiguri	Do	1.592[c]	21[x]	.3957*	.147	.1248[d]
TFG-Jalpaiguri	Do	1.704[c]	21[x]	.5340*	.132	.0409[d]
AC-Darjeeling	Do	1.946[c]	18[x]	.8924*	.1062	.0052[d]
AC-Coochbehar	Do	1.810[c]	21[x]	.8215*	0.1036	-.1137[d]
TNFG-Hooghly	Parabola	1.871[c]	15[x]	.9644*	24.332	-.0359[d]
AC-24-Parganas	Do	1.819[c]	21[x]	.8423*	16.253	.0389[d]
TNFG-Nadia	Do	1.755[c]	21[x]	.8698*	29.323	-.2003[d]
TNFG-Murshidabad	Do	1.977[c]	21[x]	.8495*	22.614	-.0605[d]
AC-Jalpaiguri	Do	2.084[c]	20[x]	.9215*	5.888	-.0406[d]
TNFG-Darjeeling	Do	1.712[c]	16[x]	.8241*	14.813	-.2517[d]
TNFG-Coochbehar	Do	1.883[c]	22[x]	.5148*	16.586	-.1238[d]
Rice-Purulia	Do	1.833[c]	21[x]	.5591*	57.869	-.0673[d]
TFG-Purulia	Do	1.773[c]	21[x]	.5506*	61.643	-.0651[d]
AC-Purulia	Do	1.736[c]	19[x]	.5601*	20.025	-.0384[d]
Barley-W.B.	Log-quadratic	1.849[c]	18[x]	.8543*	.2251	-.2093[d]
Rice-Birbhum	Do	1.694[c]	22[x]	.6866*	.1528	.0507[d]
Rice-24-Parganas	Do	1.808[c]	21[x]	.7887*	.1487	-.1432[d]
TFG-24-Parganas	Do	1.839[c]	21[x]	.7718*	.1482	-.1542[d]

Notes : Same as those in Table 4.16.

Table 4.21: Estimated Growth Rates in Crop Output in West Bengal, 1957-58 to 1994-95

(Growth rates in per cent per annum)

Name of the Crop/Crop -group	Unadjusted		Adjusted	
	Growth rate[a]	$\hat{C}$	Growth rate[a]	$\hat{C}$
Total Foodgrains	2.567*	0.00022	2.515*	0.00024*
	(0.0018)	(0.00018)	(0.0006)	(0.000064)
Total Non-Foodgrains[k]	4.269*	0.1693*	4.351*	0.1785*
	(0.0012)	(0.0169)	(0.0011)	(0.0157)
All Crops (combined)	2.821*	0.00039*	2.684*	0.00045*
	(0.0014)	(0.00014)	(0.0005)	(0.00005)

Notes : Significant at 1% level. S.E.s are within parentheses.

[a]Growth rate = $\hat{b}$ x 100 or $c\hat{b}$ x 100)/HM; [k] $\hat{c}$ is corresponding to parabolic trend and log-quadratic trend.

HM - Harmonic Mean.

Table 4.22: Estimated Growth Rates in Foodgrains Productions, Crop-wise Disaggregation, West Bengal, 1957-58 to 1994-95

(Growth rates in per cent per annum)

Name of the Crop/Crop-group	Unadjusted		Adjusted	
	Growth rate[a]	$\hat{C}$	Growth rate[a]	$\hat{C}$
Aus rice	1.776*	—	1.654*	—
	(.0030)		(.0011)	
Aman rice[k]	1.903*	5.3611*	1.961*	4. 1461*
	(.0029)	(1.4729)	(.0012)	(.6268)
Boro rice	14.657*	-.00380*	14.942*	-.00408*
	(.0074)	(.00076)	(.0034)	(.00034)
Total rice	2.627*	0.00055*	2.553*	0.00062*
	(.0018)	(.00018)	(.00061)	(.00006)
Wheat	8.670*	-.00654*	8.900*	-.00659*
	(.0088)	(.00088)	(.0038)	(.00038)
Barley[+]	-4.228*	-.000375*	-4.228*	-.00375*
	(.0035)	(.00035)	(.0035)	(.00035)
Other cereals	3.182*	—	3.167*	—
	(.0047)		(.0018)	
Total cereals	2.756*	0.00023	2.711*	.00024*
	(.0018)	(.00018)	(.0006)	(.00006)
Gram	-4.869*	-.00211*	-4.801*	-.00199*
	(.0035)	(.00035)	(.00295)	(.00030)
Other pulses	-1.435*	-.00097*	-1.625*	-.00068*
	(.0026)	(.00026)	(.00101)	(.00011)
Total pulses	2.061*	-.00112*	-2.267*	-.00083*
	(.0019)	(.000197)	(.0009)	(.00009)
Total foodgrains	2.567*	0.00022	2.515*	.00024*
	(.0018)	(.00018)	(.0006)	(.000064)

Notes : Same as those in Table 4.21.

[+] No adjustment in the data.

Table 4.23: Estimated Growth Rate in Total Non-foodgrains Production, Crop-wise Disaggregation, West Bengal, 1957-58 to 1994-95

(Growth rates in per cent per annum)

Name of the Crop/Crop -group	Unadjusted		Adjusted	
	Growth rate[a]	Ĉ	Growth rate[a]	Ĉ
Mustard	7.173*	.0027*	7.241*	.0031*
	(.0044)	(.0004)	(.0031)	(.0003)
Linseed [k]	-2.632*	-.0364*	-3.083*	-.0401*
	(.0068)	(.0049)	(.0019)	(.0018)
Sesamum	12.714*	——	12.958*	——
(or Til)	(.0053)		(.0030)	
Total oil seeds	7.334*	.0017*	7.404*	.0019*
	(.0032)	(.0003)	(.0024)	(.0002)
Jute	2.537*	——	2.540*	——
	(.0031)		(.0015)	
Sugarcane	-2.075*	-.0018*	-2.213*	-.0016*
	(.0035)	(.0003)	(.0030)	(.0003)
Potatoes	7.085*	——	7.128*	——
	(.0026)		(.0008)	
Tobacco	0.587***	——	0.713*	——
	(.0031)		(.0023)	
Total non-	4.269*	.1693*	4.351*	.1785*
foodgrains [k]	(.0012)	(.0169)	(.0011)	(.0157)

Notes : Same as those in Table 4.21.

*** Significant at 10% level.

Table 4.24 : Adjusted Growth Rates in the Production of Major Crop-groups, West Bengal by Districts, 1957-58 to 1994-95

(Growth rates in per cent per annum)

Name of the District	Rice: Growth Rate[a]	Rice: $\hat{C}$	Total foodgrains: Growth rate[a]	Total foodgrains: $\hat{C}$	Total non-foodgrains: Growth rate[a]	Total non-foodgrains: $\hat{C}$	All crops: Growth rate[a]	All crops: $\hat{C}$
Burdwan	2.730*	.0005*	2.565*	.0003*	6.148*	—	3.286*	.0004*
	(.0006)	(.0001)	(.0006)	(.0001)	(.0011)		(.0005)	(.0001)
Birbhum	2.085*	.0004***	1. 893*	—	5.092*	.0009***	2.205 *	—
	(.0023)	(.0002)	(.0010)		(.0047)	(.0005)	(.0009)	
Bankura	2.160*	.0009*	2.170*	.0007*	7.079*	.0012*	2.636*	.0010*
	(.0011)	(.0001)	(.0011)	(.0001)	(.0012)	(.0001)	(.0010)	(.0001)
Midnapore	2.587*	.0007*	2.989*	1.3791*k	7.420*	.0014*	3.166*	.0009*
	(.0021)	(.0002)	(.0023)	(.2610)	(.0009)	(.001)	(.0020)	(.0002)
Howrah	3.681*	—	3.414*	—	4.681*	-.0002	4.166*	.0001
	(.0016)		(.0014)		(.0022)	(.0002)	(.0021)	(.0002)
Hooghly	3.449*	-0.0002**	3.435*	-.0341k	5.640*	. 4155*k	4.476*	
	(.0010)	(.0001)	(.0009)	(.0354)	(.0028)	(.0378)	(.0012)	—
24-Parganas	2.629*	.0005**	2.110*	.0004***	3.632*	—	2.282*	.1008*
	(.0022)	(.0002)	(.0023)	(.0002)	(.0014)		(.0022)	(.0252)
Nadia	4.373*	.0012*	3.875*		4.769*	. 1972*k	4.163*	
	(.0011)	(.0001)	(.0009)	—	(.0042)	(.0455)	(.0008)	—

(Contd.)

Table 4.24 : *(Contd.)*

Name of the District	Rice: Growth Rate[a]	Rice: Ĉ	Total foodgrains: Growth rate[a]	Total foodgrains: Ĉ	Total non-foodgrains: Growth rate[a]	Total non-foodgrains: Ĉ	All crops: Growth rate[a]	All crops: Ĉ
Murshidabad	2.995*	.0008*	2.925*	—	4.511	.1215*k	3.237*	—
	(.0013)	(.0001)	(.0010)		(.0033)	(.0351)	(.0018)	
West	2.068*	.0008*	2.341*	—	2.070*	. 0731*k	2.267*	.0764*k
Dinajpur	(.0008)	(.0001)	(.0019)		(.0009)	(.0108)	(.0637)	(.0067)
Malda	3.774*	—	3.538*	.0002**	1.822*	—	3.144*	.1086*k
	(.0006)		(.0009)	(.0001)	(.0015)		(.0013)	(.0160)
Jalpaiguri	1.107*	—	1.789*	—	1.931	-.0587*k	1.797*	-.0017*k
	(.0022)		(.0021)		(.0012)	(.01456)	(.0009)	(.0091)
Darjeeling	1.669*	—	3.304*	.0005*	2.442*	-.1117*k	2.870*	—
	(.0009)		(.0012)	(.0001)	(.0018)	(.0230)	(.0016)	
Coochbehar	2.399*	—	2.598*	—	1.041	.1116*k	2.032*	—
	(.0008)		(.0008)		(.0022)	(.0258)	(.0016)	
Purulia	1.857*	.2498*k	1.830*	.2437*k	3.375*	-.0022*	2.007*	.0793**k
	(.0028)	(.0898)	(.0029)	(.0956)	(.0018)	(.0002)	(.0013)	(.0311)
West Bengal	2.553*	.0006*	2.515*	.00024*	4.351*	1785*k	2.684*	.0005*k
	(.0006)	(.0001)	(.0006)	(.0001)	(.0011)	(.0157)	(.0005)	(.0001)

Notes : Same as those in Table 4.21; ** Significant at 5 % level; *** Significant at 10% level.

Table 4.25 : Sub-period Growth Rates in Crop Output in West Bengal, 1957-58 to 1976-77 and 1977-78 to 1994-95

Name of the Crop/Crop-group	Whole period's growth rate	Sub-period growth Rates[a]		Trend break[b]	Kinked Trend Equation		
		1st sub-period (1957-58 to 1976-77)	2nd sub-period (1977-78 to 1994-95)		$\bar{R}2$	DW	Runs
Aus rice	1.654* (0.0011)	1.579* (.0023)	1.742* (.0026)	0.163 (.0043)	.8698*	2.199[c]	18[+]
Aman rice	1.961* (.0012)	0.673** (.0039)	3.666* (.0024)	2.993* (.0049)	.9131*	1.565[x]	18[+]
Boro rice	14.942* (.0034)	22.080* (.0080)	6.582* (.0091)	-15.498* (.0153)	.9779*	1.795[c]	19[+]
Total rice	2.553* (.0006)	1.536* (.0013)	3.805* (.0015)	2.269* (.0025)	.9813*	1.692[c]	20[+]
Wheat	8.900* (.0038)	21.927* (.0090)	-3.497* (.0103)	-25.424* (.0173)	.9548*	1.682[c]	19[+]
Barley	-4.228* (.0035)	3.150* (0078)	-11.242* (.0088)	-14.392* (.0148)	.8301*	1.641[c]	16[+]
Other cereals	3.167* (.0018)	2.780* (0037)	3.623* (.0042)	0.843 (.0070)	.9038*	2.375[c]	23[+]
Total cereals	2.711* (.0006)	2.332* (.0013)	3.727* (.0015)	1.395* (.0024)	.9832*	1.688[c]	20[+]
Gram	-4.801* (.0029)	-1.258** (.0064)	-8.952* (.0072)	-7.694* (.0122)	.8864*	1.612[c]	20[+]

(Contd.)

Table 4.25 : *(Contd.)*

Name of the Crop/Crop-group	Whole period's growth rate	Sub-period growth Rates[a] 1st sub-period (1957-58 to 1976-77)	Sub-period growth Rates[a] 2nd sub-period (1977-78 to 1994-95)	Trend break[b]	Kinked Trend Equation $\bar{R}2$	Kinked Trend Equation DW	Kinked Trend Equation Runs
Other pulses	-1.625* (.0010)	-.662* (.0022)	-3.058* (.0025)	-2.396* (.0042)	.8997*	1.929[c]	17[+]
Total pulses	-2.267* (.0009)	-1.071* (.0020)	-4.044* (.0023)	-2.973* (.0039)	.9511*	1.534[x]	17[+]
Total food-grains	2.515* (.0006)	2.058* (.0013)	3.141* (.0014)	1.083* (.0024)	.9815*	1.694[c]	20[+]
Mustard	7.241* (.0031)	1.541** (.0061)	13.917* (.0070)	12.376* (.0117)	.9515*	2.387[c]	19[+]
Linseed	-3.083* (.0019)	5.112* (.0037)	-12.484* (.0064)	-17.596* (.0088)	.9161*	1.599[c]	15[+]
Sesamum	12.958* (.0030)	13.862* (.0061)	11.900* (.0069)	-1.962*** (.0116)	.9819*	1.880[c]	17[+]
Total oil seeds	7.404* (.0024)	3.799* (.0048)	11.626* (.0055)	7.827* (.0093)	.9676*	2.300[c]	19[+]
Jute	2.540* (.0015)	2.784* (.0039)	2.391* (.0028)	-0.393 (.0056)	.8967*	1.709[c]	17[+]
Sugercane	-2.213* (.0030)	0.698* (.0063)	-5.624* (.0072)	-6.322* (.0121)	.6825*	1.779[c]	14***

(Contd.)

Table 4.25 : *(Contd.)*

Name of the Crop/Crop-group	Whole period's growth rate	*Sub-period growth Rates*[a] 1st sub-period (1957-58 to 1976-77)	2nd sub-period (1977-78 to 1994-95)	Trend break[b]	Kinked Trend Equation $\bar{R}2$	DW	Runs
Potatoes	7.218* (.0008)	7.108* (.0018)	7.151* (.0021)	0.043 (.0034)	.9949*	2.196[c]	22[+]
Tobacco	0.713* (.0023)	.887*** (.0048)	0.510 (.0054)	-0.377 (.0092)	.1773*	1.802[c]	14***
Total non-foodgrains	4.351* (.0011)	3.341* (.0011)	5.619* (.0019)	2.278* (.0048)	.9749*	1.945[c]	21[+]
All crops (combined)	2.684* (.0005)	2.081* (.0011)	3.649* (.0012)	1.568* (.0021)	.9882*	1.675[c]	18[+]

Notes: * Significant at 1% level; ** Significant at 5% level; *** Significant at 10% level.

[c] indicates presence of no autocorrelation, [A] implies inconclusive at 5% level and non-autocorrelated at 1% level. [+] random disturbance term. Terms within parentheses are S.Es.

[a] Estimated from equation: $\ln q_t$ or $q_t = a+b_1D_1t+b_2D_2t$,t = actual time - break point. Here $D_1 = 1$ and $D_2 = 0$ for first sub-period and $D_1 = 0$ & $D_2 = 1$ for 2nd sub-period. Growth rate (1957-58 - 1976-77) = $\hat{b}_1 \times 100$ or $(\hat{b}_1/HM) \times 100$ and growth rate (1977-78 - 1994-95) = $\hat{b}_2 \times 100$ or $(\hat{b}_2/HM) \times 100$;

[b] Estimating equation: $\ln q_t$ or $q_t = a+b_1t+b_2D_2t$, Trend break calculated from $\hat{b}_2$

Table 4.26 : Sub-Period Growth Rates in All Crop-output, West Bengal by Districts, 1957-58 to 1976-77 and 1977-78 to 1994-95
(Growth rates in per cent per annum)

Name of the District	Whole period's growth rate	Sub-period growth Rates[a]		Trend break[b]	Kinked Trend Equation		
		1st sub-period (1957-58 to 1976-77)	2nd sub-period (1977-78 to 1994-95)		$\bar{R}2$	DW	Runs
Burdwan	3.286* (.0005)	2.786* (.0011)	3.837* (.0013)	1.051* (.0022)	.9912*	1.515[c]	15[+]
Birbhum	2.205* (.0009)	2.086* (.0020)	2.435* (.0023)	0.349 (.0039)	.9357*	2.258[c]	26***
Bankura	2.636* (.0010)	1.269* (.0023)	4.807* (.0026)	3.538* (.0044)	.9569*	1.838[c]	17[+]
Midnapore	3.166* (.0020)	1.537* (.0043)	5.074* (.0048)	3.537* (.0082)	.8753*	2.025[c]	21[+]
Howrah	4.166* (.0021)	4.049* (.0044)	4.220* (.0049)	0.171 (.0084)	.9151*	1.789[c]	17[+]
Hooghly	4.476* (.0012)	4.261* (.0025)	4.727* (.0028)	0.466 (.0048)	.9746*	2.010[c]	22[+]

(Contd.)

Table 4.26 : (Contd.)

Name of the District	Whole period's growth rate	Sub-period growth Rates[a] 1st sub-period (1957-58 to 1976-77)	2nd sub-period (1977-78 to 1994-95)	Trend break[b]	Kinked Trend Equation $\bar{R}2$	DW	Runs
24-Parganas	2.282* (.0022)	2.007* (.0027)	2.702* (.0031)	.695** (.0039)	.8391*	1.785[c]	19[+]
Nadia	4.163* (.0008)	3.911* (.0016)	4.461* (.0018)	0.550*** (.0031)	.9884*	2.509[c]	23[+]
Murshidabad	3.237* (.0018)	3.068* (.0038)	3.583* (.0043)	0.515 (.0072)	.8973*	1.641[c]	16[+]
West Dinajpur	2.267* (.0637)	1.948* (.0017)	3.144* (.0013)	1.196* (.0027)	.9754*	1.558[c]	17[+]
Malda	3.144* (.0013)	2.742* (.0039)	3.740* (.0025)	0.998* (.0044)	.9392*	1.777[c]	20[+]
Jalpaiguri	1.797* (.0009)	2.050* (.0021)	1.484* (.0017)	-0.566 (.0034)	.9214*	2.082[c]	20[+]

(Contd.)

Table 4.26 : *(Contd.)*

Name of the District	Whole period's growth rate	Sub-period growth Rates[a]		Trend break[b]	Kinked Trend Equation		
		1st sub-period (1957-58 to 1976-77)	2nd sub-period (1977-78 to 1994-95)		$\bar{R}2$	DW	Runs
Darjeeling	2.870* (.0016)	2.513* (.0034)	3.325* (.0041)	0.812 (.0067)	.8939*	2.031[c]	16[+]
Coochbehar	2.032* (.0016)	1.792* (.0033)	2.314* (.0037)	0.522 (.0063)	.8199*	1.851[c]	21[+]
Purulia	2.007* (.0031)	0.687 (.0071)	3.325* (.0065)	2.638* (.0112)	.5622*	1.752[c]	21[+]
West Bengal	2.684* (.0005)	2.081* (.0011)	3.649* (.0012)	1.568* (.0021)	.9882*	1.675[c]	18[+]

Notes: Same as those in Table 4.25.

Table 4.27 : Sub-Period Growth Rates in Rice Production, West Bengal by Districts, 1957 -58 to 1976-77 and 1977-78 to 1994-95
(Growth rates in per cent per annum)

Name of the District	Whole period's growth rate	Sub-period growth Rates[a] 1st sub-period (1957-58 to 1976-77)	2nd sub-period (1977-78 to 1994-95)	Trend break[b]	Kinked Trend Equation $\bar{R}2$	DW	Runs
Burdwan	2.730* (.0006)	2.022* (.0014)	3.782* (.0016)	1.760* (.0026)	.9823*	1.686[c]	19[+]
Birbhum	2.085* (.0023)	1.379* (.0049)	2.912* (.0055)	1.533* (.0093)	.6895*	1.712[c]	22[+]
Bankura	2.160* (.0011)	0.780* (.0023)	4.195* (.0027)	3.415* (.0045)	.9336*	1.922[c]	19[+]
Midnapore	2.587* (.0021)	1.257* (.0045)	4.145* (.0051)	2.888* (.0086)	.8091*	2.030[c]	23[+]
Howrah	3.681* (.0016)	3.173* (.0033)	4.282* (.0038)	1.109*** (.0063)	.9402*	1.664[c]	21[+]
Hooghly	3.449* (.0010)	3.750* (.0020)	3.001* (.0023)	-0.749*** (.0038)	.9732*	2.028[c]	21[+]

(Contd.)

Table 4.27: *(Contd.)*

Name of the District	Whole period's growth rate	Sub-period growth Rates[a]		Trend break[b]	Kinked Trend Equation		
		1st sub-period (1957-58 to 1976-77)	2nd sub-period (1977-78 to 1994-95)		$\bar{R}2$	DW	Runs
24-Parganas	2.629* (.0023)	1.414* (.0048)	3.700* (.0054)	2.286* (.0091)	.7874*	1.797[c]	21[+]
Nadia	4.373* (.0011)	2.584* (.0023)	7.016* (.0027)	4.432* (.0045)	.9813*	1.818[c]	14[+]
Murshidabad	2.995* (.0013)	1.785* (.0027)	4.789* (.0031)	3.004* (.0052)	.9483*	1.719[c]	20[+]
West Dinajpur	2.068* (.0008)	0.849* (.0016)	3.858* (.0018)	3.009* (.0030)	.9664*	1.856[c]	17[+]
Malda	3.774* (.0006)	3.589* (.0013)	3.994* (.0015)	0.405 (.0026)	.9902*	1.996[c]	15[+]
Jalpaiguri	1.107* (.0022)	1.564* (.0046)	0.572 (.0052)	-0.992 (.0088)	.4001*	1.642[c]	21[+]

(Contd.)

Table 4.27: *(Contd.)*

Name of the District	Whole period's growth rate	Sub-period growth Rates[a] 1st sub-period (1957-58 to 1976-77)	2nd sub-period (1977-78 to 1994-95)	Trend break[b]	Kinked Trend Equation $\bar{R}2$	DW	Runs
Darjeeling	1.669* (.0009)	1.580* (.0019)	1.773* (.0021)	0.193 (.0036)	.9091*	2.114[c]	19[+]
Coochbehar	2.399* (.0008)	2.309* (.0018)	2.506* (.0020)	0.197 (.0034)	.9577*	1.838[c]	19[+]
Purulia	1.857* (.0028)	0.425 (.0074)	3.526* (.0067)	3.101* (.0127)	.5614*	1.851[c]	21[+]
West Bengal	2.553* (.0006)	1.536* (.0013)	3.805* (.0015)	2.269* (.0025)	.9813*	1.692[c]	20[+]

Notes: Same as those in Table 4.25.

Table 4.28: Sub-Period Growth Rates in the Production of Total Foodgrains, West Bengal by Districts, 1957 -58 to 1976-77 and 1977-78 to 1994-95.

(Growth rates in per cent per annum)

Name of the District	Whole period's growth rate	Sub-period growth Rates[a] 1st sub-period (1957-58 to 1976-77)	Sub-period growth Rates[a] 2nd sub-period (1977-78 to 1994-95)	Trend break[b]	Kinked Trend Equation $\bar{R}^2$	Kinked Trend Equation DW	Kinked Trend Equation Runs
Burdwan	2.565* (.0006)	2.101* (.0013)	3.257* (.0015)	1.156* (.0026)	.9798*	1.753[c]	19[+]
Birbhum	1.893* (.0010)	1.869* (.0022)	1.921* (.0025)	0.052 (.0043)	.9006*	2.163[c]	20[+]
Bankura	2.170* (.0011)	1.092* (.0023)	3.765* (.0027)	2.673* (.0045)	.9296*	1.940[c]	17[+]
Midnapore	2.989* (.0023)	0.857 (.0054)	4.346* (.0043)	3.489* (.0081)	.8365*	2.039[c]	21[+]
Howrah	3.414* (.0014)	3.063* (.0030)	4.030* (.0034)	0.967 (.0058)	.9421*	2.407[c]	19[+]
Hooghly	3.435* (.0009)	3.860* (.0023)	3.087* (.0015)	-0.373 (.0086)	.9759*	2.078[c]	18[+]

(Contd.)

Table 4.28 : *(Contd.)*

Name of the District	Whole period's growth rate	Sub-period growth Rates[a] 1st sub-period (1957-58 to 1976-77)	2nd sub-period (1977-78 to 1994-95)	Trend break[b]	Kinked Trend Equation $\bar{R}^2$	DW	Runs
24-Parganas	2.110* (.0013)	1.757* (.0018)	2.392* (.0024)	0.635** (.0031)	.7689*	1.818[c]	21[+]
Nadia	3.875* (.0009)	3.721* (.0019)	4.057* (.0022)	0.336 (.0036)	.9815*	2.073[c]	21[+]
Murshidabad	2.925* (.0010)	2.915* (.0022)	2.937* (.0025)	0.022 (.0042)	.9578*	1.874[c]	20[+]
West Dinajpur	2.341* (.0019)	2.084* (.0041)	2.843* (.0046)	0.759 (.0078)	.8219*	1.567[c]	16[+]
Malda	3.538* (.0009)	3.035* (.0019)	3.983* (.0022)	0.948* (.0036)	.9785*	1.733[c]	13[***]
Jalpaiguri	1.789* (.0021)	1.977* (.0042)	1.648* (.0047)	-0.329 (.0080)	.5897*	1.750[c]	25[+]
Darjeeling	3.304* (.0012)	2.056* (.0025)	4.015* (.0029)	1.959* (.0048)	.9588*	1.672[c]	16[+]

(Contd.)

Table 4.28 : *(Contd.)*

Name of the District	Whole period's growth rate	Sub-period growth Rates[a]		Trend break[b]	Kinked Trend Equation		
		1st sub-period (1957-58 to 1976-77)	2nd sub-period (1977-78 to 1994-95)		$\bar{R}^2$	DW	Runs
Coochbehar	2.598* (.0008)	2.771* (.0016)	2.394* (.0019)	-0.377 (.0031)	.9694*	1.931[c]	14[+]
Purulia	1.830* (.0029)	0.501 (.0073)	3.431* (.0067)	2.930* (.0127)	.5506*	1.787[c]	21[+]
West Bengal	2.515* (.0006)	2.058* (.0013)	3.141* (.0014)	1.083* (.0024)	.9815*	1.694[c]	20[+]

Notes: Same as those in Table 4.25

Table 4.29 : Sub-Period Growth Rates in the Production of Total Non-Foodgrains, West Bengal by Districts, 1957-58 to 1976-77 and 1977-78 to 1994-95

(Growth rates in per cent per annum)

Name of the District	Whole period's growth rate	Sub-period growth Rates[a]		Trend break[b]	Kinked Trend Equation		
		1st sub-period (1957-58 to 1976-77)	2nd sub-period (1977-78 to 1994-95)		$\bar{R}^2$	DW	Runs
Burdwan	6.148* (.0011)	6.299* (.0022)	5.969* (.0025)	-0.330 (.0043)	.9897*	2.221[c]	15[+]
Birbhum	5.092* (.0047)	3.350* (.0099)	7.132* (.0112)	3.782** (.0189)	.7630*	2.059[c]	17[+]
Bankura	7.079* (.0012)	5.393* (.0027)	9.585* (.0031)	4.192* (.0053)	.9891*	1.810[c]	14[+]
Midnapore	7.420* (.0009)	5.327* (.0020)	10.513* (.0023)	5.186* (.0039)	.9946*	1.569[c]	14[+]
Howrah	4.681* (.0022)	4.853* (.0045)	4.443* (.0051)	-0.410 (.0086)	.9249*	1.608[c]	17[+]
Hooghly	5.640* (.0028)	4.748* (.0084)	6.954* (.0031)	2.206* (.0108)	.9654*	1.899[c]	19[+]

(Contd.)

Table 4.29 : (Contd.)

Name of the District	Whole period's growth rate	Sub-period growth Rates[a] 1st sub-period (1957-58 to 1976-77)	2nd sub-period (1977-78 to 1994-95)	Trend break[b]	Kinked Trend Equation $\bar{R}^2$	DW	Runs
24-Parganas	3.632* (.0014)	3.254* (.0028)	4.079* (.0032)	0.825 (.0055)	.9536*	1.923[c]	15[+]
Nadia	4.769* (.0042)	4.049* (.0122)	5.569* (.0052)	1.520* (.0063)	.8758*	1.833[c]	21[+]
Murshidabad	4.511* (.0033)	3.429* (.0087)	4.831* (.0048)	1.402* (.0068)	.8544*	2.041[c]	21[+]
West Dinajpur	2.070* (.0009)	1.891* (.0022)	3.381* (.0018)	1.490* (.0037)	.9364*	2.347[c]	17[+]
Malda	1.822* (.0015)	1.596* (.0031)	2.090* (.0035)	0.494 (.0059)	.9144*	2.235[c]	21[+]
Jalpaiguri	1.931* (.0012)	2.249* (.0028)	0.275 (.0026)	-1.974* (.0048)	.7693*	1.953[c]	17[+]

(Contd.)

Table 4.29 : (Contd.)

Name of the District	Whole period's growth rate	Sub-period growth Rates[a]		Trend break[b]	Kinked Trend Equation		
		1st sub-period (1957-58 to 1976-77)	2nd sub-period (1977-78 to 1994-95)		$\bar{R}^2$	DW	Runs
Darjeeling	2.442* (.0018)	3.319* (.0055)	2.043* (.0036)	-1.276* (.0031)	.8231*	1.615[c]	16[+]
Coochbehar	1.041* (.0022)	0.526 (.0052)	1.688* (.0045)	1.162** (.0068)	.5185*	1.892[c]	20[+]
Purulia	3.375* (.0018)	5.453* (.0045)	1.211** (.0051)	-4.242* (.0085)	.9236*	1.475[c]	13[***]
West Bengal	4.351* (.0011)	3.341* (.0011)	5.619* (.0019)	2.278* (.0048)	.9749*	1.945[c]	21[+]

Notes: Same as those in Table 4.25

Table 4.30: Decomposition of the Output Growth Rates of Crops/Crop-groups in West Bengal, 1957-58—1994-95

Name of the crop / crop-group	Whole period (1957-58 - 1994-95)			First Sub-period (1957-58-1976-77)			Second Sub-period (1977-78 - 1994-95)		
	Growth rates			Growth rates			Growth rates		
	Output	Area	Productivity	Output	Area	Productivity	Output	Area	Productivity
Aus-rice	1.654	-.257	1.911	1.579	-.106	1.685	1.742	-.408	2.150
		(-15.5)	(115.5)		(-6.7)	(106.7)		(-23.4)	(123.4)
Aman-rice	1.961	.252	1.709	0.673	.214	.459	3.666	.306	3.360
		(12.9)	(87.1)		(31.8)	(68.2)		(8.3)	(91.7)
Boro-rice	14.942	11.966	2.976	22.080	18.848	3.232	6.582	3.863	2.719
		(80.1)	(19.9)		(85.4)	(14.6)		(58.7)	(41.3)
Total rice	2.553	.750	1.803	1.536	.786	.750	3.805	.712	3.093
		(29.4)	(70.6)		(51.2)	(48.8)		(18.7)	(81.3)
Wheat	8.900	5.447	.3453	21.927	17.981	3.946	-3.497	-6.457	2.960
		(61.2)	(38.8)		(82.0)	(18.0)		(-184.6)	(84.6)
Barley	-4.228	-5.354	1.126	3.150	1.919	1.231	-11.242	-12.026	.784
		(-126.6)	(26.6)		(60.9)	(39.1)		(-107.0)	(7.0)
Other cereals	3.167	.211	2.956	2.780	.150	2.630	3.623	.271	3.352
		(6.7)	(93.3)		(5.4)	(94.6)		(7.5)	(92.5)

(Contd.)

Table 4.30 : *(Contd.)*

Name of the crop / crop-group	Whole period (1957-58—1994-95) Growth rates			First Sub-period (1957-58—1976-77) Growth rates			Second Sub-period (1977-78—1994-95) Growth rates		
	Output	Area	Productivity	Output	Area	Productivity	Output	Area	Productivity
Total cereals	2.711	.707 (26.1)	2.004 (73.9)	2.332	.952 (40.8)	1.380 (59.2)	3.727	.592 (15.9)	3.135 (84.1)
Gram	-4.801	-5.809 (-121.0)	1.008 (21.0)	-1.258	-2.545 (-202.3)	1.287 (102.3)	-8.952	-9.487 (106.0)	.535 (6.0)
Other pulses	-1.625	-2.348 (-144.5)	.723 (44.5)	-.662	-1.517 (-229.2)	.855 (129.2)	-3.058	-3.633 (-118.8)	.575 (18.8)
Total pulses	-2.267	-3.180 (-140.3)	.913 (40.3)	-1.071	-2.291 (-213.9)	1.220 (113.9)	-4.004	-4.586 (-113.4)	.542 (13.4)
Total food-grains	2.515	.528 (21.0)	1.987 (79.0)	2.058	.740 (36.0)	1.318 (64.0)	3.141	.309 (9.8)	2.832 (90.2)
Mustard	7.241	4.068 (56.2)	3.173 (43.8)	1.541	.429 (27.8)	1.112 (72.2)	13.917	8.252 (59.3)	5.665 (40.7)
Linseed	-3.083	-3.518 (-114.1)	.435 (14.1)	5.112	4.418 (86.4)	.694 (13.6)	-12.484	-12.637 (-101.2)	.153 (1.2)
Sesamum	12.958	10.134 (78.2)	2.824 (21.8)	13.862	12.212 (88.1)	1.650 (11.9)	11.900	7.801 (65.6)	4.099 (34.4)

(Contd.)

Table 4.30: *(Contd.)*

Name of the crop / crop-group	Whole period (1957-58 - 1994-95)			First Sub-period (1957-58—1976-77)			Second Sub-period (1977-78—1994-95)		
		Growth rates			Growth rates			Growth rates	
	Output	Area	Productivity	Output	Area	Productivity	Output	Area	Productivity
Total oil seeds	7.404	4.901 (66.2)	2.503 (33.8)	3.799	2.556 (67.3)	1.243 (32.7)	11.626	7.529 (64.8)	4.097 (34.2)
Jute	2.540	1.111 (43.7)	1.429 (56.3)	2.784	1.477 (53.1)	1.307 (46.9)	2.391	.829 (34.7)	1.562 (65.3)
Sugarcane	-2.213	-3.081 (-139.2)	.868 (39.2)	.698	-.197 (-28.2)	.895 (128.2)	-5.624	-6.465 (-115.0)	.841 (15.0)
Potatoes	7.128	4.217 (59.2)	2.911 (40.9)	7.108	4.056 (57.1)	3.052 (42.9)	7.151	4.378 (61.2)	2.773 (38.8)
Tobacco	.713	-.074 (-10.4)	.787 (110.4)	.887	-.232 (-26.2)	1.119 (126.2)	.510	.083 (16.3)	.427 (83.7)
Total non-foodgrains	4.351	2.376 (54.6)	1.975 (45.4)	3.341	2.159 (64.6)	1.182 (35.4)	5.619	2.504 (44.6)	3.115 (55.4)
All crops	2.684	.701 (26.1)	1.983 (73.9)	2.081	.851 (40.9)	1.230 (59.1)	3.649	.612 (16.8)	3.037 (83.2)

Note: Figures in Parentheses indicate percentage shares; for easy interpretation only magnitudes have been considered for negative output growth rates.

Table 4.31: Decomposition of Growth Rates of Output of All Crops Combined, West Bengal and Districts, 1957-58—1994-95

Districts	Whole period (1957-58—1994-95)			First Sub-period (1957-58—1976-77)			Second Sub-period (1977-78—1994-95)		
		Growth rates			Growth rates			Growth rates	
	Output	Area	Productivity	Output	Area	Productivity	Output	Area	Productivity
Burdwan	3.286	1.291 (39.3)	1.995 (60.7)	2.786	1.328 (47.7)	1.458 (52.3)	3.837	1.200 (31.3)	2.637 (68.7)
Birbhum	2.205	.579 (26.3)	1.626 (73.7)	2.086	.956 (45.8)	1.130 (54.2)	2.435	.168 (6.9)	2.267 (93.1)
Bankura	2.636	.837 (31.8)	1.799 (68.2)	1.269	.971 (76.5)	.298 (23.5)	4.807	.680 (14.1)	4.127 (85.9)
Midnapore	3.166	.718 (22.7)	2.448 (77.3)	1.537	.637 (41.4)	.900 (58.6)	5.074	.835 (16.5)	4.239 (83.5)
Howrah	4.166	.834 (20.0)	3.332 (80.0)	4.049	.797 (19.7)	3.252 (80.3)	4.220	.870 (20.6)	3.350 (79.4)
Hooghly	4.476	1.537 (34.3)	2.939 (65.7)	4.261	1.421 (33.3)	2.840 (66.7)	4.727	1.636 (34.6)	3.091 (65.4)
24-Parganas	2.282	.294 (12.9)	1.988 (87.1)	2.007	.208 (10.4)	1.799 (89.6)	2.702	.402 (14.9)	2.300 (85.1)

(Contd.)

Table 4.31: *(Contd.)*

Districts	Whole period (1957-58—1994-95) Growth rates			First Sub-period (1957-58—1976-77) Growth rates			Second Sub-period (1977-78—1994-95) Growth rates		
	Output	Area	Productivity	Output	Area	Productivity	Output	Area	Productivity
Nadia	4.163	.476	3.687	3.911	.605	3.306	4.461	.217	4.244
		(11.4)	(88.6)		(15.5)	(84.5)		(4.9)	(95.1)
Murshidabad	3.237	.517	2.720	3.068	1.122	1.946	3.583	-.078	3.661
		(16.0)	(84.0)		(36.6)	(63.4)		(-2.2)	(102.2)
West Dinajpur	2.267	.757	1.510	1.948	.941	1.007	3.144	.613	2.531
		(33.4)	(66.6)		(48.3)	(51.7)		(19.5)	(80.5)
Malda	3.144	.375	2.769	2.742	.361	2.321	3.740	.395	3.345
		(11.9)	(88.1)		(13.2)	(86.8)		(10.6)	(89.4)
Jalpaiguri	1.797	.975	.822	2.050	1.965	.085	1.484	-.185	1.669
		(54.3)	(45.7)		(95.9)	(4.1)		(-12.5)	(112.5)
Darjeeling	2.870	.739	2.131	2.513	.756	1.757	3.325	.718	2.607
		(25.7)	(74.3)		(30.1)	(69.9)		(21.6)	(78.4)
Coochbehar	2.032	1.192	.840	1.792	1.454	.338	2.314	.705	1.609
		(58.7)	(41.3)		(81.1)	(18.9)		(30.5)	(69.5)
Purulia	2.007	.519	1.488	.687	.478	.209	3.325	.589	2.736
		(25.9)	(74.1)		(69.6)	(30.4)		(17.7)	(82.3)
West Bengal	2.684	.701	1.983	2.081	.851	1.230	3.649	.612	3.037
		(26.1)	(73.9)		(40.9)	(59.1)		(16.8)	(83.2)

Note : Figures in the parentheses indicate percentage shares.

Table 4.32 : Decomposition of Growth Rates of Total Rice Production, West Bengal and Districts, 1957-58—1994-95

Districts	Whole period (1957-58—1994-95)			First Sub-period (1957-58—1976-77)			Second Sub-period (1977-78—1994-95)		
	Growth rates			Growth rates			Growth rates		
	Output	Area	Productivity	Output	Area	Productivity	Output	Area	Productivity
Burdwan	2.730	0.761 (27.9)	1.969 (72.1)	2.022	0.741 (36.6)	1.281 (63.4)	3.782	0.784 (20.7)	2.998 (79.3)
Birbhum	2.085	0.370 (17.7)	1.715 (82.3)	1.379	0.533 (38.7)	0.846 (61.3)	2.912	0.179 (6.1)	2.733 (93.9)
Bankura	2.160	0.629 (29.1)	1.531 (70.9)	0.780	0.601 (77.1)	0.179 (22.9)	4.195	0.660 (15.7)	3.535 (84.3)
Midnapore	2.587	0.667 (25.8)	1.920 (74.2)	1.257	0.596 (47.3)	0.661 (52.6)	4.145	0.744 (17.9)	3.401 (82.1)
Howrah	3.681	1.328 (36.1)	2.353 (63.9)	3.173	1.118 (35.2)	2.055 (64.8)	4.282	1.556 (36.3)	2.726 (63.7)
Hooghly	3.449	1.219 (35.3)	2.230 (64.7)	3.750	1.617 (43.1)	2.133 (56.9)	3.001	0.686 (22.9)	2.315 (77.1)
24- Parganas	2.629	0.467 (17.8)	2.162 (82.2)	1.414	0.391 (27.7)	1.023 (72.3)	3.700	0.522 (14.1)	3.178 (85.9)

(Contd.)

Table 4.32 : (Contd.)

Districts	Whole period (1957-58—1994-95)			First Sub-period (1957-58—1976-77)			Second Sub-period (1977-78—1994-95)		
		Growth rates			Growth rates			Growth rates	
	Output	Area	Productivity	Output	Area	Productivity	Output	Area	Productivity
Nadia	4.373	1.059	3.314	2.584	0.821	1.763	7.016	1.230	5.786
		(24.2)	(75.8)		(31.8)	(68.2)		(17.5)	(82.5)
Murshidabad	2.995	0.697	2.298	1.785	0.492	1.293	4.789	0.921	3.868
		(23.3)	(76.7)		(27.6)	(72.4)		(19.2)	(80.8)
West Dinajpur	2.068	0.602	1.466	0.849	1.053	-0.204	3.858	0.074	3.784
		(29.1)	(70.9)		(124.0)	(-24.0)		(1.19)	(98.1)
Malda	3.774	1.009	2.765	3.589	1.151	2.438	3.994	0.866	3.128
		(26.7)	(73.3)		(32.1)	(67.9)		(21.7)	(78.3)
Jalpaiguri	1.107	0.916	0.191	1.564	2.265	- 0.701	0.572	-0.665	1.237
		(82.7)	(17.3)		(144.8)	(-44.8)		(-116.3)	(216.3)
Darjeeling	1.669	1.067	0.602	1.580	1.213	0.367	1.773	0.909	0.864
		(63.9)	(36.1)		(76.8)	(23.2)		(51.3)	(48.7)
Coochbehar	2.399	1.159	1.240	2.309	1.571	0.738	2.506	0.712	1.794
		(48.3)	(51.7)		(68.0)	(32.0)		(28.4)	(71.6)
Purulia	1.857	0.517	1.340	0.425	0.342	0.083	3.526	0.707	2.819
		(27.8)	(72.2)		(80.5)	(19.5)		(20.1)	(79.9)
West Bengal	2.553	0.750	1.803	1.536	0.786	0.750	3.805	0.712	3.093
		(29.4)	(70.6)		(51.2)	(48.8)		(18.7)	(81.3)

Notes: Figures in the parentheses indicate percentage shares.

Table 4.33 : Decomposion of Growth Rates of Total Foodgrains, West Bengal and Districts, 1957-58 — 1994-95

Districts	Whole period (1957-58—1994-95)			First Sub-period (1957-58—1976-77)			Second Sub-period (1977-78—1994-95)		
	Growth rates			Growth rates			Growth rates		
	Output	Area	Productivity	Output	Area	Productivity	Output	Area	Productivity
Burdwan	2.565	.484 (18.9)	2.081 (81.1)	2.101	.665 (31.7)	1.436 (68.3)	3.257	.307 (9.4)	2.950 (90.6)
Birbhum	1.893	-.015 (-0.8)	1.908 (100.8)	1.869	.270 (14.4)	1.599 (85.6)	1.921	-.296 (-15.4)	2.217 (115.4)
Bankura	2.170	.558 (25.7)	1.612 (74.3)	1.092	.607 (55.6)	.485 (44.4)	3.765	.419 (11.1)	3.346 (88.9)
Midnapore	2.989	.438 (14.7)	2.551 (85.3)	.857	.390 (45.5)	.467 (54.5)	4.364	.490 (11.3)	3.856 (88.7)
Howrah	3.414	.890 (26.1)	2.524 (73.9)	3.063	.812 (26.5)	2.251 (73.5)	4.030	.999 (24.8)	3.031 (75.2)
Hooghly	3.435	.811 (23.6)	2.624 (76.4)	3.860	1.081 (28.0)	2.779 (72.0)	3.087	.519 (16.8)	2.568 (83.2)
24-Parganas	2.110	.319 (15.1)	1.791 (84.9)	1.757	.330 (18.8)	1.427 (81.2)	2.392	.307 (12.8)	2.085 (87.2)

(Contd.)

Table 4.33 : (Contd.)

Districts	Whole period (1957-58—1994-95)			First Sub-period (1957-58—1976-77)			Second Sub-period (1977-78—1994-95)		
	Output	Growth rates		Output	Growth rates		Output	Growth rates	
		Area	Productivity		Area	Productivity		Area	Productivity
Nadia	3.875	.187 (4.8)	3.688 (95.2)	3.721	.166 (4.5)	3:555 (95.5)	4.057	.210 (5.2)	3.847 (94.8)
Murshidabad	2.925	.237 (8.1)	2.688 (91.9)	2.915	1.135 (38.9)	1.780 (61.1)	2.937	-.815 (-27.7)	3.752 (127.7)
West Dinajpur	2.341	.585 (25.0)	1.756 (75.0)	2.084	.814 (39.1)	1.270 (60.9)	2.843	.237 (8.3)	2.606 (91.7)
Malda	3.538	.530 (15.0)	3.008 (85.0)	3:035	.631 (20.8)	2.404 (79.2)	3.983	.387 (9.7)	3.596 (90.3)
Jalpaiguri	1.789	1.127 (63.0)	.662 (37.0)	1.977	2.704 (136.8)	-.727 (-36.8)	1.648	-.721 (-43.8)	2.369 (143.8)
Darjeeling	3.304	.804 (24.3)	2.500 (75.7)	2.056	.799 (38.9)	1.257 (61.1)	4.015	.808 (20.1)	3.207 (79.9)
Coochbehar	2.598	1.272 (49.0)	1.326 (51.0)	2.771	1.704 (61.5)	1.067 (38.5)	2.394	.704 (29.4)	1.690 (70.6)
Purulia	1.830	.517 (28.3)	1.313 (71.7)	.501	.437 (87.2)	.064 (12.8)	3.431	.603 (17.6)	2.828 (82.4)
West Bengal	2.515	.528 (21.0)	1.987 (79.0)	2.058	.740 (36.0)	1.318 (64.0)	3.141	.309 (9.8)	2.832 (90.2)

Note : Figures in the parentheses indicate percentage shares.

Table 4.34 : Decomposition of Growth Rates of Total Non-foodgrains Production, West Bengal and Districts, 1957-58—1994-95

Districts	Whole period (1957-58—1994-95)			First Sub-period (1957-58—1976-77)			Second Sub-period (1977-78—1994-95)		
		Growth rates			Growth rates			Growth rates	
	Output	Area	Productivity	Output	Area	Productivity	Output	Area	Productivity
Burdwan	6.148	4.698 (76.4)	1.450 (23.6)	6.299	4.664 (74.0)	1.635 (26.0)	5.969	4.732 (79.3)	1.237 (20.7)
Birbhum	5.092	4.854 (95.3)	0.238 (4.7)	3.350	4.856 (145.0)	-1.506 (-45.0)	7.132	4.852 (68.0)	2.280 (32.0)
Bankura	7.079	5.088 (71.9)	1.991 (28.1)	5.393	5.122 (95.0)	0.271 (5.0)	9.585	5.054 (52.7)	4.531 (47.3)
Midnapore	7.420	5.273 (71.1)	2.147 (28.9)	5.327	4.012 (75.3)	1.315 (24.7)	10.513	5.533 (52.6)	4.980 (47.4)
Howrah	4.681	0.588 (12.6)	4.093 (87.4)	4.853	0.747 (15.4)	4.106 (84.6)	4.443	0.351 (7.9)	4.092 (92.1)
Hooghly	5.640	2.587 (45.9)	3.053 (54.1)	4.748	1.850 (39.0)	2.898 (61.0)	6.954	3.147 (45.3)	3.807 (54.7)
24-Parganas	3.632	0.166 (4.6)	3.466 (95.4)	3.254	-0.103 (-3.2)	3.357 (103.2)	4.079	0.505 (12.4)	3.574 (87.6)
Nadia	4.769	1.082 (22.7)	3.687 (77.3)	4.049	1.410 (34.8)	2.639 (65.2)	5.569	0.829 (14.9)	4.740 (85.1)

(Contd.)

Table 4.34 : (Contd.)

Districts	Whole period (1957-58—1994-95)			First Sub-period (1957-58—1976-77)			Second Sub-period (1977-78—1994-95)		
		Growth rates			Growth rates			Growth rates	
	Output	Area	Productivity	Output	Area	Productivity	Output	Area	Productivity
Murshidabad	4.511	1.770 (39.2)	2.741 (60.8)	3.429	1.107 (32.3)	2.322 (67.7)	4.831	2.056 (42.6)	2.775 (57.4)
West Dinajpur	2.070	1.661 (80.2)	0.409 (19.8)	1.891	1.521 (80.4)	0.370 (19.6)	3.381	2.021 (59.8)	1.360 (40.2)
Malda	1.822	0.191 (10.5)	1.631 (89.5)	1.596	-0.008 (-0.5)	1.604 (100.5)	2.090	0.433 (20.7)	1.657 (79.3)
Jalpaiguri	1.931	0.490 (25.4)	1.441 (74.6)	2.249	0.398 (17.7)	1.851 (82.3)	0.275	0.590 (214.5)	-0.315 (-114.5)
Darjeeling	2.442	0.431 (17.6)	2.011 (82.4)	3.319	0.685 (20.6)	2.634 (79.4)	2.043	0.156 (7.6)	1.887 (92.4)
Coochbehar	1.041	0.491 (47.2)	0.550 (52.8)	0.526	0.217 (51.5)	0.255 (48.5)	1.688	0.711 (42.1)	0.977 (57.9)
Purulia	3.375	1.513 (44.8)	1.862 (55.2)	5.453	2.864 (52.5)	2.589 (47.5)	1.211	0.075 (6.2)	1.136 (93.8)
West Bengal	4.351	2.376 (54.6)	1.975 (45.4)	3.341	2.159 (64.6)	1.182 (35.4)	5.619	2.504 (44.6)	3.115 (55.4)

Note : Percentage Shares in the parentheses.

Table 4.35 : Cross-sectional Correlation Co-efficients among Growth Components over Major Crop-groups, West Bengal by Districts, 1957-58—1994-95

(Probability levels in parentheses)

Crop Groups	Whole period (1957-58—1994-95)		First Sub-period (1957-58—1976-77)		Second Sub-period (1977-78—1994-95)	
	Area	Productivity	Area	Productivity	Area	Productivity
Total Rice Production	0.4548 (0.088)	0.9431 (0.000)	0.4399 (0.101)	0.8466 (0.000)	0.6453 (0.009)	0.9429 (0.000)
Total Foodgrains Production	0.0138 (0.961)	0.8907 (0.000)	0.1622 (0.564)	0.8090 (0.000)	0.5974 (0.019)	0.7892 (0.000)
Total Non-foodgrains Production	0.8196 (0.000)	0.3338 (0.224)	0.6680 (0.006)	0.2846 (0.304)	0.8652 (0.000)	0.7762 (0.001)
Production of all crops combined	0.2323 (0.405)	0.9144 (0.000)	0.1233 (0.661)	0.9050 (0.000)	0.5545 (0.032)	0.8908 (0.000)

Note : Area, productivity and production imply their respective growth rates.

Table 4.36 : Period-wise Measured Growth Weights of Major Crops of West Bengal, 1957-58—1994-95

Name of the Crop	Price [a] weight	Whole period, 1957-58–1994-95		1st sub-period, 1957-58–1976-77		2nd sub-period, 1977-78–1994-95	
		Standardised[b] average production	Growth[b] weight	Standardised[b] average production	Growth[b] weight	Standardised[b] average production	Growth [b] weight
(1)	(2)	(3)	(4)	(5)	(6)	(7)	(8)
Aus rice	0.937	626.95	.0680	574.42	.0823	688.74	.0580
Aman rice	1.000	5282.36	.5726	4495.80	.6441	6207.74	.5230
Boro rice	1.123	1087.29	.1178	356.91	.0511	1946.57	.1640
Total rice	—	6996.60	.7585	5427.13	.7776	8843.05	.7450
Wheat	0.652	332.03	.0360	238.27	.0341	442.34	.0373
Barley	0.472	12.15	.0013	16.34	.0023	8.89	.0007
Gram	0.687	47.66	.0053	63.00	.0090	29.60	.0025
Other pulses	0.696	161.94	.0176	188.61	.0270	130.57	.0110
Mustard	1.046	112.90	.0122	39.61	.0057	199.12	.0168
Linseed	0.769	8.59	.0009	8.96	.0013	8.15	0007
Sesamum	1.153	37.90	.0041	7.00	.0010	74.24	.0062

(Contd.)

Table 4.36 : (Contd.)

Name of the Crop	Price [a] weight	Whole period, 1957-58–1994-95		1st sub-period, 1957-58–1976-77		2nd sub-period, 1977-78–1994-95	
		Standardised[b] average production	Growth[b] weight	Standardised[b] average production	Growth[b] weight	Standardised[b] average production	Growth [b] weight
(1)	(2)	(3)	(4)	(5)	(6)	(7)	(8)
Jute	0.166	617.70	.0669	473.16	.0678	787.75	.0664
Sugarcane	1.031	142.41	.0154	167.52	.0240	112.87	.0095
Patatoes	0.346	694.54	.0752	295.42	.0423	1164.09	.0981
Tobacco	4.857	61.34	.0066	54.84	.0079	69.00	.0058
Total	——	9225.76 (9436.56)	1.0000	6979.86 (7095.17)	1.0000	11869.67 (12038.02)	1.0000

Notes : [a]Calculated from Boyce (1987), Page 64, Table 3.4, Column 5.
[b] For estimation see text [Chapter 3, Section 3.4].
Terms within parentheses are corresponding to average indices for total agricultural production in West Bengal.

Table 4.37 : Period-wise Measured Growth Weights of Total Agricultural Production, West Bengal by Districts, 1957-58—1994-95

Name of the District	Standardised Production per unit of index No.	Whole period, 1957-58–1994-95		1st sub-period, 1957-58–1976-77		2nd sub-period, 1977-78–1994-95	
		Standardised average production	Growth weight	Standardised average production	Growth weight	Standardised average production	Growth weight
(1)	(2)	(3)	(4)	(5)	(6)	(7)	(8)
Burdwan	9.296	1174.46 (126.34)	.1245	809.40 (87.07)	.1151	1603.84 (172.53)	.1309
Birbhum	6.255	623.50 (99.68)	.0661	508.03 (81.22)	.0722	759.36 (121.40)	.0620
Bankura	5.848	616.85 (105.48)	.0654	464.68 (79.46)	.0661	795.85 (136.09)	.0650
Midnapore	12.957	1497.44 (115.57)	.1588	1105.10 (85.29)	.1571	1958.97 (151.19)	.1598
Howrah	1.702	182.66 (107.32)	.0194	121.47 (71.37)	.0173	254.64 (149.61)	.0208

(Contd.)

Table 4.37 : (Contd.)

Name of the District	Standardised Production per unit of index No.	Whole period, 1957-58–1994-95		1st sub-period, 1957-58–1976-77		2nd sub-period, 1977-78–1994-95	
		Standardised average production	Growth weight	Standardised average production	Growth weight	Standardised average production	Growth weight
(1)	(2)	(3)	(4)	(5)	(6)	(7)	(8)
Hooghly	5.770	770.18 (133.48)	.0817	489.07 (84.76)	.0695	1100.92 (190.80)	.0899
24-Parganas	8.541	1044.31 (122.27)	.1107	812.16 (95.09)	.1155	1317.45 (154.25)	.1075
Nadia	4.607	608.68 (132.12)	.0645	392.29 (85.15)	.0558	863.21 (187.37)	.0705
Murshidabad	6.645	751.08 (113.03)	.0796	565.56 (85.11)	.0804	969.31 (145.87)	.0791
West Dinajpur	5.897	613.05 (103.96)	.0650	499.65 (84.73)	.0710	746.50 (126.59)	.0609
Malda	3.111	376.34 (120.97)	.0399	276.60 (88.91)	.0393	493.68 (158.69)	.0403

(Contd.)

Table 4.37 : (Contd.)

Name of the District	Standardised Production per unit of index No.	Whole period, 1957-58–1994-95		1st sub-period, 1957-58–1976-77		2nd sub-period, 1977-78—1994-95	
		Standardised average production	Growth weight	Standardised average production	Growth weight	Standardised average production	Growth weight
(1)	(2)	(3)	(4)	(5)	(6)	(7)	(8)
Jalpaiguri	3.543	385.90 (108.92)	.0409	330.42 (93.26)	.0470	451.17 (127.34)	.0368
Darjeeling	0.567	69.46 (122.51)	.0074	52.12 (91.93)	.0074	89.86 (158.48)	.0073
Coochbehar	4.076	408.50 (100.22)	.0433	337.86 (82.89)	.0480	491.57 (120.60)	.0401
Purulia	2.898	309.48 (106.79)	.0328	269.54 (93.01)	.0383	356.45 (123.00)	.0291
West Bengal	72.444	9436.56 (130.26)	——	7095.17 (97.94)	——	12038.02 (166.17)	——
Dist.Total	81.713	9431.89	1.0000	7033.95	1.0000	12252.78	1.0000

Notes : Terms within parentheses imply average index No. in the respective periods.

For estimation procedure see text [Chapter 3, Section 3.4].

Table 4.38 : Contribution of the Crops to Agricultural Output Growth Rates in West Bengal, 1957-58—1994-95

Name of the Crop	Whole period (1957-58–1994-95)			1st Sub-period (1957-58–1976-77)			2nd Sub-period (1977-78–1994-95)		
	Output growth rate	Weighted output growth rate	Crop's[a] contribution	Output growth rate	Weighted output growth rate	Crop's[a] contribution	Output growth rate	Weighted output growth rate	Crop's[a] contribution
(1)	(2)	(3)	(4)	(5)	(6)	(7)	(8)	(9)	(10)
Aus rice	1.654	0.1130	3.75	1.579	.130	5.28	1.742	.101	2.69
Aman rice	1.961	1.1230	39.29	.673	.433	17.57	3.666	1.917	51.11
Boro rice	14.942	1.760 (.700)	23.24	22.080	1.128 (.631)	25.61	6.582	1.079 (.817)	21.78
Total rice	2.553	1.936	64.28	1.536	1.194	48.46	3.805	2.835	75.58
Wheat	8.900	.320	10.62	21.927	.748	30.36	-3.497	-.130	-3.47
Barley	-4.228	-.005	-0.17	3.150	.007	0.28	-11.242	-.008	-0.21
Gram	-4.801	-.025	-0.83	-1.258	-.011	-0.45	-8.952	-.022	-0.59
Other pulses	-1.625	-.029	-0.96	-.662	-.018	-0.73	-3.058	-.034	-0.91
Mustard	7.241	.088	2.92	1.541	.009	0.37	13.917	.234	6.24

(Contd.)

Table 4.38 : *(Contd.)*

Name of the Crop	Whole period (1957-58-1994-95)			1st Sub-period (1957-58-1976-77)			2nd Sub-period (1977-78-1994-95)		
	Output growth rate	Weighted output growth rate	Crop's[a] contribution	Output growth rate	Weighted output growth rate	Crop's[a] contribution	Output growth rate	Weighted output growth rate	Crop's[a] contribution
(1)	(2)	(3)	(4)	(5)	(6)	(7)	(8)	(9)	(10)
Linseed	-3.083	-.003	-0.10	5.112	.007	0.28	-12.484	-.009	-0.24
Sesamum	12.958	.053	1.76	13.862	.014	0.57	11.900	.074	1.97
Jute	•2.540	.170	5.64	2.784	.189	7.67	2.391	.159	4.24
Sugarcane	-2.213	-.034	-1.13	.698	.017	0.69	-5.624	-0.053	-1.41
Potatoes	7.128	.536	17.80	7.108	.301	12.22	7.151	.702	18.72
Tobacco	.713	.005	0.17	.887	.007	0.28	.510	.003	0.08
All Crops (Combined)	2.684	4.072 (3.012)	100.00	2.081	2.961 (2.464)	100.00	3.649	4.013 (3.751)	100.00

Notes : [a] Contributions are in percentage form. Terms within parentheses represent measurement by residual method.
For estimation: (i) Weighted output growth rate = (Output growth rate) x Growth weight (Table 4.36).
(ii) Crop's contribution = (Weighted output growth rate of the crop) ÷ (Sum of weighted output growth rates of all crops) x 100.

Table 4.39 : Contribution of the Districts to Agricultural Output Growth Rates in West Bengal, 1957-58—1994-95

Name of the District	Whole period (1957-58—1994-95)			1st Sub-period (1957-58—1976-77)			2nd Sub-period (1977-78—1994-95)		
	Output growth rate	Weighted output growth rate	District's[a] contribution	Output growth rate	Weighted output growth rate	District's[a] contribution	Output growth rate	Weighted output growth rate	District's[a] contribution
(1)	(2)	(3)	(4)	(5)	(6)	(7)	(8)	(9)	(10)
Burdwan	3.286	.409	13.74	2.786	.321	13.76	3.837	.502	13.25
Birbhum	2.205	.146	4.90	2.086	.151	6.47	2.435	.151	3.99
Bankura	2.636	.172	5.78	1.269	.084	3.60	4.807	.312	8.24
Midnapore	3.166	.503	16.89	1.537	.241	10.33	5.074	.811	21.41
Howrah	4.166	.081	2.72	4.049	.070	3.00	4.220	.088	2.32
Hooghly	4.476	.366	12.29	4.261	.296	12.69	4.727	.425	11.22
24-Parganas	2.282	.253	8.50	2.007	.232	9.94	2.702	.290	7.66
Nadia	4.163	.269	9.03	3.911	.218	9.34	4.461	.315	8.32
Murshidabad	3.237	.258	8.67	3.068	.247	10.59	3.583	.283	7.47

(Contd.)

Table 4.39 : (*Contd.*)

Name of the District	Whole period (1957-58–1994-95)			1st Sub-period (1957-58–1976-77)			2nd Sub-period (1977-78–1994-95)		
	Output growth rate	Weighted output growth rate	District's[a] contribu-tion	Output growth rate	Weighted output growth rate	District's[a] contribu-tion	Output growth rate	Weighted output growth rate	District's[a] contribu-tion
(1)	(2)	(3)	(4)	(5)	(6)	(7)	(8)	(9)	(10)
West Dinajpur	2.267	.147	4.94	1.948	.138	5.92	3.144	.191	5.04
Malda	3.144	.125	4.20	2.742	.108	4.63	3.740	.151	3.99
Jalpaiguri	1.797	.073	2.45	2.050	.096	4.12	1.484	.055	1.45
Darjeeling	2.870	.021	0.71	2.513	.019	0.81	3.325	.024	0.63
Coochbehar	2.032	.088	2.96	1.792	.086	3.69	2.314	.093	2.45
Purulia	2.007	.066	2.22	.687	.026	1.11	3.325	.097	2.56
West Bengal	2.684	2.977	100.00	2.081	2.333	100.00	3.649	3.788	100.00

Notes : [a] Contributions are in percentage form.

For estimation : (i) Weighted output growth rate = (Output growth rate) x Growth weight (Table 4.37).

(ii) District's contribution = (Weighted output growth rate of the district) ÷ (Sum of weighted output growth rates over all districts) x 100.

5

Measurement of Agricultural Stability

As stability in agricultural production is as much desirable as growth, it has drawn the attention of quite a number of scholars in this field. Stability in agriculture is customarily measured from its counterpart concept, i.e., instability. To know the state of stability in West Bengal agriculture (a) the nature of instability (i.e., the nature of the movement of instability over time), (b) the extent (i.e., level or magnitude) of instability and (c) the decomposition of output instability into its components have been measured in subsequent first three sections of this chapter. In the last section an attempt is made to enquire into the relationship between growth and instability in West Bengal agriculture.

5.1 Nature of Instability in Agricultural Production: Crop-wise and District-wise Disaggregation

The results on the nature of instability in the production of different crops and crop-groups in West Bengal, obtained by applying the techniques described in section 3.5 of Chapter 3, are presented in Table 5.1. It is observed from Table 5.1 that both $\hat{b}$ (estimated trend coefficient by fitting linear trend equation on the detrend statistic, $|X_t|$) and rank correlation coefficient (between $|X_t|$ and t) are of same sign in all cases. Among 22 crops and crop-groups of West Bengal, in 15 cases both rank correlations and $\hat{b}$s are negative and only in 7 cases they are positive. Further both $\hat{b}$s and ranks are statistically significant in 8 cases (having negative signs in 5 cases and positive signs in other 3 cases) and in 4 cases only rank correlations are statistically significant (with negative signs). Therefore, altogether in 9 cases [*viz.*, boro rice, total rice, wheat, total cereals, total foodgrains, jute, potatoes, total non-foodgrains and all crops] the instability in production is declining significantly over time (i.e., during LFG regime compared to pre-LFG regime). Only for three crops (*viz.*, gram, sesamum and sugarcane) the output instability is increasing significantly over time. For the remaining crops and crop-groups no significant change in the instability in production over time is obtained. However, the output instability is positive, though not statistically significant, only for mustard, linseed, total oilseeds and tobacco. It is further observed that the increasing instabilities in the output of some individual crops, with the exception of total oilseeds, were not strong enough to reverse the declining trend in instability in the total agricultural production and in their respective crop-group's production. From these observations it can be said that the instability in agricultural production of West Bengal during LFG regime (compared to pre-LFG regime) has declined significantly so far as the important crops and major crop-groups are concerned.

In order to measure the spatial difference in output instability in West Bengal agriculture the nature of instability in the production of four major crop-groups—total rice, total foodgrains, total non-foodgrains and total agricultural production — in the districts have been measured. The results are given in Table 5.2

An analysis of the results given in Table 5.2 shows that there has been significant decline in the instability in the production of total rice in the State and only in three districts in the State, namely, Howrah, Hooghly, and Nadia. There has been no statistically significant change in the instability in the total rice production in any of the remaining twelve districts. In other words, it has remained the same as before.

So far as the instability in total foodgrains is concerned, it is found to have significant (statistically) decline in the State and in the districts of Howrah, Hooghly, Murshidabad, Darjeeling and Coochbehar. There has been no significant change in instability in the total foodgrains production in the other ten districts. In the case of total non-foodgrains instability has declined in the State and in six districts of Burdwan, Bankura, Midnapore, Howrah, 24-Parganas and West Dinajpur. There is no significant change in instability in the remaining nine districts. In the case of the total agricultural production (all crops) only in three districts—Howrah, Hooghly and Nadia—the instability has declined significantly. It has also declined in the State. In a nutshell, in only two districts, namely, Howrah and Hooghly, the instability in the production of all the crop-groups such as total rice, total foodgrains, total non-foodgrains and total agricultural production, has declined significantly over the period under study. In only five districts of Burdwan, Bankura, Midnapore, 24-Parganas and West Dinajpur, significant decline in the instability has occurred in the case of total non-foodgrains only. In Murshidabad, Malda, Darjeeling and Coochbehar, instability has declined significantly in the production of total foodgrains only. In Nadia the instability is seen to have declined significantly in the production of total rice and in total agricultural production. However, as a result of the combined effect of the performance of some districts in respect of different crop-groups, the State has experienced significant decline in respect of the major crop-groups and the total agricultural production.

A few important points need to be mentioned here about the results, statistical and otherwise, presented in Tables 5.1 and 5.2. First, the nature of instability (for statistical significance) is reflected more by $\hat{b}$s than rank correlations for the crop-groups of the districts and the reverse is true for the crop-groups of the State. However, the disagreements between the values of $\hat{b}$ and rank correlation in respect of their levels of statistical significance (not in signs) have occurred only at higher levels of probability (mostly at 10% level). Second, from the measurement of nature of production instability it is seen that altogether in 30 cases (out of total 82 crops/crop-groups) instabilities are statistically significant. And from the heteroscedasticity measures [Tables 4.12 to 4.15, section 4.2, Chapter 4] on the residuals (derived from the fitted regression equation on the original series, not on the adjusted series as in case of instability measures) it is seen that 19 crops/crop-groups were characterized by significant long-term variations in their production series. Now if we compare these two measures over the crops, it is seen that for 14 crops/crop-groups the $\hat{b}$s and rank correlations have the same sign and level of significance. In other 21 cases [16 crops/crop-groups for instability measure and 5 crops/crop-groups for heteroscedasticity measure] there is dissimilarity between these two measures, specially in respect of levels of statistical significance. As both these measures heavily depend upon the specification of growth trend equation one should place more confidence on the instability measure of this study (based on the correct specification of trend equation on the adjusted series) than the heteroscedasticity measure (i.e., the instability measure of the earlier studies, based on the specification of trend equation on the original series with a number of other econometric problems).

5.2 Levels of Instability in Crop Output: Disaggregation by Crops and Districts

It is desirable that both the level of instability, i.e., the magnitude of instability and the changing pattern of instability over time are to be considered for a proper measurement of the state of stability in crop output. For example, a crop even with increasing nature of instability is more stable in its production if its levels of output instability remain at low level for all the periods than another crop whose levels of instability remain always high though the nature of instability is declining or unchanging over time. Crop-wise levels of instability in the agricultural production of West Bengal for the whole period (i.e., 1957-58 to 1994-95) and also for two sub-periods (i.e., 1957-58 to 1976-77 and 1977-78 to 1994-95) are presented in Table 5.3. From Table 5.3 it is observed that the patterns of changes in the levels of instability (measured by the summary measure i.e., mean of $|Z_t|$, as explained in section 3.5, Chapter 3) over the two sub-periods maintain consistency in regard to the nature of instability, as explained in the previous section (Table 5.1) for all 22 crops and crop-groups of West Bengal. But the magnitude of these changes in the levels of instability between two sub-periods does not maintain consistency in all cases. The limitations of the summary measure might have contributed to this inconsistency.

However, replacing the strong outliers by the next higher value in the summary measures it is seen that the differences in the magnitudes of the levels of instability between two sub-periods become consistent with the nature of instability in all the cases, which was absent in the earlier summary measures. From the estimated results of Table 5.3 it is seen that the levels of instability are much higher in the cases of individual crops than the crop-groups. In all the crop-groups (except total oil seeds) the levels of instability for all the periods are low (i.e., less than 10%). Among the individual crops, levels of instability are very high (≥20%) for all the periods in case of wheat, boro rice, linseed, sesamum and other cereals; only for the first sub-period in case of barley and jute and only for the second sub-period in case of sugarcane. In the case of mustard this is high for both the whole period and the second sub-period. The other crops in which the levels of instability in the production are high (lies in between 20% and 15%) for the whole period and at least for one of the sub-periods are aus rice, gram and tobacco (along with total oilseeds) and only for first sub-period in case of potatoes. For the other two crops, namely, aman rice and other pulses, the levels of instability for all three periods (whole period and two sub-periods) are moderate, lying between 15% and 10%. Only for three crops such as, potatoes, other pulses and jute, the levels of instability have changed from high/moderate to low value and that too for the second sub-period.

To enquire into the district-wise variations in the levels of instability in the productions of four major crop-groups such as, total rice, TFG, TNFG and all crops, the summary measures have been calculated for these crop-groups over the districts for all three periods. The results are given in Tables 5.4 to 5.7. From these four tables it is seen that the changes in the levels of instability over two sub-periods maintain consistency with the movements of the nature of instability over time (Table 5.2) in all cases, except for TFG in Burdwan, West Dinajpur and Jalpaiguri and total rice and TFG in 24-Parganas. Out of these five exceptional cases (out of total 60 cases), in first three cases, that is, for TFG in Burdwan, West Dinajpur and Jalpaiguri, the movements of the nature of instability over time, as measured by both $\hat{b}$s and rank values, are very close to zero whereas in these cases levels of instability, measured by summary measures, give slightly different results perhaps because of the presence of some moderate outliers in the data set. But in the case of productions of total rice and TFG in 24-Parganas the situation is different. In these two cases the $\hat{b}$s are statistically insignificant, but the rank coefficients are not so. Further in these two cases instabilities are very sensitive to the trend specification. As it has been mentioned in section 4.1, Chapter 4, (Table 4.7), the parabolic trend implies increasing instability over time while the log-quadratic trend implies decreasing instability. The presence of these inconsistencies in respect of the nature of instabilities is confirmed by the estimated values of $\hat{b}$s and ranks (based on the log-quadratic trend). Similarly, the inconsistency in respect of the level of instability is confirmed by the values of summary measures. On these grounds, one can consider the summary measures as more reliable in these two cases.

So far as the detection of strong outliers are concerned, the year 1982-83 acts as the year of strong outliers in most cases of the productions of total rice, TFG and all crops. Actually 1982-83 was a bad crop year in West Bengal agriculture. However, the values of the strong outliers are not same in all cases; they depend on the nature of the series and also on the values of the neighbouring years. With the replacement of strong outliers by next higher value, it is also observed here that the values of the summary measures are consistent with the nature of instability in all cases.

From the district-wise estimated results on the level of instability in crop-output the following observations can be made.

(i) In case of total rice production (Table 5.4) it is seen that the levels of instability are relatively high in the districts of Howrah, Purulia and Bankura for all periods, in Hooghly and Nadia for the first sub-period and in 24-Parganas for the second sub-period, compared to those in other districts. In the cases of Burdwan, Birbhum, Midnapore, West Dinajpur, Jalpaiguri and Malda the levels of instability are relatively lower than those in other districts for all the three periods.

(ii) In case of total foodgrains (Table 5.5) the levels of production instability remain more or less same as those in rice production. The positions of the districts also remain more or less the same, only exception being Malda where instability in the production of TFG during second sub-period has increased remarkably compared to that in the rice production.

(iii) As regards the output instability in total non-foodgrains (Table 5.6) the situations over the districts are altogether different from those in either rice production or TFG production. In general, here levels of instability over the districts are much higher in comparison to the other two crop-groups. In respect of the inter-district variations in the output instability in TNFG, it is seen that with the exception of Midnapore, West Dinajpur, Jalpaiguri, Darjeeling and Coochbehar, the levels of instability are high in all other districts specially for the whole period and for the first sub-period. In the second sub-period the instability has declined remarkably in 24-Parganas and Nadia. The levels of instability in the production of TNFG remain relatively high for all the periods in Purulia, Howrah, Birbhum, Malda, Burdwan and Bankura; they remain relatively low all through the periods in Darjeeling, Jalpaiguri and Coochbehar.

(iv) The magnitudes of output instability in all crops combined (Table 5.7) over the districts are, in general, much lower compared to other three crop-groups. So far as the inter-district variations are concerned, production instabilities in all crops are relatively high in Howrah, Purulia and Bankura and they are very low in Jalpaiguri, Darjeeling, Burdwan, Malda and West Dinajpur compared to other districts.

Therefore, from the observations on the levels of instability over the districts in the productions of four major crop-groups it can be concluded that the extent of instability is lower in case of all crops combined than other crop-groups and it is higher in case of TNFG. Further, the levels of instability are generally high for all crop-groups over all the periods in three districts, namely, Bankura, Howrah and Purulia. On the other hand, the instabilities in the northern districts are generally lower. However, it should be mentioned that though in the case of Howrah, instability has declined significantly (negative and significant $\hat{b}$ and rank correlation) over the whole period, the levels of instability have remained quite high. Actually, the instability has declined over time mainly in those districts where levels are high.

5.3 Decomposition of Output Instability in Agriculture

The instability in agricultural production of West Bengal has been decomposed in terms of individual crops and districts. While decomposing the instability in the total production of the State by the districts, only all crops combined is considered. Methodology used for this decomposition has been discussed in section 3.5 of

Chapter 3. For instability decomposition, output instability has been alternatively defined as the variability in point-to-point (i.e., annual) growth rates of the original production series, measured around the estimated trend growth rate [and denoted as var(G_o)]. But this measure of instability is not strictly comparable with the earlier two measures of nature and level of instability because point-to-point growth series (on which this new measure depends) are more liable to volatile fluctuations than the original series (on which earlier two measures depend). The estimated var(G_o)s of the crops in the State and all crops combined of the districts for different sub-periods are presented in Tables 5.8 and 5.9 respectively. From these two Tables it is observed that the years which have been deleted in the calculation of var(G_o)s for containing strong outliers, remain same as those mentioned earlier (Tables 5.3 and 5.7) in the cases of measuring levels of instability [however, with one period lead or lag because of the consideration of first difference of production series in calculating var(G_o)s]. In addition, some more years have also been detected in some cases for containing extremely high or low values whose occurrence may be attributed to the volatile nature of the point-to-point growth series. Comparing var(G_o)s with the corresponding levels of instability (Tables 5.3 and 5.7) it is seen that though exact rankings differ slightly, the broad classifications of the crops and districts on the basis of the magnitude of instability do not differ for any of the periods. Further, comparing the movements of the estimated var (G_o) over two sub-periods with the natures of instability for the corresponding crops and districts (i.e., comparing cols. 4 and 5 with col. 2 of Tables 5.8 and 5.9) it is seen that there is consistency in the results in all crops, excepting three major crops in the State (*viz.*, aman rice, total rice and all crops) and in all districts, excepting two districts for all crops (*viz.*, Bankura and Midnapore). Actually, as indicated above, this variance measure of instability is non-comparable with the trend measure of instability and the achievement of consistency, between them, wherever it has occurred, is a matter of coincidence. However, considering the broad spectrum of instability measures and also considering the coincidence of consistency in most cases this new measure of instability has been used (as in the earlier studies but with certain improvements) in the decomposition model to determine the relative shares of its components.

It has been mentioned in Chapter 3, section 3.5, that an error term is very likely to appear in the instability (variance) decomposition model,

$$Var(G_o) = Var(G_a) + Var(G_p) + 2Cov(G_a, G_p),$$

particularly because of varying nature of output, area and productivity data and consequent presence of different trends in these data. The necessary adjustments in the data and estimation of variances around trend growth rate instead of mean growth rate have contributed to substantial reduction in the errors in all the cases, particularly with respect to their absolute and relative contributions to the overall output instability. Further in comparison with the synchronous movements of area and productivity variabilities [i.e., $2Cov(G_a, G_p)$] it is seen from Table 5.12 that almost in all cases the error term is found to be too weak to have any significant effect upon the output instability, the only exception being Howrah (for first sub-period only), Nadia and Coochbehar (for whole period only) and 24-Parganas (for the whole period and second sub-period). These few exceptional cases (out of total 48 estimated values of error term) also arise mainly because of relatively lower value of $2Cov(G_a, G_p)$ than the low value of the error term. No definite pattern is obtained on the nature of error term. In other words, nothing can be said conclusively as regards the overestimation or underestimation in the components' contributions towards output instability. [In 17 cases error term is negative, in 28 cases it is positive and in 7 cases (4 of them having negative sign) it is zero or very close to zero]. Therefore, it can be said that the instability decomposition model used in our study is more reliable and the residual effect is dominated mainly by the synchronous movement of area and productivity variabilities, i.e., $2Cov(G_a,G_p)$.

The final estimated results on the instability decomposition analysis for crops/crop-groups of the State and all crops combined for the districts have been reported in Tables 5.10 and 5.11 respectively. From Table 5.10 it is seen that productivity variability acts as the main source of output instability for major crops/crop-groups in the State, namely, aman rice, total rice, other cereals, total cereals, total foodgrains, mustard and all crops for all the periods (i.e., for whole period and two sub-periods). Likewise, the variability in area acts as the dominant source of output instability for all the periods in the cases of boro rice, wheat, sesamum, sugarcane, tobacco and total non-foodgrains. For aus rice, linseed, total oilseeds and potatoes productivity variation was the principal contributor to output instability during the first sub-period; but during the second sub-period the area variation contributed most to the output instability. In general, it can be said that the area variability is the principal contributor to the output instability for the non-foodgrain crops and the crops whose output growth rates have declined over time; on the other hand, productivity variability is the principal contributor to the output instability for the foodgrains and also for the crops where output growth rates have increased over time with the exception of total oilseeds and TNFG.

As regards the residual effect it is seen that its contribution to the output instability is smaller in the case of crop-groups, except TNFG, in comparison to that in the case of individual crops. It is also smaller in the case of foodgrains in comparison to that in the case of non-foodgrains. It is further seen that the positive effect (negative effect) of the residual component is higher (lower) in magnitude in the first sub-period than in the second sub-period. Further, as the outcome of the movements of area and productivity variabilities in opposite direction (i.e., due to negative residual effect), the output instability has declined (i) in the cases of other cereals, gram, potatoes and TNFG during the whole period, 1957-58 — 1994-95 and its two sub-periods; (ii) in case of linseed and total oil seeds during 1957-58 — 1994-95 and during the first sub-period 1957-58 — 1976-77; (iii) in the cases of boro rice, wheat, total pulses, Jute and sugarcane during the second sub-period, 1977-78—1994-95; (iv) in the case of mustard in the first sub-period and (v) in the case of sesamum during the whole period as well as in the second sub-period. In all other cases this synchronous movement exerts positive pressure on output instability. Therefore, negative residual effect on output instability (i.e., the movement of area and productivity variabilities in opposite direction leading to lower output instability) occurs in the case of those crops where variability in area is high (except other cereals).

From the decomposition of output instability in total agricultural production (i.e., all crops) in the districts (Table 5.11) it is observed that in all the districts for all the periods the contribution of productivity variability in output instability is much higher than the contributions of other components. However, the variability in area as source of output instability is also remarkable in the cases of Burdwan, Malda and Darjeeling for all the periods and Howrah, Hooghly, Jalpaiguri and Coochbehar for the second sub-period. Likewise, the contribution of residual component is not negligible for at least one of the sub-periods in the districts other than Burdwan, 24-Parganas, Nadia, West Dinajpur and Coochbehar. As far as the periodic changes in the contributions of area and productivity variabilities towards output instability over the districts are concerned, it is observed that over these two sub-periods (i) the contribution of area variability has increased for Burdwan, Howrah and Coochbehar, (ii) the contribution of productivity variability has increased for Nadia and Purulia, (iii) the contributions of both area and productivity variabilities have increased (declined) for Hooghly, Malda and Jalpaiguri (for only Birbhum) and (iv) in all other districts the relative contributions of the components remain more or less constant. These periodic changes in the relative contributions of area and productivity components have been quite obviously reflected in the contribution of the residual component. Thus, it is seen that the contribution of the residual component has declined over these two sub-periods (i) from positive to negative in Burdwan, Hooghly, Nadia, Malda and Coochbehar, (ii) from high positive value to low positive value in Purulia and (iii) from low negative value to high negative value in Howrah; for all other districts the contribution of residual component remains more or less constant (with declining pressure), except in Birbhum. Therefore, over the districts the synchronous movements (in opposite direction) of area and productivity variabilities have exerted downward pressure on output instability more in second sub-period than in first sub-period. Further, it is observed that this downward pressure is quite strong for all the periods in those districts where area variability is also high (similar results were obtained for all crops in the State).

These diversified analyses on instability decomposition have been summarized by the correlation coefficients (Table 5.13), measured cross sectionally (taking individual crops of West Bengal and all crops of the districts), between variability in annual output growth rates and its constituent components. From Table 5.13 it is observed that all correlation coefficients are positive, implying that more variability in the component leads to more output instability and *vice versa*. From the probability levels of the statistical significance of the measured correlation coefficients the following observations can be made. In the case of individual crops in the State variability in area plays significant role in explaining output instability during first sub-period; but both area and productivity variabilities turn out to be the significant sources of output instability during second sub-period. No significant systematic relationship between output instability and the synchronous movements of area and productivity variabilities could be observed for all of these periods. But in the case of all crops in the districts all the three components are equally powerful in explaining output instability during 1957-58 — 1994-95 and 1957-58 — 1976-77. The area component was, however, insignificant for the second sub-period.

We have also measured cross sectionally the inter-period correlation coefficients of the instability components to know the nature of their movements over two sub-periods (Table 5.14). From this table it is seen that for individual crops in the State all the inter-period correlation coefficients are statistically insignificant, excepting those in productivity variability. This implies that there is no systematic relation in the period-wise movements of the components, except for productivity, and for productivity, the crops having higher/lower productivity variability in first sub-period face similar pattern in the second sub-period. But in case of all crops of the districts all these correlation coefficients are positive and statistically significant. This implies that over two sub-periods the positions of the districts in regard to the relative strength of the instability components remain on the whole unchanged.

5.4 Association between Growth and Instability in Crop Output

In this section an attempt is made to determine the nature of association between growth and instability. In other words, in this section a hypothesis that there is direct association between growth and instability is tested statistically. For this purpose the correlation coefficients between the growth rates and the levels of instability are estimated taking individual crops in the State and each crop-group in the districts for different periods into consideration. The estimated correlation coefficients are presented in cols.2, 3 and 4 of Table 5.15. The correlation coefficients are statistically significant only in three cases, that is, in the case of the individual crops of the State (i) for the whole period and (ii) for the first sub-period and (iii) in the case of total non-foodgrains of the districts for the first sub-period. In these three cases the correlation coefficients are also positive, implying that there are positive association between growth and instability in these three cases. As all other estimated correlation coefficients are statistically insignificant no systematic relationship between growth and instability is obtained in the other cases.

To enquire into the nature of association between change of growth rate and change of instability over time, the correlation coefficients between the trend breaks and the values of rank correlation (i.e., the nature of instability) over these cross sections have been measured. The estimated correlation coefficients are presented in the last column of Table 5.15. Here also all the estimated correlations are not statistically significant. From these results it can be said that there is practically no association between the change in growth rates, measured by trend breaks, and the change in instability.

Through these summary measures (i.e., measured correlation coefficients), however, no specific conclusion can be drawn on the relation between growth and stability in the State or in the districts, since the sign of the correlation coefficients do not indicate any particular type of association between growth and instability. For example, from positive correlation one may infer that higher growth is associated with higher instability or lower growth with lower instability, while from negative correlation one may infer that higher growth is associated with lower instability or lower growth with higher instability. Therefore, to get out of this impasse and to draw some meaningful inference on the association between crop-wise changing nature of growth and instability in agricultural production in West Bengal all the 82 crops and crop-groups in the State and the districts have been arranged on the basis of the following four classifications of trend breaks and the nature of instability. These classifications have been made into (i) positive and statistically significant coefficients, (ii) positive and statistically insignificant coefficients, (iii) negative and statistically significant coefficients and (iv) negative and statistically insignificant coefficients. The results are presented in Tables 5.16 and 5.17. In each of the tables there are 16 cells, each indicating the crops with changing growth rates and instability. These 16 cells have been classified into four broad groups consisting of four cells each. The left hand bottom group contains the crops for which high growth is associated with low instability, whereas the right hand top group contains the crops that have declining growth but higher instability. Thus, these eight cells clearly indicate the crops for which there is negative association between growth and instability. On the other hand, for the crops in the other eight cells (left top and right bottom), there is positive association between growth rate and instability, i.e., high growth - high instability and low growth - low instability. For example, for mustard and total oilseeds high growth rate is attained at the cost of high instability and for barley, other pulses, total pulses, boro rice and wheat low growth is associated with low instability. However, from this table it is clear that in West Bengal during 1957-58—1994-95 nine crops or crop-groups have registered high growth rate with declining instability. However, the most undesirable performance is encountered in the case of tobacco, gram, sesamum and sugarcane that fall in the right-top group in Table 5.16.

The results presented in Table 5.16 may be further interpreted to indicate the performance of the crops in West Bengal during the pre-LFG (1957-58—1976-77) and LFG period (1977-78—1994-95), in respect of their growth and instability. There is one trend break between these two sub-periods. The positive and statistically significant trend break, therefore, mean significantly higher growth rate during the second sub-period than during the first sub-period. Implication is just the reverse when the trend break is negative and statistically significant. If the breaks are not statistically significant, whether they are positive or negative, it means that there is no statistically significant change in growth rates. The positive and statistically significant trend breaks with statistically significantly negative instability have occurred in the case of total agricultural production and four crop-groups, namely, total rice, total cereals, total foodgrains and total non-foodgrains. This means that during the LFG period there has been higher growth of these crop-groups and total agricultural production and instability in the production of these crop-groups and total production has significantly declined. This is undoubtedly an achievement during the LFG regime. Along side this, mustard, total oilseeds and aman rice have registered higher growth rate during the LFG period without any significant increase or decrease in instability. There is not a single crop that has registered higher growth with increasing instability (instability positive and significant).

But, during the LFG regime three crops—gram, sesamum and sugarcane—have registered declining growth but increasing instability and another four crops and crop-groups, namely, linseed, barley, other pulses and total pulses that have declines in growth rate and no significant increase or decrease in instability. In the case of boro-rice and wheat the significantly declining growth rate has been associated with significantly declining instability. Thus, to conclude our assessment of the performance of agriculture in West Bengal, during the LFG period with respect to growth and instability of the crops and crop-groups, it can be said that while during the LFG period the performance of certain crop-groups such as total rice, total cereals, total foodgrains, total non-foodgrains and total agricultural production has been highly satisfactory in terms of higher growth and declining instability compared to the pre-LFG period, the performance of the crops such as gram, sesamum and sugarcane has been highly distressing in terms of lower growth and higher instability in comparison to the pre-LFG period. Not less unsatisfactory is the performance of the crops like linseed, barley, other pulses and total pulses which have experienced lower growth rates and almost the same magnitude of instability (not statistically significantly different from zero) during the LFG period. As regards the relation between growth and instability in crop output in the State there is enough evidence to accept the hypothesis of negative association between growth and instability in the State, particularly in relation to major crop groups (Cell 4,1 of Table 5.16).

Table 5.17 presents a classification of the districts in the State on the basis of varying nature of relation between growth and instability. The presentation of the results in this table is same as that in Table 5.16, and therefore, the interpretation runs on almost the same line. It is seen from the left bottom cell [4,1] of the table that only three crop-groups, including total agricultural production, in five districts have registered significantly higher growth rate (trend break -positive and statistically significant) with significantly declining instability during the second sub-period. This means that during the LFG period (1977-78—1994-95) only five districts, namely, Howrah, Nadia, Bankura, Midnapore and West Dinajpur could boast of significantly higher growth rate with significantly declining instability in only three crop groups such as total rice, total non-foodgrains and total agricultural production. It is seen from Table cells (2,1) and (3,1) of Table 5.17 that total rice in the districts of Burdwan, Birbhum, Bankura, Midnapore, 24-Parganas, Murshidabad, West Dinajpur and Purulia; total foodgrains production in the districts of Burdwan, Bankura, Midnapore, 24-Parganas and Purulia; the total non-foodgrains in Hooghly, Nadia, Murshidabad and Coochbehar and the total agricultural production (AC) in the districts of Bankura, Burdwan, Midnapore, 24-Parganas, Malda, West Dinajpur and Purulia—have experienced significantly higher growth rate with no significant increase or decrease in the output instability during the second sub-period.

During the second sub-period the only crop-group of total foodgrains in the districts of Malda and Darjeeling (Cell 1,1) registered significantly higher growth rate, but not without significantly increasing output instability. As against this it is to be seen that the total rice in only one district, Hooghly, has experienced significant fall in the growth rate along with significant fall in output instability. Thus, the evidence [only three data points—two in cell (1,1) and one in cell (4,4)] is not sufficient to indicate any positive relation between the growth and instability (higher growth - higher instability or lower growth - lower instability). Similarly, only six data points relating to six districts in the left bottom corner of the table indicating negative association between growth and instability are insufficient to establish negative association between growth and instability.

Considering the information given in cells (2,4) and (3,4), it can be said that the growth rates of total non-foodgrains production in the districts of Purulia, Darjeeling and Jalpaiguri have significantly declined but no change has taken place in the instability during the second sub-period.

The information in cell (4,2) reveals that the growth rates of the total foodgrains, total non-foodgrains and total agricultural production in the districts of Howrah, Murshidabad, 24-Parganas and Hooghly during the second sub-period have remained the same (in statistical sense) as those in the first sub-period, but the instability has declined significantly in the second period.

Thus, no definite relationship could be established between growth and instability in the districts of the State. This result is in contrast with that for the crops and crop-groups in the State. The negative correlation between growth and instability in the output of the crop-groups, particularly between significantly higher growth and significantly declining instability in the State, may be ascribed to the aggregation at the State level of output of different crops in the crop-groups in the districts, that wipe off a significant part of variations in individual crop output.

Table 5.1 : Trend Measures of Instability in Crop-Output, West Bengal, 1957-58—1994-95

Crop/Crop-group	Nature of Instability $\hat{b}$ (Parametric)	Nature of Instability Rank Correlation (Non-parametric)	Crop/Crop-group	Nature of Instability $\hat{b}$ (Parametric)	Nature of Instability Rank Correlation (Non-parametric)
Aus rice	-00035 (-1.023)	-.1707	Total food-grains	-.00010 (-.692)	-.3107***
Aman rice	-.00264 (-1.075)	-.2651	Mustard	.00050 (.753)	.1907
Boro rice	-.00497* (-5.828)	-.7313*	Linseed	.00602 (1.259)	.2659
Total rice	-.00013 (-.827)	-.3331**	Sesamum	0.00629* (2.853)	0.4844*
Wheat	-.00437* (-4.796)	-.5735*	Total oil seeds	.00021 (.528)	.0708
Barley	-.00077 (-1.088)	-.2093	Jute	-.00454*** (-1.955)	-.3712**
Other cereals	-.00104 (-1.256)	-.2252	Sugarcane	.00074*** (1.906)	.3184***
Total cereals	-.00012 (-.797)	-.3344**	Potatoes	-.00086* (-3.804)	-.5351*
Gram	.00142** (2.540)	.3359**	Tobacco	.00059 (.795)	.0839
Other pulses	-.00009 (.395)	-.0301	Total non-foodgrains	-.00196* (-2.820)	-.4481*
Total pulses	-.00004 (.218)	-.0358	All crops	-.00021 (-1.024)	-.3915**

Notes : * Significant at 1% level, ** Significant at 5% level,
*** Significant at 10% level.
$\hat{b}$ is the parameter estimated from $|X_t| = a + bt$,
Rank correlation is between $|X_t|$ and t,
For X_t see text (section 3.5, Chapter 3),
t ratios in the parentheses.

Table 5.2 : Trend Measures of Instability in Agricultural Production, Major Crop-groups, West Bengal by Districts, 1957-58 — 1994-95

District	Total Rice		Total foodgrains		Total non-foodgrains		All Crops	
	Nature of Instability		Nature of Instability		Nature of Instability		Nature of Instability	
	$\hat{b}$ (Parametric)	r (Non parametric)	$\hat{b}$ (Parametric)	r (Non-parametric)	$\hat{b}$ (Parametric)	r (Non-parametric)	$\hat{b}$ (Parametric)	r (Non-parametric)
1	2	3	4	5	6	7	8	9
Burdwan	-.00004 (-.213)	-.1959	-.00001 (-.029)	-.0646	.00074** (-2.003)	-.2250	0.00003 (.163)	.0808
Birbhum	.00003 (.106)	.0507	-.00001 (-.044)	-.0638	.00052 (.688)	.0165	.00020 (.483)	.0278
Bankura	-.00032 (-.863)	-.2553	-.00029 (-.801)	-.2618	-.00061*** (-1.679)	-.2250	-.00036 (-.777)	-.2399
Midnapore	-.00012 (-.566)	-.2543	-.00150 (-1.049)	-.1959	-.00061*** (-1.871)	-.1979	-.00019 (-.578)	-.2227
Howrah	-.00169* (-2.891)	-.4003**	-.00166* (-3.231)	-.4281*	-.00117*** (-1.936)	-.2700	-.00206* (-4.036)	-.5650*
Hooghly	-.00051*** (-1.813)	-.3223***	-.00273*** (-1.758)	-.3439**	-.00056 (-0.714)	-.0359	-.00086* (-3.067)	-.5503*
24-Parganas	-.00009 (-.455)	-.1432	-.00013 (-.673)	-.1542	-.00095** (-2.593)	-.3786**	.00068 (0.356)	.0389
Nadia	-.00061** (-2.278)	-.3647**	-.00021 (-.916)	-.0656	-.00121 (-.696)	-.2003	-.00043*** (-1.803)	-.2412

(Contd.)

Table 5.2 : *(Contd.)*

District	Total Rice		Total foodgrains		Total non-foodgrains		All Crops	
	Nature of Instability		Nature of Instability		Nature of Instability		Nature of Instability	
	$\hat{b}$ (Parametric)	r (Non parametric)	$\hat{b}$ (Parametric)	r (Non-parametric)	$\hat{b}$ (Parametric)	r (Non-parametric)	$\hat{b}$ (Parametric)	r (Non-parametric)
1	2	3	4	5	6	7	8	9
Murshidabad	-.00016 (-.515)	-.0803	-.00033 (-1.382)	-.3048***	-.00215 (-1.174)	-.0605	-.00017 (-.768)	-.0445
West Dinajpur	-.00009 (-.438)	-.0510	-.00005 (-.024)	-.0026	-.00372** (-2.112)	-.3086***	-.00093 (-.767)	-.1279
Malda	-.00033 (-1.562)	-.1653	.00047*** (1.906)	.1869	-.00015 (-.324)	-.0129	.00006 (.098)	.0832
Jalpaiguri	.00022 (.764)	.1248	.00010 (.390)	.0409	-.00092 (-.404)	-.0260	-.00026 (-.485)	-.0406
Darjeeling	-.00032 (-.866)	-.1372	.00064*** (1.669)	.2010	-.00157 (-1.142)	-.2517	.00028 (.575)	.0052
Coochbehar	-.00032 (-1.181)	-.2185	-.00047** (-2.385)	-.3959**	-.00083 (-.501)	-.1238	-.00025 (-1.293)	-.1137
Purulia	-.00183 (-.934)	-.0673	-.00147 (-.746)	-.0651	.00019 (.258)	.0015	-.00136 (-.695)	-.0384
West Bengal	-.0013 (-.827)	-.3331**	-.00010 (-.692)	-.3107***	-.00196* (-2.820)	-.4481*	-.00021 (-1.024)	-.3915**

Notes : Same as those in Table 5.1. Here r implies rank correlation.

Table 5.3 : Summary Measure on Instability in Agricultural Production, West Bengal by Crops, 1957-58—1994-95 and Sub-Periods 1957-58 — 1976-77 and 1977-78 — 1994-95

Crop/Crop-group	Strong Outlier(s)		Value of the summary measure, mean of $\|Z_t\|$		
	Year(s)	Extreme value(s)	Whole period (1957-58–1994-95)	First sub-period (1957-58–1976-77)	Second sub-period (1977-78–1994-95)
Aus rice	—	—	0.1701	.1969	.1401
Aman rice	—	—	0.1361	.1370	.1350
Boro rice	1971-72	2.01	0.4041 (.4224)	.5856 (.6203)	.2014
Total rice	1982-83	0.35	.0792 (.0834)	.0844	.0735 (.0823)
Wheat	1970-71	1.99	.4607 (.4776)	.6591 (.6912)	.2389
Barley	—	—	.1762	.2356	.1098
Other cereals	—	—	.2677	.3164	.2133
Total cereals	1982-83	0.37	.0813 (.0846)	.0881	.0736 (.0807)
Gram	—	—	.1797	.1627	0.1987
Other pulses	—	—	.1131	.1244	.1005
Total pulses	—	—	.0898	.0945	.0845
Total food-grains	1982-83	0.34	.0794 (.0822)	.0841	.0741 (.0800)
Mustard	—	—	.2124	.1694	.2604
Linseed	—	—	.3127	.2540	.3784
Sesamum	—	—	.3012	.2054	.4083
Total oil seeds	1987-88	0.81	.1508 (.1564)	.1306	.1733 (.1851)
Jute	1968-69 1985-86	0.56 0.58	.1558 (.1625)	.2021 (.2079)	.1040 (.1117)
Sugarcane	—	—	.1813	.1435	.2236
Potatoes	—	—	.1410	.1788	.0986
Tobacco	—	—	.1672	.1624	.1725
Total non-foodgrains	—	—	.0633	.0779	.0471
All crops	1982-83	0.27	.0622 (.0647)	.0656	.0584 (.0637)

Notes : Values not adjusted by the deletion of strong outlier(s) have been presented in the parentheses. For $|Z_t|$ see text (section 3.5, Chapter 3).

Table 5.4 : Levels of Instability in Rice Productions, West Bengal by Districts, Whole Period and Two Sub-periods

District	Strong Outlier(s)		Value of the summary measure, mean of $\|Z_t\|$		
	Year(s)	Extreme value(s)	Whole period (1957-58–1994-95)	First sub-period (1957-58–1976-77)	Second sub-period (1977-78–1994-95)
Burdwan	—	—	.0829	.0935	.0735
Birbhum	1982-83	0.45	.0979	.0898	.1073
			(.1010)		(.1135)
Bankura	1982-83	0.48	.1316	.1399	.1224
			(.1341)		(.1277)
Midnapore	1982-83	0.39	.1022	.1079	.0959
			(.1047)		(.1012)
Howrah	1960-61	0.97	.1901	.2145	.1629
			(.2032)	(.2392)	
Hooghly	—	—	.1273	.1482	.1039
24-Parganas	—	—	.1230	.1091	.1386
Nadia	—	—	.1175	.1388	.0935
Murshidabad	1982-83	0.42	.1207	.1262	.1146
			(1243)		(.1222)
West Dinajpur	1971-72	0.45	.0994	.1019	.0965
			(.1030)	(.1087)	
Malda	—	—	.1008	.1070	.0939
Jalpaiguri	1971-72	0.53	.1024	.0942	.1117
			(.1080)	(.1047)	
Darjeeling	—	—	.1248	.1290	.1202
Coochbehar	—	—	.10[illegible]3	.1226	.0943
Purulia	1979-80	0.50	.1406	.1517	.1281
	1982-83	0.58	(.1506)		(.1493)
West Bengal	1982-83	0.35	.0792	.0844	.0735
			(.0834)		(.0823)

Notes : Values not adjusted by the deletion of strong outlier(s) have been presented in the parentheses. For $|Z_t|$ see text (section 3.5, Chapter 3).

Table 5.5: Levels of Instability in the Production of Total Foodgrains, West Bengal by Districts, Whole Period and Two Sub-periods

District	Strong Outlier(s)		Value of the summary measure, mean of $\|Z_t\|$		
	Year(s)	Extreme value(s)	Whole period (1957-58–1994-95)	First sub-period (1957-58–1976-77)	Second sub-period (1977-78–1994-95)
Burdwan	—	—	.0827	.0746	.0918
Birbhum	1982-83	0.48	.1094	.1099	.1089
			(.1133)		(.1171)
Bankura	1982-83	0.48	.1301	.1399	.1192
			(.1332)		(.1257)
Midnapore	1982-83	0.36	.1074	.1192	0941
			(.1096)		(.0988)
Howrah	1960-61	0.93	.1807	.2091	.1488
			(.1940)	(.2344)	
Hooghly	—	—	.1321	.1571	.1041
24-Parganas	—	—	.1231	.1124	.1352
Nadia	—	—	.1060	.1123	.0990
Murshidabad	1970-71	0.42	.1106	.1220	.0979
			(.1137)	(.1278)	
West Dinajpur	—	—	.0950	.0945	.0956
Malda	1990-91	0.61	.1115	.0866	.1393
	1991-92	0.51	(.1137)		(.1511)
Jalpaiguri	1971-72	0.44	.0987	.1007	.0964
			(.1026)	(.1081)	
Darjeeling	1986-87	0.98	.1169	.0935	.1429
			(.1352)		(.1817)
Coochbehar	—	—	.1019	.1214	.0800
Purulia	1979-80	0.51	.1347	.1470	.1211
	1982-83	0.59	(.1464)		(.1458)
West Bengal	1982-83	0.34	.0794	.0841	.0741
			(.0822)		(.0800)

Notes : Values not adjusted by the deletion of strong outlier(s) have been presented in the parentheses. For $|Z_t|$ see text (section 3.5, Chapter 3).

Table 5.6: Levels of Instability in the Production of Total Non-foodgrains, West Bengal by Districts, Whole Period and Two Sub-periods

District	Strong Outlier(s)		Value of the summary measure, mean of $\|Z_t\|$		
	Year(s)	Extreme value(s)	Whole period (1957-58–1994-95)	First sub-period (1957-58–1976-77)	Second sub-period (1977-78–1994-95)
Burdwan	—	—	.1652	.1845	.1437
Birbhum	—	—	.1746	.1596	.1915
Bankura	—	—	.1629	.1795	.1443
Midnapore	—	—	.1289	.1375	.1192
Howrah	1962-63	0.86	.2067	.2182	.1938
			(.2150)	(.2340)	
Hooghly	1961-62	0.50	.1535	.1735	.1312
	1962-63	0.57	(.1616)	(.1888)	
24-Parganas	—	—	.1513	.1864	.1121
Nadia	—	—	.1794	.2252	.1282
Murshidabad	—	—	.1407	.1669	.1215
West Dinajpur	1959-60	0.50	.1281	.1399	.1150
			(.1320)	(.1473)	
Malda	—	—	.1729	.1793	.1658
Jalpaiguri	—	—	.0994	.1088	.0889
Darjeeling	1964-65	0.44	.0896	.1043	.0731
	1965-66	0.49	(.0988)	(.1217)	
Coochbehar	—	—	.1051	.1066	.1033
Purulia	—	—	.3236	.3103	.3385
West Bengal	—	—	.0633	.0779	.0471

Notes : Values not adjusted by the deletion of strong outlier(s) have been presented in the parentheses. For $|Z_t|$ see text (section 3.5, Chapter 3).

Table 5.7: Levels of Instability in the Production of All Crops, West Bengal by Districts, Whole Period and Two Sub-periods

District	Strong Outlier(s)		Value of the summary measure, mean of $\|Z_t\|$		
	Year(s)	Extreme value(s)	Whole period (1957-58–1994-95)	First sub-period (1957-58–1976-77)	Second sub-period (1977-78–1994-95)
Burdwan	—	—	.0696	.0608	.0794
Birbhum	1982-83	0.44	.1006	.0915	.1107
			(.1053)		(.1207)
Bankura	1982-83	0.48	.1252	.1310	.1188
			(.1294)		(.1276)
Midnapore	1982-83	0.38	.0961	.0976	.0945
			(.0986)		(.0998)
Howrah	—	—	.1723	.2235	.1151
Hooghly	—	—	.0862	.1103	.0593
24-Parganas	—	—	.1128	.1115	.1143
Nadia	—	—	.1083	.1170	.0987
Murshidabad	—	—	.0958	.0999	.0907
West Dinajpur	—	—	.0762	.0803	.0715
Malda	—	—	.0669	.0640	.0702
Jalpaiguri	—	—	.0435	.0455	.0413
Darjeeling	1964-65	0.33	.0661	.0625	.0701
			(.0700)	(.0699)	
Coochbehar	—	—	.0802	.0887	.0708
Purulia	1979-80	0.50	.1295	.1390	.1189
	1982-83	0.58	(.1406)		(.1424)
West Bengal	1982-83	0.27	.0622	.0656	.0584
			(.0647)		(.0637)

Notes : Values not adjusted by the deletion of strong outlier(s) have been presented in the parentheses. For $|Z_t|$ see text (section 3.5, Chapter 3).

Table 5.8 : Movements of Variability in Annual Output Growth Rates over Two Sub-periods, West Bengal by Crops, 1957-58—1976-77 and 1977-78—1994-95

Crop/Crop-group	Nature of Instability overtime[a]	Strong Outlier Year	First Sub-period, 1957-58–1976-77, Var(Go)[b]	Second Sub-period, 1977-78–1994-95, Var (Go)[b]
Aus-rice	Declining	1988-89	0.0407	0.0226
Aman-rice	Declining	1983-84	0.0131	0,0252
Boro-rice	Declining +	1967-68	0.1762	0.0601
Total rice	Declining +	1983-84	0.0132	0.0173
Wheat	Declining +	1968-69	0.0981	0.0426
Barley	Declining		0.2015	0.0630
Other cereals	Declining		0.1369	0.1182
Total cereals	Declining +	1983-84	0.0136	0.0132
Gram	Increasing +		0.0482	0.0514
Other pulses	Declining	1958-59	0.0283	0.0277
Total pulses	Declining		0.0319	0.0203
Total foodgrains	Declining +	1983-84	0.0123	0.0120
Mustard	Increasing	1958-59	0.0665	0.0867
Linseed	Increasing	1958-59	0.0807	0.1384
Sesamum	Increasing +		0.0740	0.0874
Total oil seeds	Increasing		0.0505	0.0743
Jute	Declining +	1969-70	0.1058	0.0494
Sugarcane	Increasing +		0.0582	0.0877
Potatoes	Declining +		0.0647	0.0176
Tobacco	Increasing		0.0420	0.0490
Total non-food-grains	Declining +		0.0143	0.0055
All crops	Declining +	1983-84	0.0081	0.0094

Notes : a- reference to Table 5.1 + statistically significant, b- variance in annual output growth rates.

Table 5.9 : Movements of Variability in Annual Output Growth Rates over Two Sub-periods, All Crops West Benpal by Districts, 1957-58—1976-77 and 1977-78—1994-95

District	Nature of Instability overtime[a]	Strong Outlier Year(s)	First Sub-period, 1957-58–1976-77, Var(Go)[b]	Second Sub-period, 1977-78–1994-95, Var (Go)[b]
Burdwan	Increasing	—	0.0093	0.0158
Birbhum	Increasing	1983-84	0.0198	0.0309
Bankura	Decreasing	1980-81 1983-84	0.0377	0.0387
Midnapore	Decreasing	1983-84	0.0214	0.0292
Howrah	Decreasing +	1960-61 1969-70	0.0461	0.0311
Hooghly	Decreasing +	—	0.0407	0.0134
24-Parganas	Increasing	1960-61, 1971-72	0.0143	0.0225
Nadia	Decreasing +	—	0.0293	0.0170
Murshidabad	Decreasing	—	0.0305	0.0228
West Dinajpur	Decreasing	1958-59	0.0167	0.0164
Malda	Increasing	1958-59	0.0113	0.0165
Jalpaiguri	Decreasing	—	0.0087	0.0051
Darjeeling	Increasing	1964-65,	0.0066	0.0147
Coochbehar	Decreasing	1958-59	0.0184	0.0138
Purulia	Decreasing	1980-81, 1983-84	0.0604	0.0569
West Bengal	Decreasing +	1983-84	0.0081	0.0094

Notes: a-reference to Table 5.2, Others as in Table 5.8.

Table 5.10 : Decomposition of Instability in Annual Crop Output Growth Rates, West Bengal by Crops, 1957-58—1976-77 and 1977-78—1994-95

Crop/Crop-group	Whole period (1957-58—1994-95)				First sub-period (1957-58—1976-77)				Second sub-period (1977-78—1994-95)			
	Var.in output growth	Var.in area growth	Var.in productivity growth	Residual effect	Var.in output growth	Var.in area growth	Var. in productivity growth	Residual effect	Var.in output growth	Var.in area growth	Var. in productivity growth	Residual effect
Aus-rice	0.0321	0.0128	0.0134	0.0059	0.0407	0.0095	0.0206	0.0106	0.0226	0.0164	0.0054	0.0008
		(39.9)	(41.7)	(18.4)		(23.3)	(50.6)	(26.1)		(72.6)	(23.9)	(3.5)
Aman-rice	0.0186	0.0017	0.0150	0.0019	0.0131	0.0006	0.0102	0.0023	0.0252	0.0029	0.0205	0.0018
		(9.1)	(80.6)	(10.3)		(4.6)	(77.9)	(17.5)		(11.5)	(81.3)	(7.2)
Boro-rice	0.1210	0.0960	0.0084	0.0166	0.1762	0.1395	0.0123	0.0244	0.0601	0.0567	0.0042	-0.0008
		(79..3)	(6.9)	(13.8)		(79.2)	(7.0)	(13.8)		(94.3)	(7.0)	(-1.3)
Total rice	0.0150	0.0013	0.0109	0.0028	0.0132	0.0007	0.0094	0.0031	0.0173	0.0020	0.0127	0.0026
		(8.7)	(72.7)	(18.6)		(5.3)	(71.2)	(23.5)		(11.6)	(73.4)	(15.0)
Wheat	0.0794	0.0636	0.0153	0.0005	0.0981	0.0781	0.0111	0.0089	0.0426	0.0365	0.0194	-0.0133
		(80.1)	(19.3)	(0.6)		(79.6)	(11.3)	(9.1)		(85.7)	(45.50)	(-31.20)
Barley	0.1409	0.0514	0.0572	0.0323	0.2015	0.0789	0.0660	0.0566	0.0630	0.0144	0.0479	0.0007
		(36.5)	(40.6)	(22.9)		(39.2)	(32.8)	(28.0)		(22.9)	(76.0)	(1.1)

(Contd.)

Table 5.10 : (*Contd.*)

Crop/Crop-group	Whole period (1957-58—1994-95)				First sub-period (1957-58—1976-77)				Second sub -period (1977-78—1994-95)			
	Var.in output growth	Var.in area growth	Var.in product-ivity growth	Residual effect	Var.in output growth	Var.in area growth	Var. in product-ivity growth	Residual effect	Var.in output growth	Var.in area growth	Var. in product-ivity growth	Residual effect
Other cereals	0.1279	0.0126	0.1272	-0.0119	0.1369	0.0086	0.1383	-0.0100	0.1182	0.0169	0.1152	-0.0139
		(9.9)	(99.5)	(-9.4)		(6.3)	(101.0)	(-7.3)		(14.3)	(97.5)	(-11.8)
Total cereals	0.0132	0.0011	0.0101	0.002	0.0136	0.0007	0.0095	0.0034	0.0132	0.0015	0.0110	0.0007
		(8.3)	(76.5)	(15.2)		(5.1)	(69.9)	(25.0)		(11.4)	(83.3)	(5.3)
Gram	0.0500	0.0330	0.0454	-0.0284	0.0482	0,0280	0.0324	-0.0122	0.0514	0.0381	0.0592	-0.0459
		(66.0)	(90.8)	(-56.8)		(58.1)	(67.2)	(-25.3)		(74.1)	(115.2)	(-89.3)
Other pulses	0.0280	0.0107	0.0119	0.0054	0.0283	0.0130	0.0076	0.0077	0.0277	0.0076	0.0172	0.0029
		(38.2)	(42.5)	(19.3)		(45.9)	(26.9)	(27.2)		(27.4)	(62.1)	(10.5)
Total pulses	0.0270	0.0111	0.0110	0.0049	0.0319	0.0137	0.0081	0.0101	0.0203	0.0073	0.0139	-0.0009
		(41.1)	(40.7)	(18.2)		(42.9)	(25.4)	(31.7)		(36.0)	(68.5)	(-4.5)
Total	0.0120	0.0009	0.0093	0.0018	0.0123	0.0004	0.0086	0.0033	0.0120	0.0013	0.0101	0.0006
foodgrains		(7.5)	(77.5)	(15.0)		(3.3)	(69.9)	(26.8)		(10.8)	(84.2)	(5.0)
Mustard	0.0766	0.0231	0.0519	0.0016	0.0665	0.0289	0.0547	-0.0171	0.0867	0.0223	0.0435	0.0209
		(30.2)	(67.8)	(2.0)		(43.5)	(82.3)	(-25.8)		(25.7)	(50.2)	(24.1)

(*Contd.*)

Table 5.10 : *(Contd.)*

Crop/Crop-group	Whole period (1957-58—1994-95)				First sub-period (1957-58—1976-77)				Second sub -period (1977-78—1994-95)			
	Var.in output growth	Var.in area growth	Var.in product-ivity growth	Residual effect	Var.in output growth	Var.in area growth	Var. in product-ivity growth	Residual effect	Var.in output growth	Var.in area growth	Var. in product-ivity growth	Residual effect
Linseed	0.1043	0.0692	0.0645	-0.0294	0.0807	0.0482	0.0993	-0.0668	0.1384	0.1000	0.0296	0.0088
		(66.3)	(61.8)	(-28.1)		(59.7)	(123.0)	(-82.7)		(72.3)	(21.4)	(6.3)
Sesamum	0.0810	0.0637	0.0261	-0.0088	0.0740	0.0473	0.0060	0.0207	0.0874	0.0802	0.0476	-0.0404
		(78.6)	(32.2)	(-10.8)		(63.9)	(8.1)	(28.0)		(91.8)	(54.5)	(-46.3)
Total oil	0.0642	0.0357	0.0341	-0.0056	0.0505	0.0251	0.0546	-0.0292	0.0743	0.0449	0.0125	0.0169
seeds		(55.6)	(53.1)	(-8.7)		(49.7)	(108.1)	(-57.8)		(60.4)	(16.8)	(22.8)
Jute	0.0776	0.0367	0.0258	0.0151	0.1058	0.0474	0.0278	0.0306	0.0494	0.0261	0.0238	-0.0005
		(47.3)	(33.2)	(19.5)		(44.8)	(26.3)	(28.9)		(52.8)	(48.2)	(-1.0)
Sugarcane	0.0739	0.0550	0.0142	0.0047	0.0582	0.0215	0.0154	0.0213	0.0877	0.0904	0.0129	-0.0156
		(74.4)	(19.2)	(6.4)		(36.9)	(26.5)	(36.6)		(103.1)	(14.7)	(-17.8)
Potatoes	0.0418	0.0189	0.0286	-0.0057	0.0647	0.0186	0.0496	-0.0035	0.0176	0.1092	0.0063	-0.0079
		(45.2)	(68.4)	(-13.6)		(28.7)	(76.7)	(-5.4)		(109.1)	(35.8)	(-44.9)

(Contd.)

Table 5.10 : *(Contd.)*

Crop/Crop-group	Whole period (1957-58—1994-95)				First sub-period (1957-58—1976-77)				Second sub -period (1977-78—1994-95)			
	Var.in output growth	Var.in area growth	Var.in productivity growth	Residual effect	Var.in output growth	Var.in area growth	Var. in productivity growth	Residual effect	Var.in output growth	Var.in area growth	Var. in productivity growth	Residual effect
Tobacco	0.4560	0.0277	0.0074	0.0105	0.0420	0.0197	0.0096	0.0127	0.0490	0.0377	0.0040	0.0073
		(60.7)	(16.2)	(23.1)		(46.9)	(22.9)	(30.2)		(76.9)	(8.2)	(14.9)
Total non-	0.0099	0.0133	0.0030	-0.0064	0.0143	0.0207	0.0036	-0.0100	0.0055	0.0054	0.0024	-0.0023
foodgrains		(134.3)	(30.3)	(-64.6)		(144.8)	(25.2)	(-70.0)		(98.2)	(43.6)	(-41.8)
All crops	0.0086	0.0007	0.0056	0.0023	0.0081	0.0006	0.0050	0.0025	0.0094	0.0008	0.0064	0.0022
		(8.1)	(65.1)	(26.8)		(7.4)	(61.7)	(30.9)		(8.5)	(68.1)	(23.4)

Key: Var = Variance.

Notes : Percentage shares of the components in variability in annual output growth rates are within the parentheses. Annual growth rate is $[(y_t - Y_{t-1}) / y_{t-1}]$ (calculated from original series). For calculation of variances, see text [section 3.5, Chapter-3].

Table 5.11 : Decomposition of Instability in Annual Output Growth Rates of Agriculture (All crops), West Bengal by Districts, 1957-58—1994-95, 1957-58—1976 and 1977-78—1994-95

District	Whole period (1957-58—1994-95)				First sub-period (1957-58—1976-77)				Second sub -period (1977-78—1994-95)			
	Var.in output growth	Var.in area growth	Var.in productivity growth	Residual effect	Var.in output growth	Var.in area growth	Var. in productivity growth	Residual effect	Var.in output growth	Var.in area growth	Var. in productivity growth	Residual effect
Burdwan	0.0125	0.0045	0.0083	-0.0003	0.0093	0.0021	0.0060	0.0012	0.0158	0.0071	0.0105	-0.0018
		(36.0)	(66.4)	(-2.4)		(22.6)	(64.5)	(12.9)		(44.9)	(66.4)	(-11.3)
Birbhum	0.0252	0.0049	0.0139	0.0064	0.0198	0.0044	0.0128	0.0026	0.0309	0.0054	0.0149	0.0106
		(19.4)	(55.2)	(25.4)		(22.2)	(64.6)	(13.2)		(17.5)	(48.2)	(34.3)
Bankura	0.0381	0.0056	0.0190	0.0135	0.0377	0.0051	0.0196	0.0130	0.0387	0.0063	0.0183	0.0141
		(14.7)	(49.9)	(35.4)		(13.5)	(52.0)	(34.5)		(16.3)	(47.3)	(36.4)
Midnapore	0.0249	0.0013	0.0183	0.0053	0.0214	0.0015	0.0145	0.0054	0.0292	0.0012	0.0229	0.0051
		(5.2)	(73.5)	(21.3)		(7.0)	(67.8)	(25.2)		(4.1)	(78.4)	(17.5)
Howrah	0.0385	0.0095	0.0341	-0.0051	0.0461	0.0075	0.0389	-0.0003	0.0311	0.0115	0.0294	-0.0098
		(24.7)	(88.6)	(-13.3)		(16.3)	(84.4)	(-0.7)		(37.0)	(94.5)	(-31.5)
Hooghly	0.0274	0.0056	0.0165	0.0053	0.0407	0.0034	0.0216	0.0157	0.0134	0.0080	0.0111	-0.0057
		(20.4)	(60.2)	(19.4)		(8.4)	(53.1)	(38.5)		(59.7)	(82.8)	(-42.5)

(Contd.)

Table 5.11 : (Contd.)

District	Whole period (1957-58—1994-95)				First sub-period (1957-58—1976-77)				Second sub -period (1977-78—1994-95)			
	Var.in output growth	Var.in area growth	Var.in productivity growth	Residual effect	Var.in output growth	Var.in area growth	Var. in productivity growth	Residual effect	Var.in output growth	Var.in area growth	Var. in productivity growth	Residual effect
24-Parganas	0.0185	0.0009	0.0169	0.0007	0.0143	0.0007	0.0128	0.0008	0.0225	0.0011	0.0208	0.0006
		(4.9)	(91.4)	(3.7)		(4.9)	(89.5)	(5.6)		(4.9)	(92.4)	(2.7)
Nadia	0.0233	0.0035	0.0198	0.000	0.0293	0.0049	0.0219	0.0025	0.0170	0.0021	0.0174	-0.0025
		(15.0)	(85.0)	(0.0)		(16.7)	(74.7)	(8.6)		(12.4)	(102.4)	(-14.6)
Murshidabad	0.0267	0.0037	0.0175	0.0055	0.0305	0.0032	0.0201	0.0072	0.0228	0.0040	0.0147	0.0041
		(13.9)	(65.5)	(20.6)		(10.5)	(65.9)	(23.6)		(17.5)	(64.5)	(18.0)
West Dinajpur	0.0166	0.0013	0.0163	-0.0010	0.0167	0.0008	0.0168	-0.0009	0.0164	0.0017	0.0159	-0.0012
		(7.8)	(98.2)	(-6.0)		(4.8)	(100.6)	(-5.4)		(10.4)	(96.9)	(-7.3)
Malda	0.0139	0.0046	0.0121	-0.0028	0.0113	0.0023	0.0089	0.0001	0.0165	0.0069	0.0153	-0.0057
		(33.1)	(87.1)	(-20.2)		(20.4)	(78.8)	(0.8)		(41.8)	(92.7)	(-34.5)
Jalpaiguri	0.0070	0.0016	0.0057	-0.0003	0.0087	0.0011	0.0052	0.0024	0.0051	0.0018	0.0061	-0.0028
		(22.9)	(81.4)	(-4.3)		(12.6)	(59.8)	(27.6)		(35.3)	(119.6)	(-54.9)

(Contd.)

Table 5.11 : *(Contd.)*

District	Whole period (1957-58—1994-95)				First sub-period (1957-58—1976-77)				Second sub-period (1977-78—1994-95)			
	Var.in output growth	Var.in area growth	Var.in productivity growth	Residual effect	Var.in output growth	Var.in area growth	Var. in productivity growth	Residual effect	Var.in output growth	Var.in area growth	Var. in productivity growth	Residual effect
Darjeeling	0.0107	0.0063	0.0092	-0.0048	0.0066	0.0030	0.0064	-0.0028	0.0147	0.0096	0.0119	-0.0068
		(58.9)	(86.0)	(-44.9)		(45.5)	(97.0)	(-42.5)		(65.3)	(80.9)	(-46.2)
Coochbehar	0.0161	0.0038	0.0126	-0.0003	0.0184	0.0023	0.0139	0.0022	0.0138	0.0051	0.0111	-0.0024
		(23.6)	(78.3)	(-1.9)		(12.5)	(75.5)	(12.0)		(37.0)	(80.4)	(-17.4)
Purulia	0.0580	0.0085	0.0345	0.0150	0.0604	0.0084	0.0306	0.0214	0.0569	0.0086	0..0406	0.0077
		(14.7)	(59.5)	(25.8)		(13.9)	(50.7)	(35.4)		(51.1)	(71.4)	(13.5)
West Bengal	0.0086	0.0007	0.0056	0.0023	0.0081	0.0006	0.0050	0.0025	0.0094	0.0008	0.0064	0.0022
		(8.1)	(65.1)	(26.8)		(7.4)	(61.7)	(30.9)		(8.5)	(68.1)	(23.4)

Key : Var = Variance.

Notes : Same as those in Table 5.10.

Table 5.12 : Classified Results on the Residual Effect of Instability Decomposition, All Crops, West Bengal by Districts, Whole Period and Two Sub-periods

District	Whole Period (1957-58—1994-95)			1st Sub-Period (1957-58—1976-77)			2nd Sub-period (1977-78—1994-95)		
	Residual effect[a]	2 Cov (Ga, Gp)	Error	Residual effect[a]	2 Cov (Ga, Gp)	Error	Residual effect[a]	2 Cov (Ga, Gp)	Error
Burdwan	-0.0003	-0.0008	0.0005	0.0012	0.0010	0.0002	-0.0018	-0.0028	0.0010
	(-2.4)	(-6.4)	(4.0)	(12.9)	(10.8)	(2.1)	(-11.3)	(-17.7)	(6.4)
Birbhum	0.0064	0.0056	0.0008	0.0026	0.0032	-0.0006	0.0106	0.0082	0.0024
	(25.4)	(22.2)	(3.2)	(13.2)	(16.2)	(-3.0)	(34.3)	(26.5)	(7.8)
Bankura	0.0135	0.0126	0.0009	0.0130	0.0104	0.0026	0.0141	0.0144	-0.0003
	(35.4)	(33.1)	(2.3)	(34.5)	(27.6)	(6.9)	(36.4)	(37.2)	(-0.8)
Midnapore	0.0053	0.0036	0.0017	0.0054	0.0036	0.0018	0.0051	0.0038	0.0013
	(21.3)	(14.5)	(6.8)	(25.2)	(16.8)	(8.4)	(17.5)	(13.0)	(4.5)
Howrah	-0.0051	-0.0050	-0.0001	-0.0003	0.0009	-0.0012	-0.0098	-0.0104	0.0006
	(-13.3)	(-13.0)	(-0.3)	(-0.7)	(1..9)	(-2.6)	(-31.5)	(-33.4)	(1.9)
Hooghly	0.0053	0.0032	0.0021	0.0157	0.0122	0.0035	-0.0057	-0.0062	0.0005
	(19.4)	(11.7)	(7.7)	(38.5)	(30.0)	(8.5)	(-42.5)	(-46.3)	(3.8)
24-Parganas	0.0007	0.0002	0.0005	0.0008	0.0006	0.0002	0.0006	-0.0002	0.0008
	(3.7)	(1.1)	(2.6)	(5.6)	(4.2)	(1.4)	(2.7)	(-0.9)	(3.6)
Nadia	0.0000	-0.0006	0.0006	0.0025	0.0014	0.0011	-0.0025	-0.0024	-0.0001
	(0.0)	(-2.6)	(2.6)	(3.6)	(4.8)	(3.8)	(-14.6)	(-14.1)	(-0.5)

(Contd.)

Table 5.12 : *(Contd.)*

District	Whole Period (1957-58—1994-95)			1st Sub-period (1957-58—1976-77)			2nd Sub-period (1977-78—1994-95)		
	Residual effect[a]	2 Cov (Ga, Gp)	Error	Residual effect[a]	2 Cov (Ga, Gp)	Error	Residual effect[a]	2 Cov (Ga, Gp)	Error
Murshidabad	0.0055	0.0036	0.0019	0.0072	0.0054	0.0018	0.0041	0.0022	0.0019
	(20.6)	(13.5)	(7.1)	(23.6)	(17.7)	(5.9)	(18.0)	(9.6)	(8.4)
West Dinajpur	-0.0010	-0.0007	-0.0003	-0.0009	-0.0008	-0.0001	-0.0012	-0.0006	-0.0006
	(-6.0)	(-4.2)	(-1.8)	(-5.4)	(-4.8)	(-0.6)	(-7.3)	(-3.7)	(-3.6)
Maida	-0.0028	-0.0026	-0.0002	0.0001	0.0001	0.0000	-0.0057	-0.0053	-0.0004
	(-20.2)	(-18.7)	(-1.5)	(0.8)	(0.8)	(0.0)	(-34.5)	(-32.1)	(-2.4)
Jalpaiguri	-0.0003	-0.0004	0.0001	0.0024	0.0022	0.0002	-0.0028	-0.0028	0.0000
	(-4.3)	(-5.7)	(1.4)	(27.6)	(25.3)	(2.3)	(-54.9)	(-54.9)	(0.0)
Darjeeling	-0.0048	-0.0040	-0.0008	-0.0028	-0.0022	-0.0006	-0.0068	-0.0056	-0.0012
	(-44.9)	(-37.4)	(-7.5)	(-42.5)	(-33.3)	(-9.2)	(-46.2)	(-38.1)	(-8.1)
Coochbehar	-0.0003	0.0006	-0.0009	0.0022	0.0038	-0.0016	-0.0024	-0.0024	0.0000
	(-1.9)	(3.7)	(-5.6)	(12.0)	(20.6)	(-8.6)	(-17.4)	(-17.4)	(0.0)
Purulia	0.0150	0.0164	-0.0014	0.0214	0.0196	0.0018	0.0077	0.0128	-0.0051
	(25.8)	(28.3)	(-2.5)	(35.4)	(32.5)	(2.9)	(13.5)	(22.5)	(-9.0)
West Bengal	0.0023	0.0016	0.0007	0.0025	0.0018	0.0007	0.0022	0.0016	0.0006
	(26.8)	(18.6)	(8.2)	(30.9)	(22.2)	(8.7)	(23.4)	(17.0)	(6.4)

Notes : Terms within the parentheses denote percentage shares in the variability of annual output growth rates. [a] for residual effect see Table 5.11.

Key : Cov = Covariance, Ga = annual growth rate in area, Gp = annual growth rate in productivity.

Table 5.13 : Cross Sectional Correlation Coefficients between Variability in Annual Output Growth Rates and its Components, Crops and Districts of West Bengal, 1957-58—1994-95 and 1957-58—1976-77 and 1977-78—1994-95

Cross sectional base	Correlation Coefficient								
	Whole Period (1957-58—1994-95)			Ist Sub-period (1957-58—1976-77)			2nd Sub-period (1977-78—1994-95)		
[Variability in annual output growth rates of]	Variability in area growth rates	Variability in productivity growth rates	Synchronous movement of area and productivity	Variability in area growth rates	Variability in productivity growth rates	Synchronous movement of area and productivity	Variability in area growth rates	Variability in productivity growth rates	Synchronous movement of area and productivity
15 individual crops of West Bengal+	0.6228 (.013)	0.5845 (.022)	0.0036 (.990)	0.7525 (.001)	0.4004 (.139)	0.1259 (.655)	0.6409 (.010)	0.5770 (.024)	0.1660 (.554)
All crops (combined) of the Districts	0.6324 (.011)	0.8985 (.000)	0.7361 (.002)	0.8432 (.000)	0.9050 (.000)	0.7997 .(000)	0.2995 (.278)	0.8771 (.000)	0.7255 (.002)

Notes : Probability values in parentheses.

+15 individual crops are aus-rice, aman-rice, boro-rice, wheat, barley other cereals, gram, other pulses, mustard, linseed, sesamum, jute, sugarcane, potatoes and tobacco.

Table 5.14 : Inter-period Correlation Coefficients of the components of Output Instability Cross-sectionally, Crops and Districts of West Bengal, Two Sub-periods: 1957-58—1976-77 and 1977-78—1994-95

Cross Section	Var(Go) t_1 and t_2	Var(Ga) t_1 and t_2	Var(Gp) t_1 and t_2	2 Cov(Ga,Gp) t_1 and t_2
15 individual crops of West Bengal +	0.3395 (.216)	0.3123 (0.257)	0.7210 (.002)	-0.3202 (.245)
All crops (combined) of the Districts	0.7430 (.002)	0.6528 (.008)	0.7424 (.002)	0.6084 (.016)

Notes : t_1 implies first sub-period (1957-58 to 1976-77) and t_2 implies second sub-period (1977-78 to 1994-95)

Var implies variability in annual growth rates and Cov implies synchronous movement (i.e., covariance).

Go = annual growth rate in output,

Ga = annual growth rate in area,

Gp = annual growth rate in productivity; probability levels in parentheses.

+ same as in Table 5.13.

Table 5.15 : Cross Sectionally Measured Correlation Coefficients between Growth and Instability in Agricultural Production, West Bengal by Crops and Districts by Crop-groups, 1957-58—1994-95

Cross sectional base	Correlation coefficient between growth rate[a] and level of instability[b]			Correlation coefficient between trend break[c] and nature of instability[d] (i.e, the value of rank correlation coefficient)
	Whole period (1957-8–1994-95)	Ist sub-period (1957-58–1976-77)	2nd sub-period (1977-78–1994-95)	
(1)	(2)	(3)	(4)	(5)
15 individual crops of West Bengal +	0.5869 (.021)	0.8825 (.000)	0.0858 (.761)	0.2614 (.347)
Total rice production of the Districts	-0.0696 (.805)	0.3223 (.241)	-0.0705 (.803)	-0.1946 (.488)
Total foodgrains production of the Districts	0.2077 (.458)	0.1227 (.663)	0.3804 (.162)	0.1587 (.572)
Total non-foodgrains production of the Districts	0.1862 (.506)	0.5441 (.036)	-0.0779 (.783)	-0.1384 (.623)
All crops production of the Districts	0.2985 (.280)	0.2377 (.394)	0.3176 (.249)	-0.0255 (.928)

Notes: Probability levels in the parentheses, [a] output growth rates from Tables 4.25 to 4.29, [b] levels of instability from Tables 5.3 to 5.7, [c] trend breaks (difference between sub-period growth rates) from Tables 4.25 to 4.29, [d] rank correlation between $|X_t|$ and t (see Tables 5. 1 and 5.2), [+] as in Table 5.13.

Table 5.16 : Classification of Crops /Crop-groups by Trend Break[a] and Nature of Instability[b] in Agricultural Production, West Bengal, 1957-58—1994-95

Trend Break → / Nature of Instability ↓	Positive and Significant (1)	Positive and Insignificant (2)	Negative and Insignificant (3)	Negative and Significant (4)
Positive and Significant (1)	—	—	—	Gram, Sesamum, Sugarcane
Positive and Insignificant (2)	Mustard, Total oil seeds	—	Tobacco	Linseed
Negative and Insignificant (3)	Aman-rice	Aus-rice, Other cereals	—	Barley, Other pulses, Total pulses
Negative and Significant (4)	Total rice, Total cereals, Total food-grains, Total non-foodgrains, All crops	Potatoes	Jute	Boro-rice, Wheat

Notes: [a] For trend break (between two sub-periods) see Table 4.25,
[b] for nature of instability ($\hat{b}$/rank correlation) see Table 5.1.

Table 5.17 : Classification of Crop-groups by Trend Break[a] and Nature of Instability[b] in Agricultural Production, West Bengal by Districts, 1957-58—1994-95

Trend Break → Nature of ↓ Instability	Positive and Significant (1)	Positive and Insignificant (2)	Negative and Insignificant (3)	Negative and Significant (4)
Positive and Significant (1)	Malda-TFG, Darjeeling-TFG	—	—	—
Positive and Insignificant (2)	Birbhum-TR, Birbhum -TNFG, Burdwan - AC, 24-Parganas-AC, Malda - AC.	Birbhum-AC, Darjeeling-AC.	Jalpaiguri-TR, Jalpaiguri-TFG.	Purulia - TNFG.
Negative and Insignificant (3)	Burdwan-TR, Bankura-TR, Midnapur-TR, 24-Parganas-TR, Murshidabad-TR-West Dinajpur-TR, Purulia-TR, Burdwan-TFG, Bankura-TFG-Midnapur-TFG, 24-Parganas-TFG- Purulia-TFG, Hooghly-TNFG, Nadia-TNFG, Murshidabad-TNFG,	Malda-TR, Darjeeling -TR, Coochbehar-TR, Birbhum-TFG, Nadia-TFG, West Dinajpur-TFG, Malda-TNFG, Murshidabad-AC, Coochbehar-AC.	Jalpaiguri-AC.	Jalpaiguri-TNFG, Darjeeling-TNFG

(Contd.)

Table 5.17 : *(Contd.)*

Trend Break → Nature of ↓ Instability	Positive and Significant (1)	Positive and Insignificant (2)	Negative and Insignificant (3)	Negative and Significant (4)
(3)	Coochbehar-TNFG, Bankura-AC, Midnapur-AC, West Dinajpur-AC, Purulia-AC			
Negative and Significant (4)	Howrah-TR, Nadia-TR, Bankura-TNFG Midnapur-TNFG, West Dinajpur-TNFG Nadia-AC.	Howrah-TFG, Murshidabad -TFG, 24-Parganas-TNFG, Howrah-AC, Hooghly-AC.	Hooghly-TFG, Coochbehar-TFG, Burdwan-TNFG, Howrah - TNFG.	Hooghly - TR

Notes: [a] for trend break (between two sub-periods) see Tables 4.26 to 4.29, [b] for nature of instability ($\hat{b}$ / rank correlation) see Table 5.2.

Key : TR = total rice, TFG = total foodgrains, TNFG = total non-foodgrains and AC = all crops.

6

Growth and Stability Contributing Factors in Agriculture

For sustained growth in agricultural production, the growth in productivity is a necessary prerequisite as the scope of area expansion is dwindling day by day. Also from the decomposition of output growth in West Bengal agriculture for the whole period (i.e., 1957-58—1994-95) it is observed that productivity plays a significant role (section 4.5, Chapter 4). In consideration of that, in the first section of this chapter an analysis is made, both from time series and cross section studies, of different factors that determine productivity (more specifically, land productivity, i.e., total agricultural production divided by total cropped area) in West Bengal agriculture. Further, from the decomposition of the variability in annual output growth rates (section 5.3, Chapter 5) it has been observed that all the constituent components (i.e., area and productivity variabilities and their synchronous movements) are equally powerful (measured by correlation coefficients) in explaining output instability in West Bengal agriculture during the period 1957-58—1994-95. As a result, instead of identifying the factors that explain the variabilities in the components, an attempt is made here to identify the factors that explain the variabilities in the output as a whole in West Bengal agriculture. In section 3, the effects of rainfall on the growth and stability in agricultural production of West Bengal is measured. Lastly, an attempt is made to enquire into the dependence of agricultural output growth on the growth of rural population in West Bengal, cross sectionally over the districts.

6.1 Factors Affecting Agricultural Productivity

6.1.1 *Time Series Analysis*

To measure the influence of technological factors on agricultural productivity in West Bengal during LFG regime [i.e., from 1977-78 to 1994-95] the productivity index (PI) of all crops in West Bengal is regressed on three technological factors, namely, fertilizer intensity (FI), HYV intensity (HYVI) and irrigation intensity from government canals (II) for the period 1977-78 to 1994-95. The regression results are as follows :

$$PI_t = \underset{(0.420)}{7.099} - \underset{(-0.633)}{0.053FI_t} + \underset{(8.300)}{196.254^{*} HYVI_t} + \underset{(3.110)}{395.714^{*} II_t} \quad \text{..........(6.1)}$$

$DW = 2.240^{+}$, $t = 1997/8........1994/5$, $\bar{R}^2 = 0.8970$ $[F = 50.337^{*}]$

[Notes : t values in the parentheses; * implies significant at 1% level, $^{+}$ presence of non-autocorrelated disturbance term. Along with these symbols, the other symbols which has been used through out this chapter for the equations are :
**implies significant at 5% level and *** implies significant at 10% level].

From the above regression results of equation (6.1) we see that though $\bar{R}^2$ is significantly high (i.e., major portion of sampling variance in PI is explained by the explanatory variables) with no autocorrelation problem in the disturbance term and though the signs of the estimated coefficients of HYVI and II are in conformity with our expectations, the estimated coefficient of FI (along with the magnitude of the intercept term) is found to be insignificant (in statistical sense), implying thereby that as the intensity of fertilizer application increases, the productivity index of all crops decline. This is contrary to our general expectation about the relationship between these two variables. The cause of this may be traced in the multicollinearity between FI and HYVI, as is evident from the following simple correlation matrix.

Correlation Matrix

	FI	HYVI	II
FI	1	0.7042 (.000)	0.2296 (.359)
HYVI	0.7042 (.000)	1	0.2354 (.347)
II	0.2296 (.359)	0.2354 (.347)	1

[Note : Probability values in the parentheses].

To avoid the problem of multicollinearity principal component analysis is made for FI and HYVI and the results of this principal component analysis are reported below:

Factor	Eigen value	Percent of Variance
1	1.70423	85.2
2	0.29577	14.8

Bartlett test of sphericity = 10.619^{*}

From this data reduction we have chosen (following Kaiser's criterion, i.e., for the chosen principal component, Eigen value ≥ 1) the first principal component (P_1) which accounts for 85.2% of the total variance in sample sets of FI and HYVI. Actually, this P_1 will give the joint effect of fertilizer and HYV seeds on the productivity index. Now, regressing PI on P_1 and II we get the following results:

$$PI_t = 75.844^{**} + 17.024^{*} P_{1t} + 400.575^{***} II_t \ldots\ldots (6.2)$$
$$(2.781) \quad (6.116) \quad (1.986)$$

$DW = 2.022^{+}, \bar{R}^2 = 0.7412$ [$F = 25.346^{*}$]

From these regression results it is seen that the equation (6.2) is an improvement over the equation (6.1). Here the estimated parameters satisfy the theoretical expectations that the technological factors in combination or individually usually exert significantly positive influence on agricultural productivity. But the strength of the dependence of productivity on irrigation here appears significant (with positive value) only at 10% level. Though the magnitude of the estimated parameter of II is abnormally high [like equation (6.1)], it also contains high value of standard error. Moreover, contrary to the earlier findings on West Bengal agriculture [from cross section studies of Desai and Singh (1973), Boyce (1987), etc.] it is seen here that irrigation (II) is uncorrelated with other technological factors, which is again contrary to our general perception about the relationship among irrigation, use of fertilizer, application of HYV, etc. This unseemly behaviour may be attributed to the inadequacy of irrigation data used in the above analysis. Actually, irrigation from government canals covers only a part of total irrigation in West Bengal. To cover the whole part of the irrigation a separate estimation of the irrigation variable has been made (see section 3.2, Table 3.1, Chapter 3).

Replacing II by the estimated irrigation intensity (EII) strong correlations are obtained among the technological factors. The correlation results are presented in the following correlation matrix:

Correlation Matrix

	FI	**HYVI**	**EII**
FI	1	0.7042 (.001)	0.6126 (.007)
HYVI	0.7042 (.001)	1	0.8401 (.000)
EII	0.6126 (.007)	0.8401 (.000)	1

[Probability levels in the parentheses].

Under the circumstances, for getting reliable estimate of the regression parameters the principal component analysis has been carried out on all these three explanatory variables, the results of which are given below :

Factor	**Eigen Value**	**Percent of Variance**
1	2.44207	81.4
2	0.41033	13.7
3	0.14760	4.9

Bertlett test of sphericity = 28.987*

From the above mentioned results it is observed that here factor analysis is not an inappropriate exercise (reference to Bertlett test of sphericity, that tests whether the correlation matrix is identity matrix, which would indicate that the factor model is inappropriate). Applying Kaiser's criterion we have chosen the first principal component (P^m_1) which accounts for 81.4% of the sampling variances of the explanatory variables and this P^m_1 may be considered as the joint effect of the technological factors.

Using P^m_1 as the regressor in the regression equation of productivity index we get the following final results:

$$PI_t = \underset{(60.147)}{129.768^*} + \underset{(8.984)}{19.944^*}\, P^m_{1t}\text{..........} \quad (6.3)$$

$$DW = 2.006^+,\ \bar{R}^2 = 0.8242\ [F = 80.706^*]$$

In comparison to earlier results [of regression equations (6.1) and (6.2)] here the value of F statistic for $\bar{R}^2$ and the values of t-statistic of the estimated parameters have improved very significantly. It is to be noted that contrary to the conventional measurement of factor elasticities through the estimation of regression equation in log-log configuration, regression equation (6.3) is fitted on the original scale because some of the observations of P^m_1 are negative whose $\log_e$ values are undefined. However, in the construction of principal components, as the explanatory variables have been standardized [i.e., $\{y_t - E(Y)\} \div \sigma_y$, for any Y] so that the inference can be drawn from equation (6.3), that one standard unit change in the technological factors in composite form leads to change in productivity index in West Bengal agriculture by 19.94. This change is positive with high statistical significance. So technological factors jointly have played a very significant role in bringing about the growth in productivity which, in turn, has led to the growth in the agricultural production in West Bengal during LFG regime.

6.1.2 *Cross Section Study*

In time series analysis the influence on agricultural productivity of technological factors only has been considered and estimated, omitting other important set of explanatory variables, namely, the institutional factors. This omission of relevant variables leads to mis-specification of the regression model which, in turn, leads to inefficient (also very likely biased and inconsistent) estimates of the parameters [Kmenta (1971), Pp. 442-446]. To take care of other relevant explanatory variables as determinants of agricultural productivity, a cross section study is made here. In the cross section study total eight explanatory variables are considered. These are : (i) % of GCA irrigated (T_1), (ii) fertilizer intensity (T_2), (iii) % of GCA under HYV (T_3), (iv) agricultural machinery intensity (T_4), (v) average size of operational holding (I_1), (vi) % of agricultural labourers in total agricultural work force (I_2), (vii) % of net cropped area (NCA) under tenancy (I_3) and (viii) % of NCA under surplus land distributed (I_4) [for detailed discussion please refer to section 3.2, Chapter 3, Tables 3.2-3.4]. The productivity indices of the districts [reported in the same Tables 3.2-3.4] have been regressed on these eight variables at three different time points — 1970-71(i.e., a year representing the levels of the above variables in the pre-LFG regime), 1980-81 (i.e., a year representing the above in the first few years of LFG regime) and 1990-91 (i.e., a year representing the above nearly after a decade and a half of the LFG rule in West Bengal). These eight explanatory variables may be classified broadly into two categories: first four (T_1 to T_4) as technological factors and last four (I_1 to I_4) as institutional factors. On the basis of these two classifications the explanatory variables are clubbed into two groups by applying principal component analysis in order to avoid the problem of multicollinearity.

From the measured correlation coefficients among four technological factors (reported in Table 6.1) it is observed that the technological factors are in general highly correlated with each other at least for most of the time points. But from Table 6.2 it is seen that, in general, the institutional factors are not significantly correlated with each other, except I_2 and I_4 for all the three time points and I_2 and I_3 for 1990-91. Actually, the correlation coefficient between I_2 and I_4 is significantly negative for all three time points, which implies that the surplus land has been distributed less in those districts where the number of agricultural labourers is high and *vice versa*. The average size of operational holdings, I_1, is found to have no significant correlation (also in respect of multiple correlation) with other institutional factors in any of the years under study. Therefore, to avoid the problem of multicollinearity among some of the explanatory variables and to facilitate comparison of the regression results at the three different time points, the principal component analysis has been used to club these technological factors and institutional factors, except I_1. The exclusion of I_1 from clubbing has been done mainly for the following reason.

If I_1 is clubbed with the other three institutional factors, and Kaiser's criterion of selection of principal components is followed, then the first two principal components are to be selected for the institutional factors. This will pose serious problem in the interpretation of the components. By excluding I_1 from the clubbing of the set of institutional factors, the only first principal component can be selected, thus avoiding the problem of interpretation of the component.

In Table 6.3 the important results on the principal component analysis for the technological factors are presented. From Bartlett test of sphericity it is seen that principal component analyses on technological factors for all three time points give meaningful results (as this statistic is significant in all cases). For each of the three time points we have chosen (applying Kaiser's criterion) the first principal component (denoted by P_1) to get the joint effect of technological factors, which accounts maximum part of the sampling variances of the data. Similar conclusions can also be drawn (from the reported results of Table 6.4) by applying factors analysis on the 'other institutional factors' (i.e., I_2, I_3 and I_4). However, the percentage of sampling variance explained by the first principal component for these 'other institutional factors' is lower than that for the technological factors.

By applying principal component analysis, therefore, the original eight explanatory variables are represented by only three variables, namely, i) composite form of the technological factors (P_1), ii) average farm size (I_1) and iii) composite form of 'other institutional factors' (F_1). It has been observed that these three variables are uncorrelated with one another for all the three time points. The estimated results of regression fitted to the cross-sectional data on productivity indices and the above three explanatory variables are presented in Table 6.5, for all three time points. It is observed that for all the years, 1970-71, 1980-81 and 1990-91, the regression equations give good fit (evident from high value of $\bar{R}^2$ along with the significant value of F statistic) in the sense that these three variables explain a significant part of the productivity variation over the districts. From the regression results it is further observed that technological factors significantly and positively affect the agricultural productivity in West Bengal for all the years. But the effect of 'other institutional factors' is mixed. For the years 1970-71 and 1980-81, 'other institutional factors' jointly affect the agricultural land productivity in West Bengal positively with statistical significance [but less strong (in statistical sense) than the joint effect of technological factors]; but for the year 1990-91 their joint effect on productivity is not only insignificant but also negative. Regarding the effect of farm size on land productivity in West Bengal no statistically significant relation is observed.

From these observations one can infer that there was no change in the pattern of causes behind productivity growth between the period before LFG rule and the initial period of the LFG rule. But the recent growth in productivity of West Bengal agriculture has been caused, as this study shows, by technological improvements (similar observation was derived from time series analysis). Moreover, contrary to the popular belief, 'other institutional factors' (i.e., the growth in labour force and land reforms) play very insignificant (rather negative) role in bringing about rapid growth in productivity in West Bengal agriculture in the recent past, as represented by the year, 1990-91.

6.2 Causes of Output Instability in Agriculture

Unlike the determination of growth contributing factors, it is difficult to form an appropriate econometric model for testing the causes of output instability. The prime factor which affects instability in agricultural production is weather determined random factor for which a random regressor model is to be formed (this has been discussed in the following section). Apart from this, there are some other factors (say, pest attack, crop disease etc.) which are difficult to measure. Moreover, in the cross section study some district-specific geo-physical characteristics might be contributing to the output instability in the districts. However, some other systematic factors (explained in section 3.2, Chapter 3) have been considered here to examine their possible effects, *ceteris paribus*, on output instability.

6.2.1 *Time Series Analysis*

Due to non-availability of data on area under current fallow for last few years, the time series has been initially confined to the period 1957-58 to 1988-89. The estimated coefficients obtained by regressing $|Z_t|$ (defined in section 3.5, Chapter 3) against percentage of NCA under current fallow (henceforth termed as only fallow), crop base, locational base [presented in Table 3.6] and proportion of GCA irrigated from govt. canals (hence forth termed as only irrigation) are reported in Table 6.6 for the whole period [where $|Z_t|$ is calculated from the whole period's trend equation] and also for two sub-periods [where $|Z_t|$ is calculated by using kinked trend equation]. From Table 6.6 it is seen that these four independent variables could not explain the output instability in West Bengal agriculture during first sub-period; during this period a major part of instability in agricultural output in West Bengal may be attributed to variations in rainfall, since in this period the proportion of area irrigated in the total cultivated area was neither significant, nor significantly varying. Some other random variable might have also contributed to output instability in this period. On the other hand, during the second sub-period some of these factors significantly explain the production instability in West Bengal agriculture. Perhaps the dominating influence of one of these factors, namely, irrigation, in the second sub-period has outweighed its insignificant influence during the first sub-period and resulted in overall significant, though at higher levels of significance, effect of this variable during the whole period. The other factors which were significant during the second sub-period, though at a very high level of significance, were not strong enough to have their influences felt on the whole period.

But, in general, whole period's regression equation (which also contains first sub-period's influences) gives poorer fit than the regression equation for the second sub-period. So these independent variables are able to explain instability in that situation when agriculture is found to have been relatively developed. However, before accepting the regression results the possibility of the presence of multicollinearity among the explanatory variables has been checked and tested.

From product moment correlation matrix among independent variables (Table 6.7) it is observed that most of the variables are significantly correlated for the whole period (more specifically, irrigation and locational base are correlated with other independent variables). And during the second sub-period the correlations between locational base & fallow and between locational base and crop base are found to be statistically significant. But the simple correlation only signifies the symptom of multicollinearity. To judge whether multicollinearity is harmful or not one should consider the multiple correlation coefficients of each of the independent variables with others. In this regard following Klein's rule that "intercorrelation of variables is not necessarily a problem unless it is high relative to the overall degree of multiple correlation" [Klein (1962), P.101], it is seen from Table 6.8 that the multiple correlations of each independent variable (with other independent variables) are higher than overall multiple correlation (of dependent variable with all independent variables) for the whole period and lower for second sub-period. So multicollinearity problem appears severe for the whole period, not for the second sub-period.To avoid multicollinearity problem here the variable which is the source of multicollinearity and also which plays very insignificant role in explaining output instability has been omitted. Following these rules the fallow and locational base variables have been omitted from the whole periods' regression equation. Similarly, fallow may be excluded form second sub-period's regression equation though multicollinearity problem is not so severe here. It should be mentioned in this connection that it is the presence of its strong multicollinearity (specially with irrigation) that might have contributed to the negative and statistically significant coefficient of fallow in the preliminary regression analysis (Table 6.6), contrary to our expectation. The re-estimated regression equations fitted to the whole period (1957-58 — 1994-95) and second sub-period (1977-78—1994-95) are presented in Table 6.6. (Because of the omission of the variable-fallow, the period is now 1957-58—1994-95, instead of 1957-58 — 1988-89). From the final estimated results, given in Table 6.9, it is seen that a negative relationship exists between output instability and irrigation for the whole period as well as for the second sub-period and , as is expected, this negative relationship is stronger in the later period (when irrigation facilities were developed over large areas) than that in the whole period. Replacing Govt. canal irrigation by the estimated total irrigation (defined in section 3.2, Chapter 3) similar result (not reported here) are obtained but with higher levels of significance. From this result it can be said that the assured irrigation (from Govt. canals) has more stabilizing effect on agricultural production of West Bengal than total irrigation, which comprise irrigation from those sources that get dried up or fail during some part of a crop year. But contrary to the popular belief that squeezing agricultural (crop, locational, etc.) base leads to greater instability, a statistically significantly negative relationship between agricultural base and output instability is obtained here during the second sub-period. This result may arise because of the cultivation of only the principal crops under favourable geo-climatic conditions.

6.2.2 *Cross Section Analysis*

In explaining output instability from cross section study total five independent variables have been identified. These are, namely, fallow, crop base and locational base (defined in Tables 3.7 to 3.9) and irrigation and holding size (defined in Tables 3.2 to 3.4) on which instability statistic $|Z_t|$ (defined earlier in sections 3.5 and 3.6, Chapter 3) has been regressed over 15 districts at three time points, 1970-71, 1980-81 and 1990-91. Before fitting regression equations, the presence of multicollinearity has been examined. In addition, considering relatively small number of observations (n=15) it is desirable that the variables that have negligible effect on the dependent variable are omitted.

From simple product moment correlation coefficients among independent variables (Table 6.10) it is observed that most of the independent variables are not intercorrelated. The significant correlations are obtained only (i) between fallow and crop base in 1970-71, (ii) between crop base and locational base in both 1980-81 and 1990-91 and (iii) between crop base and holding size in 1990-91. So, in general, multicollinearity problem is not a serious one in this cross sectional study. However, it is observed from these correlations that the nature of association between different pairs of variables remains the same over time except for irrigation and fallow, and locational base and holding size. In these two cases positive correlations in 1970-71 turn to negative correlations in 1980-81 and 1990-91. This may imply that initially the districts which had high current fallow (holding size) undertook more irrigation (contributed larger shares in the State's agricultural production); but overtime due to expansion of irrigation current fallow has come down and perhaps due to incoming of more cultivators the contribution of larger shares by the districts leads to reduction in their holding size. Regarding other correlations the following observations can be made: (i) more irrigation leads to the squeezing of crop base with higher share in the State's agricultural production (i.e., locational base), (ii) increase in the area of larger holdings under irrigation has led to fall in current fallow and (iii) larger share in State's production is associated with increasing production of principal crops (i.e., crop base) and lower level of current fallow. But all these observations are statistically not maintainable.

To detect the severity of multicollinearity problem Klien's rule has been applied as before. From Table 6.11 it is seen that multiple correlations among independent variables are lower than overall degree of multiple correlation (i.e., of dependent variable) in the years 1970-71 and 1980-81. But this is not so for the year 1990-91, when locational base is detected as the source of multicollinearity. So for the year 1990-91 locational base has been omitted from the regression analysis and to increase the degree of freedom for other years where multicollinearity is not serious, the locational base has also been omitted from the regression analysis of years 1970-71 and 1980-81 as in these two years locational base plays a very insignificant role (considering both simple and partial correlations) in explaining output instability.

In Table 6.12 the cross sectional regression results on the causes of output instability are presented. The regression equation gives very poorer fit to the data sets for the year 1990-91, compared with those for the earlier two years, to explain output instability, contrary to the findings of time series analysis. From the estimated coefficients of the independent variables it is seen that larger current fallow, squeezing crop base and larger holding size significantly contribute to greater output instability in 1970-71. But only current fallow is positively and significantly related to output instability in the year 1980-81. In the year 1990-91 large holding size significantly contributes to the stability in agricultural production. All other estimated coefficients are not statistically significant.

From the cross sectional regression analysis an unseemly result is obtained in regard to the effect of irrigation on output instability. For the initial two years (i.e., 1970-71 and 1980-81) the insignificant contribution of irrigation to the output instability may be explained in terms of its low level in most of the districts during these years. But the statistically insignificant relationship between irrigation and the output instability during 1990-91 is very difficult to explain in terms of the level of irrigation since there was significant increase in the area under irrigation during 1990-91. However, in a cross-section study involving the districts with greatly varying areas under irrigation, a result indicating no statistical significant association between output instability and irrigation is perhaps a logical outcome. To get a reliable estimate of the effect of irrigation on output instability a separate regression equation should be fitted for each group of the districts, classified on the basis of the levels of irrigation expansion. But as the number of districts for each class becomes too small (leading to small degrees of freedom), fitting separate regression equation may yield erroneous results. However, excluding low irrigated districts (having less then 35% of GCA irrigated), actually a significant negative correlation (r = -0.632*) between instability and irrigation is obtained for the year 1990-91.

Now taking both time series and cross sectional regression results on the causes of output instability together it can be concluded that expansion of irrigation has stabilizing effect on agricultural production in West Bengal. Crop base and holding size have changed their positions from factors causing instability (in the early years) to factors contributing to stability (in the later years) in agricultural production. In regard to current fallow it is observed from time series analysis that it has no significant effect on instability (perhaps because of our failure to consider its unknown lead-lag structure for a particular region). But in the cross-section analysis the current fallow is seen to have significantly positive effect upon output instability in the early years. This relationship may be explained by the increase in marginal lands but which might have remained uncultivated because of low level of irrigation and insufficient rainfall.

6.3 Effects of Rainfall on Agricultural Production

For sustained and uninterrupted growth in agricultural production, it is commonly believed that good rainfall is a necessary prerequisite, specially for a State like West Bengal where, inspite of significant increase in the area under irrigation in recent years, a major part of the cropped area is rain-fed (see Table 3.1). If we consider the pattern of fluctuations in rainfall index (measured as mean absolute percent deviation of actual rainfall from the normal level) over two sub-periods for the districts (see Table 6.13), it is seen that actual rainfall indices fluctuate less around normal level for all the districts (except Howrah) and the State during second sub-period compared to first sub-period. This lower fluctuation is also evident from the period-wise numbers of deviation of actual rainfall above normal rainfall by ± 20 per cent. Therefore, from Table 6.13 it can be concluded that rainfall situation in West Bengal was much better during the second sub-period than in the first sub-period and this may be one of the main factors behind the remarkable agricultural improvement during the LFG regime over the early period.

Conventionally, the effect of rainfall (more specifically, the elasticity of rainfall) on agricultural production is measured by the estimated coefficient of log of rainfall index, included in the exponential trend equation for agricultural production, i.e., by fitting log (production) = $\alpha + \beta t + \gamma$ log (rainfall index).

Fitting this trend equation to the West Bengal data for the whole period (1957-58—1994-95), the following trend equation is obtained :

$$\ln Prod_t = 3.3975^{*} + 0.02724^{*} + 0.1951^{**} \ln rain_t \; ..... (6.4)$$
$$\text{t value:} \quad (7.995) \quad (17.929) \quad (2.048)$$

$$\bar{R}^2 = 0.9126, F = 193.374^{*}$$

The coefficient of ln $rain_t$ is positive and statistically significant at 5% probability level, implying thereby that the elasticity of agricultural production with respect to rainfall for the whole period is 0.19.

But, interestingly, when the same trend equation is fitted to the sub-period data, the rainfall is found to have statistically insignificant effect upon agricultural production. The equations fitted to the first sub-period data (1957-58 — 1976-77) and the second-period data (1977-78 — 1994-95) are given below :

FIRST SUB-PERIOD (1957-58 — 1976-77):

$$\ln Prod_t = 3.7429^{*} + 0.02583^{*}t + 0.1243 \ln rain_t \; .....(6.5)$$
$$(9.838) \quad (9.489) \quad (1.497)$$

$$\bar{R}^2 = .8232$$
$$F = 45.244^{*},$$

SECOND SUB-PERIOD (1977-78 — 1994-95) :

$$\ln Prod_t = 2.6445^{*} + 0.03817^{*}t + 0.2851 \ln Rain_t \; .... (6.6)$$
$$(2.798) \quad (7.196) \quad (1.320)$$

$$\bar{R}^2 = .8061$$
$$F = 36.342^{*}.$$

It is seen that for both the first and the second sub-periods [equations (6.5) and (6.6) respectively] the effect of rainfall on agricultural production is not statistically significant. Saha and Swaminathan (1994) also obtained the same result implying thereby that the effect of rainfall on agricultural production was not statistically significant during the LFG period.

These results for whole period and the two sub-periods are inconsistent. While the effects of rainfall on agricultural production during the two sub-periods are found to be statistically not significant, the same during the whole period is statistically significant at 5% probability level. This inconsistency may have its origin in the mis-specification of the equations fitted to the whole period and the sub-periods' data. A mis-specified trend equation may leave a substantial part of the systematic variation in the dependent variable unexplained. Introduction of an additional variable in the equation may capture a part of the left-out unexplained variation and may turn out to be statistically significant. This is what has happened in the case of the fitted trend equation for the whole period. Exponential trend equation fitted to the whole period data left a substantial part of the systematic variation in the dependent variable unexplained. Inclusion of additional variable, rainfall, captured the left-out part of explainable variation and was found to be statistically significant. This exponential trend equation is, however, inappropriate for the whole period data. The log-quadratic equation gives a better fit in all respects. (This point has been discussed elaborately in sec. 4.1 of Chapter 4). For this equation the effect of rainfall is found to be statistically not significant for the whole period. On the other hand, the exponential trend is the most appropriate trend equation for the sub-periods as the length of the period is shorter and the observations corresponding to these sub-periods exhibit an exponential trend.

Thus, it is seen that the period-wise inconsistency arises out of the mis-specification of the trend equation. Hence, to get rid of this problem, one has to get rid of trend specification. This has been done here by estimating the absolute fluctuations of all crops, production around the appropriate trend values and the absolute fluctuations of rainfall index around the normal level, i.e., 100, and regressing the former on the latter for the whole period as well as the two sub-periods. The results are presented below.

WHOLE PERIOD (1957-58 — 1994-95) :

$$|PF_t| = 0.05001^{*} + 0.00108\,|RF_t| \quad \text{........} \quad (6.7)$$
$$(2.658) \qquad (0.969)$$

FIRST SUB-PERIOD :

$$|PF_t| = 0.05601^{*} + 0.00064\,|RF_t| \quad \ldots\ldots\ldots\ (6.8)$$
$$(3.754) \qquad (0.861)$$

SECOND SUB-PERIOD :

$$|PF_t| = 0.02737 + 0.00121\,|RF_t| \quad \ldots\ldots\ldots\ (6.9)$$
$$(0.600) \qquad (0.888)$$

Where PF denotes ln (Prod.) minus trend value and RF denotes rainfall index minus 100.

From these three equations [(6.7) to (6.9)] it is observed that for all three periods the estimated coefficient of $|RF_t|$ is not statistically significant. Thus, consistency is restored over the period-wise estimates. Another important problem emerges from the above discussion. That relates to the disagreement between the statistically insignificant rainfall (log) coefficients and our expectations about the relation between rainfall and agricultural production during the pre-LFG and the LFG regimes. The statistical results show that during each of the sub-periods the effect of rainfall on the agricultural production was insignificant. But it is very difficult to accept these results as showing the true relation between rainfall and agricultural production in West Bengal during these two sub-periods. In the first sub-period (1957-58—1976-77) the rainfall was the main source of irrigation and it was the season-wise and year-wise fluctuations in the rainfall that determined the level as well as the trend in agricultural production in a significant way. Similarly, during the second sub-period there was good rainfall in a number of years. But the effect of rainfall on agricultural production is not reflected in the estimated coefficient of rainfall. As it has been explained in sec. 3.6 of Chapter 3, the rainfall is a random regressor with measurement errors. The application of the ordinary least squares method to estimate the coefficient of such a regressor will yield both biased and inconsistent result. The Durbin's ranking method is suggested in such cases. The method, as applied to our problem, is discussed below.

First, the variables $|PF_t|$ and $|RF_t|$ have been expressed in the forms of mean deviations. Second, the ranked values (to be considered as new variable) have been calculated in descending order on the basis of the values of $|RF_t|$ in mean deviation form. Finally, the variables $|PF_t|$ and $|RF_t|$ in mean deviations have been multiplied with the ranked variables to construct the new dependent (i.e., DPF_t) and explanatory (i.e., DRF_t) variables according to Durbin's ranking method. Applying OLS method on these new set of variables for each period the following estimates are obtained.

WHOLE PERIOD :

$$DPF_t = \underset{(1.921)}{0.001634^{***}} DRF_t; \; DW = 1.870^{+} \quad \text{........ (6.10)}$$

FIRST SUB-PERIOD :

$$DPF_t = \underset{(2.193)}{.001230^{**}} DRF_t; \; DW = 1.911^{+} \quad \text{.......... (6.11)}$$

SECOND SUB-PERIOD :

$$DPF_t = \underset{(2.050)}{0.002734^{***}} DRF_t; \; DW = 1.583^{+} \quad \text{.......... (6.12)}$$

From these three final estimated equations [(6.10) to (6.12)] it is observed that the fluctuations in agricultural production in West Bengal are significantly positively determined by the fluctuations in rainfall index for all the periods. This implies that higher fluctuations in rainfall index leads to higher fluctuations in agricultural production in West Bengal and *vice versa*. Further, it is seen that the dependence of West Bengal agriculture on rainfall is relatively stronger in first sub-period than in the second sub-period. This may be due to more expansion of irrigation during second sub-period than of first sub-period [area irrigated only from government canal increased two fold between 1960-61 and 1977-78, i.e., from 473.3 thousand hectares in 1960-61 to 962.6 thousand hectares in 1977-78]. However, the dependence of agriculture on rainfall should be considered crop-wise (but could not be carried out here due to difficulties in the construction of crop-wise rainfall indices). In aggregative analysis, like this, some parts of fluctuation may have been smoothed out that, perhaps, has led to significance here only at 5 per cent or 10 per cent level.

6.4 Rural Population and Agricultural Growth

From inter-district variations in rural population growth and agricultural output growth, the demographic effect on the performance of West Bengal agriculture has been assessed with the help of Boyce's model [Boyce (1987), Pp. 149-150]:

$$GO_{Ti} = a + b\ GP_{(T\text{-}K),i} + u_i \ldots\ldots\ldots\ldots (6.13)$$

Where GO_{Ti} denotes agricultural output growth rate of ith district during the period T (here T=1957-58—1994-95 or 1957-58—1976-77 or 1977-78—1994-95), $GP_{(T\text{-}k),i}$ is rural population growth rate in ith district for the period T-k (here k being a positive integer determines the lag structure) and u_i = disturbance term.

In this model output growth in one period is regressed on the population growth in an earlier period to exclude the possibility of reverse causal relationship (i.e., agricultural production induced population growth). This model, as argued by Boyce, does not imply that past population growth is the sole determinant of subsequent agricultural performance, rather it merely tests how, *ceteris paribus*, population growth affects agricultural growth in a later period. More specifically, significantly positive (negative) $\hat{b}$ implies the existence of induced innovation (neo-Malthusian) theory and for 'surplus labour theory' $\hat{b}$ would be insignificant and near about zero. Actually, the estimated coefficient $\hat{b}$ [equation (6.13)] here provides the direct estimate of the elasticity of output (in period T) with respect to population [in period (T-k)]. It should be noted in this connection that one may get biased estimate of $\hat{b}$ in this model if one period's (say second sub-period) output growth is highly collinear with the output growth of the earlier period (say first sub-period), because in this case the population growth getting inducement from first sub-period's output growth will always positively affect the output growth of the second sub-period. But such possibility does not arise here as the correlation between two sub-periods' output growth is very weak [reference to Table 4.41 of Chapter 4].

The relationship between population growth and agricultural performance may be restated in terms of output growth per capita, GOPC, (i.e., output growth minus population growth) as dependent variable :

$$GOPC_{Ti} = Go_{Ti} - GP_{Ti}$$

$$= [a + b\, GP_{(T-k)i} + u_i] - [a'' + b''\, GP_{(T-k)i} + u''_i]$$

(assuming population growth as a function of prior population growth)

$$= (a-a'') + (b-b'')\, GP_{(T-k)i} + (u_i-u''_i)$$

$$\text{i.e., } GOPC_{Ti} = a' + b'GP_{(T-k)i} + u'_i \text{ .... (6.14)}$$

To determine the demographic effect on agricultural performance we have estimated equations (6.13) and (6.14) separately and from these, an indirect estimate of b'' has been made (to know whether the districts experiencing high population growth in the past are also experiencing high population growth in the present or not).

The rural population growth rate for a particular period has been estimated from inter-census rates of growth of rural population (presented in Table 3.5 of Chapter 3) with the help of geometric mean and per capita output growth rate for a particular period has been calculated from output growth rate minus rural population growth rate for that period. In Table 6.14 the values of the dependent variables over the districts for the whole period and for two sub-periods are presented. At the outset we can explain the results of Table 6.14. From the analysis of section 4.4 of Chapter 4 it was established earlier that output growth rate in West Bengal increased remarkably during the second sub-period compared with first sub-period. This increase is much more evident (from Table 6.14) in consideration of the per capita output growth rates. In the first sub-period per capita output growth rate in West Bengal was negative (-0.17%) and during second sub-period this turned to remarkably positive (1.46%). Nearly half the number of the districts registered negative growth in per capita output during the first sub-period while during the second sub-period all the districts (except Jalpaiguri) experienced positive and relatively higher growth rates. For the two districts, *viz.*, Midnapore and Bankura that registered highest growth rates during LFG regime, the per capita output growth rates were also negative in the previous sub-period. In general, (considering the whole period's per capita output growth) agricultural performance of the northern districts is worse in comparison to that of the central districts.

To examine the lagged effect of rural population growth on agricultural performance in West Bengal the total output growth rate and also per capita output growth rate have been regressed on the past growth of rural population. From these cross sectional regression analysis two districts — Howrah and 24-Parganas—have been excluded, as in the study of Boyce (1987), because of their proximity to the big Metropolis, since it may lead to the inclusion of semi-urban population. Further, as the exact lag structure in the relationship between rural population growth and agricultural growth is unknown, different lagged values of rural population growth (as independent variable) have been considered in the regression equation and the statistical significance of the estimated coefficients is examined. In Table 6.15 the most reliable estimates have been presented.

From Table 6.15 it is observed that the impact of rural population growth on agricultural performance of West Bengal for all periods is felt most strongly after a lag of approximately 25 years (more or less similar to Boyce's observation which established the lag structure of 30 years). Further, for all the periods the districts which experienced high past rural population growth, are also presently experiencing high rural population growth (as $b'' > 0$), however, with varying magnitude. When whole period's agricultural output growth is regressed upon 1931—71 rural population growth, the elasticity is estimated to be 1.3277 ($t = 3.449$). In other words, a 1 per cent increase in rural population for the period 1931-71 is associated with more than 1 per cent (1.33%) increase in agricultural output for the period 1957-58—1994-95. During the same periods the rural population growth elasticity of per capita output growth is also high and positive, but less than the elasticity of total output growth.

So far as the two sub-periods are concerned, the estimates of rural population elasticity of agricultural output/per capita output are strikingly different. During the first sub-period these elasticities of agricultural output and per capita output were positive and high with strong statistical significance. But for the second sub-period these elasticities were negative and statistically significant at 20% and 10% probability levels. The weak statistical significance (i.e., substantial dispersion around trend line) may be accepted as natural because past population growth could not completely explain subsequent substantially improved agricultural performance. However, due to high level of population growth (in 1951-71 compared to 1931-51) the magnitude as well as the statistical significance ot the population growth elasticity of per capita output growth are stronger than that of total output growth rate during the second sub-period. The above discussion may be summarized in the following way:

The first sub-period's total output/per capita output growth in agriculture had the strong induced effect from the past population growth (i.e., the operation of induced innovation theory) in West Bengal. On the other hand, the past population growth had detrimental effect on agricultural performance during the second sub-period (i.e.,the operation of the neo-Malthusian theory). However, this detrimental effect was too weak to outweigh the induced effect of population growth during the whole period. This detrimental effect signals that if population growth is not checked, the present level of satisfactory growth in agriculture may not be sustained in West Bengal.

Table 6.1 : Correlation Among Technological Factors; 1970-71, 1980-81 and 1990-91 (Probability levels in the parentheses)

Variables	Correlation Coefficients 1970-71	1980-81	1990-91
T_1 and T_2	0.6817 (.005)	0.4500 (.092)	0.6054 (.017)
T_1 and T_3	0.5286 (.043)	0.8005 (.000)	0.8526 (.000)
T_1 and T_4	0.1697 (.546)	0.5895 (.021)	0.6415 (.010)
T_2 and T_3	0.5627 (.029)	0.3502 (.201)	0.6035 (.017)
T_2 and T_4	0.2916 (.292)	0.6964 (.004)	0.5916 (.020)
T_3 and T_4	0.4701 (.077)	0.7207 (.002)	0.8766 (.000)

Notes : T_1, T_2, T_3 & T_4 imply % of GCA irrigated, fertiliser intensity, % of GCA under HYV and agricultural machinery intensity respectively.

Table 6.2 : Correlation Among Institutional Factors; 1970-71, 1980-81 and 1990-91 (Probability levels in the parentheses)

Variables	Correlation Coefficients 1970-71	1980-81	1990-91
I_1 and I_2	0.0834 (.767)	0.0551 (.845)	-0.0814 (.773)
I_1 and I_3	0.1467 (.602)	0.1968 (.482)	0.0152 (.957)
I_1 and I_4	0.0516 (.855)	0.0472 (.867)	.3198 (.245)
I_2 and I_3	-0.2315 (.406)	-0.0848 (.764)	0.4455 (.096)
I_2 and I_4	-0.8224 (.000)	-0.6340 (.011)	0.6219 (.013)
I_3 and I_4	0.3286 (.232)	0.4067 (.133)	-0.0725 (.797)

Notes : I_1, I_2, I_3 & I_4 imply average size of operational holding, % of agricultural labourers in total agricultural work-force, % of NCA under tenancy/registered bargadars and % of NCA under surplus land distributed respectively.

Table 6.3 : Principal Component Analysis on Four Technological Factors, 1970-71, 1980-81 and 1990-91

Component /Statistic	1970-71		1980-81		1990-91	
	Eigen value	Percent of variance accounted	Eigen value	Percent of variance accounted	Eigen value	Percent of variance accounted
Factor 1	2.39096	59.8	2.81874	70.5	3.09902	77.5
Factor 2	0.90775	22.7	0.75927	19.0	0.49032	12.3
Factor 3	0.39697	9.9	0.32280	8.1	0.35746	8.9
Factor 4	0.30432	7.6	0.09919	2.4	0.05321	1.3
Bartlett Test of Sphericity	15.84077**		31.71997*		41.93573*	
Chosen Principal component		First (PI)		First (PI)		First (PI)

Notes: * implies significant at 1% level,
** implies significant at 5% level.

Table 6.4 : Principal Component Analysis on Three Institutional Factors; 1970-71, 1980-81 and 1990-91

Component /Statistic	1970-71		1980-81		1990-91	
	Eigen value	Percent of variance accounted	Eigen value	Percent of variance accounted	Eigen value	Percent of variance accounted
Factor 1	1.98248	66.1	1.79353	59.8	1.80039	60.1
Factor 2	0.84637	28.2	0.92309	30.8	0.93144	31.0
Factor 3	0.17115	5.7	0.28338	9.4	0.26817	8.9
Bartlett Test of Sphericity		15.17982*		9.2077		9.72300**
Chosen Principal component		First (FI)		First (FI)		First (FI)

Notes: * implies significant at 1% level.
** implie significant at 5% level.

Table 6.5: Determinants of Agricultural Productivity: Cross Sectional Regression Results, 1970-71, 1980-81 and 1990-91, West Bengal by Districts

Time point	$\bar{R}^2$ [F Statistic]	Estimated Coefficient of			
		Intercept term	Techno-logical factors (PI)[a]	Average farm size (II)	Other institutional factors (FI)[a]
1970-71	.6591*	93.783*	21.603*	-13.830	8.887***
	[10.022]	(6.227)	(4.536)	(-I.289)	(1.881)
1980-81	.8809*	102.768*	28.254*	-18.245	8.178*¹
	[35.154]	(7.137)	(8.957)	(-I.574)	(2.594)
1990-91	.7673*	154.224*	38.640*	-32.079	-4.699
	[16.386]	(5.623)	(4.432)	(-I.154)	(-O.532)

Notes: Figures in the parenthes. under the estimated coefficients indicate t values and those under $\bar{R}^2$ indicate F values.

*Significant at 1% lev

**Significant at 5% level;

*** Significant at 10% level; [a]expressed in standardised form.

Table 6.6 : Regression Analysis of Instability Causing Factors in West Bengal Agriculture, 1957-58—1988-89. (A preliminary analysis)

Independent variable/$\bar{R}^2$/ DW	Estimated Coefficients Whole period (1957-58—1998-99)	First sub-period (1957-58—1976-77)	Second sub-period (1977-78—1988-89)
Irrigation	-1.5692**	-0.4350	-3.8417*
	(-2.239)	(-.278)	(-3.998)
Fallow	-0.0091⁺	-0.0004	0.0083
	(-1.637)	(-.044)	(.935)
Crop base	-0.00005	0.0024	-0.0090⁺⁺
	(-0.013)	(.280)	(-1.617)
Locational base	0.00302	0.0032	-0.0196⁺⁺
	(.475)	(.382)	(-1.625)
Intercept term	0.1622	0.1505	1.9841*
	(.623)	(.180)	(3.294)
$\bar{R}^2$	0.0626⁺⁺	-0.2410	0.7028*
DW	1.830	2.155	1.900

Notes : t values in the parentheses, * significant at 1% level, * significant at 5% level, ⁺ ⁺ significant at 20% level.

Irrigation = Proportion of GCA irrigated from Govt. canals.

Fallow = Percentage of NCA under current fallow.

Crop base = Percentage of GCA under aman rice, boro rice, jute, potatoes and mustard (i.e., 5 principal corps).

Locational base = Percentage share of 4 main districts [namely, Burdwan, Midnapore, Hooghly and 24-Parganas] in State's total agricultural production.

Table 6.7 : Product Moment Correlation Matrix of Independent Variables of Output Instability in West Bengal Agriculture, 1957-58 1988-89 and 1977-78—1988-89.

	Irrigation	Fallow	Crop base	Locational base	
Irrigation	1	-0.7229 (.000)	0.3400 (.037)	0.6775 (.000)	
Fallow	-0.0558 (.863)	1	0.1942 (.287)	-0.3708 (.037)	
Crop base	0.1746 (.448)	0.4642 (.128)	1	0.6665 (0.000)	
Locational base	0.2343 (.349)	0.6390 (.025)	0.5595 (.016)	1	↑ Whole Period

← Second Period

Notes: Probability levels within parentheses.

Correlations above diagonal represent for whole period.

Correlations below diagonal represent for second sub-period.

Table 6.8 : Selection of Independent Variables for Explaining, Instability in Agricultural Production of West Bengal, 1957-58—1988-89 and 1977-78—1988-89

Variable/ others	Whole period (1957-58—1988-89)			Second Sub-period (1977-78—1988-89)		
	Simple Correlation with dependent variable	Multiple Correlation among independent variables		Simple Correlation with dependent variable	Multiple Correlation among independent variables	
		Value	Significantly related variables		Value	Significantly related variables
Irrigation	-0.3283 (.044)	0.8171	Fallow* & Locational base*	-0.7295 (.001)	0.3492	—
Fallow	0.0011 (.995)	0.7646	Irrigation* & Crop base***	-0.1096 (.734)	0.7368	Locational base**
Crop base	-0.2371 (.152)	0.5757	Fallow*** & Locational base*	-0.4082 (.093)	0.4775	—
Locational base	-0.1832 (.213)	0.7326	Irrigation* & Crop base*	-0.4227 (.081)	0.6900	Fallow**
Multiple correlation, dependent variable with independent variables		0.4182			0.9005	
Decision on omission of independent variable(s)[a]	Fallow & Locational base			Fallow		

Notes : Probability levels in the parentheses*, ** & ***imply significant at 1%, 5% & 10 % level respectively.

[a] omitted independent variables are unimportant and also cause of multicollinearity problem with other independent variables. Klien's rule has been applied to detect multicollinearity problem.

Table 6.9 : Final Regression Results on the Determinants of Output Instability in West Bengal Agriculture, 1957-58 — 1994-95[a] and 1977-78—1994-95

Independent Variable /$\bar{R}^2$ /DW	Whole period (1957-58—1994-95)	Second Sub-period (1977-78—1994-95)
	Estimated Coefficient	Estimated Coefficient
Irrigation	-0.6246*** (-1.766)	-3.6541* (-3.921)
Crop base	-0.0019 (-.844)	-0.0096*** (-1.858)
Locational base	——	-0.0058 (-.764)
Intercept term	0.2775*** (1.751)	1.1290* (3.720)
$\bar{R}^2$	0.0956***	0.5495*
DW	1.835	1.997

Notes: t values in the parentheses.

* Significant at 1% level.

*** Significant 10% level.

[a] Time period has been extended from 1957-58 — 1988-89 to 1957-58 — 1994-95 owing to the omission of the variable, fallow (for which data are not available for the years 1989-90 to 1994-95).

Table 6.10 : Product Moment Correlation Coefficients among Independent Variables of Output Instability in Agriculture, West Bengal by Districts, 1970-71, 1980-81 and 1990-91

Independent variables	Correlation Coefficients		
	1970-71	1980-81	1990-91
Irrigation & Fallow	0.2441	-0.0741	-0.2309
	(.381)	(.793)	(.408)
Irrigation & Crop base	0.2205	0.1810	0.1861
	(.430)	(.518)	(.507)
Irrigation & Locational base	0.1918	0.2479	0.2820
	(.494)	(.373)	(.309)
Irrigation & Holding size	0.4264	0.3585	0.1801
	(.113)	(.189)	(.521)
Fallow & Crop base	0.5151	0.4321	0.0906
	(.049)	(.108)	(.748)
Fallow & Locational base	-0.2196	-0.0394	-0.4008
	(.432)	(.889)	(.139)
Fallow & Holding size	-0.3721	-0.3757	-0.1687
	(.172)	(.168)	(.548)
Crop base & Locational base	0.3307	0.5809	0.5384
	(.229)	(.023)	(.038)
Crop base & Holding size	-0.0991	-0.3590	-0.4954
	(.725)	(.189)	(.060)
Locational base & Holding size	0.4364	-0.1312	-0.1627
	(.104)	(.641)	(.562)

Note: Probability levels in the parentheses.

Table 6.11 : Selection of Independent Variables for Explaining Output Instability[s] in Agriculture, West Bengal by Districts, 1970-71, 1980-81 and 1990-91

Variable/ Others	1970-71				1980 - 81				1990 - 91			
	Correlation with Dependent Variable		Correlation among Indep. Variables		Correlation with Dependent Variable		Correlation among Indep. Variables		Correlation with Dependent Variable		Correlation among Indep. Variables	
	Simple	Partial	Multiple Corr. Value	Multiple Corr. Significantly related Vari.	Simple	Partial	Multiple Corr. Value	Multiple Corr. significantly related vari.	Simple	Partial	Multiple Corr. Value	Multiple Corr. Significantly related vari.
Ir	-.2688 (.333)	-.2377 (.482)	.6103	Hs***	-.2154 (.441)	-.2108 (.534)	.5055	—	.0974 (.730)	.2052 (.545)	.4255	—
Fl	.5300 (.049)	.5172 (.103)	.7308	Cb***	.7530 (.000)	.6102 (.046)	.5953	—	-.1774 (.527)	-.1842 (.588)	.5592	Lb***
Cb	.6213 (.013)	.4625 (.152)	.6931	Fl*** & Lb***	.5410 (.037)	.3661 (.268)	.7589	Lb**	.1788 (.524)	-.1887 (.578)	.7354	Lb**
Lb	.2040 (.466)	-.0311 (.928)	.6493	Cb***	.0587 (.835)	-.1418 (.677)	.6720	Cb**	.2222 (.426)	.0993 (.772)	.7073	Fl*** & Cb**
Hs	.2067 (.460)	.5369 (.089)	.6964	Ir***	-.5561 (.031)	-.2976 (.374)	.5890	—	-.5011 (.057)	-.5610 (.073)	.5720	—

(Contd.)

Table : 6.11 : *(Contd.)*

Variable/ Others	1970-71	1980 - 81	1990 - 91
Multiple corr. (Dep.vari. with indep. variables)	0.7884	0.8408	0.6038
Decision on omission of Independent Variable [a]	Lb	Lb	Lb

Key: Ir = Irrigation Fl = Fallow, Cb = Crop base, Lb = Locational base and Hs = Holding size.

Notes: ** Significant at 5% level, *** Significant at 10% level.

Probability levels in the parentheses.

[a] omitted variable is unimportant and source of multicollinearity problem (detected by Klien's rule).

[s] Instability statistic $|Z_t|$ is measured not from a single time point, rather it is averaged over three neighbouring time points (t-1, t, t+1) to avoid yearly volatile fluctuation. Here t = 1970-71, 1980-81 and 1990-91.

Table 6.12 : Cross Sectional Regression Analysis of Output Instability in West Bengal Agriculture, 1970-71, 1980-81 and 1990-91

Variable/ $\bar{R}^2$	Estimated Regression Coefficients		
	1970-71	1980-81	1990-91
Irrigation	-0.00059 (-.775)	-0.00042 (-.743)	0.00029 (.668)
Fallow	0.00471*** (2.071)	0.01348** (2.792)	-0.00123 (-.868)
Crop base	0.00149*** (1.882)	0.00091 (1.207)	-0.00031 (-.518)
Holding size	0.13155*** (2.091)	-0.06387 (-.950)	-0.11965*** (-2.119)
Intercept term	-0.15558++ (-1.644)	0.13157++ (1.623)	0.22605** (2.633)
$\bar{R}^2$	0.4698**	0.5812*	0.1015

Notes: * significant at 1% level,
**significant at 5% level,
***significant at 10% level,
++ significant at 20% level,
t values in the Parentheses.

Table 6.13 : Fluctuations in All Crops Rainfall Index, West Bengal By Districts, 1957-58—1994-95

District	Mean absolute percent deviation in all crops rainfall index from the normal level			Frequency of deviation of actual from normal rainfall index for all crops by ± 20% & above		
	Whole period, 1957-58–1994-95	1st sub-period, 1957-58–1976-77	2nd sub-period, 1977-78–1994-95	Whole period, 1957-58–1994-95	1st sub-period, 1957-58–1976-77	2nd sub-period, 1977-78–1994-95
Burdwan	18.555	23.265	13.322	14	11	3
Birbhum	19.226	20.695	17.594	13	8	5
Bankura	20.989	26.245	15.150	15	10	5
Midnapore	17.305	17.705	16.861	13	7	6
Howrah	20.305	19.330	21.389	15	7	8
Hooghly	18.721	21.005	16.183	14	9	5
24-Parganas	15.268	16.600	13.789	12	6	6
Nadia	19.039	23.785	13.767	15	10	5
Murshidabad	20.629	21.590	19.561	18	9	9
West Dinajpur	26.200	29.690	22.322	23	14	9
Malda	21.103	23.460	18.483	16	10	6
Jalpaiguri	19.945	21.485	18.233	17	11	6
Darjeeling	15.729	16.320	15.072	12	8	4
Coochbehar	15.447	16.475	14.306	15	9	6
Purulia	18.750	20.650	16.639	13	7	6
West Bengal	14.239	16.865	11.322	9	7	2

Table 6.14 : Growth Rates in Agricultural Output, Rural Population and Agricultural Output Per Capita, West Bengal by Districts, Whole Period and Two Sub-periods

District	Whole period (1957-58—1994-95)			First sub-period (1957-58—1976-77)			Second sub - period (1977-78—1994-95)		
	GO	GP	GOPC	GO	GP	GOPC	GO	GP	GOPC
Burdwan	3.29	1.57	1.72	2.79	1.73	1.06	3.84	1.41	2.43
Birbhum	2.21	1.99	0.22	2.09	2.02	0.07	2.44	1.95	0.49
Bankura	2.64	1.79	0.85	1.27	1.90	-0.63	4.81	1.69	3.12
Midnapore	3.17	2.18	0.99	1.54	2.26	-0.72	5.07	2.10	2.97
Howrah	4.17	1.44	2.73	4.05	1.37	2.68	4.22	1.52	2.70
Hooghly	4.48	2.10	2.38	4.26	2.30	1.96	4.73	1.90	2.83
24-Parganas	2.28	2.45	-0.17	2.01	2.29	-0.28	2.70	2.64	0.06
Nadia	4.16	2.78	1.38	3.91	2.81	1.10	4.46	2.76	1.70
Murshidabad	3.24	2.51	0.73	3.07	2.47	0.60	3.58	2.56	1.02
West Dinajpur	2.27	2.70	-0.43	1.95	2.78	-0.83	3.14	2.61	0.53
Malda	3.14	2.58	0.56	2.74	2.58	0.16	3.74	2.58	1.16
Jalpaiguri	1.80	2.32	-0.52	2.05	2.45	-0.40	1.48	2.18	-0.70
Darjeeling	2.87	2.20	0.67	2.51	2.28	0.23	3.33	2.11	1.22
Coochbehar	2.03	2.60	-0.57	1.79	3.09	-1.30	2.31	2.15	0.16
Purulia	2.01	1.61	0.40	0.69	1.45	-0.76	3.33	1.80	1.53
WEST BENGAL	2.68	2.22	0.46	2.08	2.25	-0.17	3.65	2.19	1.46

Key : GO = output growth rate (from Table 4.26), GP = rural population growth rate (calculated from Table 3.5), GOPC = per capita output growth rate.

Notes: (i) Population growth rates for the years 1991-92, 1992-93, 1993-94 and 1994-95 have been estimated from 1991 census.

(ii) GP has been calculated from inter-census rates of growth of rural population by using geometric mean.

Table 6.15 : Regression Analysis of the Impact of Lagged Rural Population Growth upon (i) Agricultural Output Growth and (ii) Agricultural Output Growth Per Capita in West Bengal by Districts [a], 1957-58—1994-95, 1957-58—1976-77 and 1977-78—1994-95

Dependent variable	Independent variable	Estimated regression coefficient $\hat{b}$	Estimated regression coefficient $\hat{b}'$	Indirect estimate $\hat{b}''$
$GO_T/GOPC_T$	GP_{31-71}	1.3277* (3.449)	0.9299** (1.810)	0.3978
$GO_1/GOPC_1$	GP_{31-51}	1.5897* (3.366)	1.4531* (3.241)	0.1366
$GO_2/GOPC_2$	GP_{51-71}	-0.6137++ (-1.088)	-1.0040*** (-1.736)	0.3903

Key : GO = output growth rate, GOPC = output growth rate per capita, GP = rural population growth rate.

Subscripts T, 1 and 2 refer to 1957-58—1994-95, 1957-58—1976-77 and 1977-78—1994-95 respectively.

Subscripts 31-71, 31-51 and 51-71 refer to 1931-71, 1931-51 and 1951-71 respectively.

$\hat{b}$ estimated from equation: $GO_t = a + b\ GP_{(T-K)}$

$\hat{b}'$ estimated from equation : $GOPC_t = a^1 + b^1 GP_{(T-K)}$

$\hat{b}' = \hat{b}-\hat{b}''$ is for the equation : $GP = a'' + b''\ GP_{(T-K)}$

(t = T, 1 and 2)

Notes: [a] Excluding Howrah and 24-Parganas.

* Significant at 1% level by one-tailed test,

**significant at 5% level by one-tailed test,

*** significant at 10% level by one-tailed test,

++ significant at 20% level by one-tailed test, t values in the parentheses.

7

Measures of Equity and its Relation with Growth and Stability in Agriculture

District-wise growth rates in money as well as real wage rates of agricultural labourers during the period 1978-79—1991-92 in West Bengal have been estimated in the first section of this chapter. Section 2 deals with the consideration of inequalities in the distributions of ownership land holding and monthly per capita consumer expenditure in rural West Bengal. The incidence of rural poverty in West Bengal during LFG regime has also been considered here. In section 3, effects of agricultural growth and stability on equity in West Bengal have been examined. The last section discusses the extent of inequality among different classes of the rural households in the distribution of land holdings, income and consumption, as measured from the data collected from primary survey of some sample villages.

7.1 Growth in Agricultural Wage Rates

Using simple log-linear trend equation, the growth rates in the wages of agricultural labourers have been estimated over the districts during LFG regime. The estimated growth rates (as well as the base levels) in money wages of male and female agricultural labourers are presented in Table 7.1. From the estimated growth rates of money wages of male agricultural labourers it is observed that in all the districts these growth rates are very high. The highest growth rate is reported from Malda district (13.491%), followed by Nadia. The other districts where growth rate is more than 10% are Burdwan,

Birbhum, Midnapore, Hooghly and Jalpaiguri. Darjeeling has registered the lowest increase (i.e., 8.17%) of money wages of male agricultural labourers. In all other districts the increase of money wages of male labourers per annum is between 9 per cent and 10 per cent. Comparing these growth rates with the base levels, it is, in general, observed that the districts with lower base level of money wages have experienced higher growth rates. This is also confirmed from the measured correlation coefficient [correlation coefficient = -0.5772; probability level of significance = 0.024] between initial levels and growth rates of money wage of male labourers over the districts. This implies that during LFG regime the inter-district differences in agricultural wages of male workers have been narrowed down. This might have happened because of the rising bargaining strength of the agricultural labourers through "Krishak Sabha" (agricultural labourers' union) movement from mid-eighties.

Growth rates of agricultural money wage rate of female labourers exhibit more or less a similar trend. Here also Malda records the highest annual growth rate (15.17%) and it is followed by Nadia, 24-Parganas and Midnapore. Here also Darjeeling is experiencing the lowest growth rate (7.851%). For all other districts growth rates of money wage of female labourers have increased by more than 10 per cent (but less than 11 per cent) per annum, except two districts—Jalpaiguri and Coochbehar—where the increases are between 9 per cent to 10 per cent. Here also it is observed that the districts with lower base levels are, in general, experiencing higher growth in money wages of female labourers and the measured product moment correlation coefficient between initial levels and growth rates is statistically significant [r = -0.4805, level of significance is 0.070]. Another interesting point to note is that, for all the districts, except Burdwan and three northern districts (Jalpaiguri, Darjeeling and Coochbehar) the increase in the wages of female labourers is more than that of male labourers for the districts. The estimated correlation coefficient between differences in the levels and differences in the growth rates of wages of male and female labourers is -0.4167 (with probability level 0.122). This lends support to the fact that there is also narrowing down of the wage gap of both types of agricultural labourers during LFG regime. The possible reason behind higher rate of growth of female labourers' money wage may be that the difference between the wages of the female labourers and the wages of the male labourers over the districts had been higher in the early years of the period under review and after 1986-87 there was an increase in the ratio of female wage to male wage in most of the districts (but in no case that ratio was greater than one). In the sphere of agricultural production the female labourers also perform the same job in West Bengal like male labourers, except ploughing and some other specific operations. The female labourers are equally competent in sowing, transplanting, harvesting operations, etc. From the basic data (not reported here) it is observed that in recent years the male and female labourers are paid equal wages in the peak agricultural season in most of the districts of West Bengal. However, in the lean season the male labourers are paid relatively higher wages. As yearly average wage rate is taken here, we get minimum gender disparity in wages, particularly after 1987-88.

From these findings it may be concluded that during LFG regime in West Bengal the differences in agricultural money wages over the districts and genders are coming down and the agricultural labourers are being paid good remuneration for their labour. But to determine the real improvement in this regard, one should consider the real wages instead of money wages. The estimated growth rates in real wage rates of agricultural labourers (male and female) along with the per capita crop-output growth rates for the districts are presented in Table 7.2. In consideration of growth rates in real wages over the districts the trend is almost the same for both male and female agricultural labourers as observed in case of money wage rate (i.e., inter-district and inter-gender variations in real wages are gradually declining). Actually, growth in real wage is equal to growth in money wage minus growth in price level. Due to non-availability of price index (CPIAL) for individual districts, the State's CPIAL has been used as a common deflator in calculating real wages of the districts. As a result, the similar pattern of changes in real and money wages over the districts and also over male and female labourers is obtained and the difference between the growth rates of money and real wages is accounted for by the growth rate of the State's CPIAL during the same period.

Now, comparing the growth of real wages (per day per labourer) with the growth of per capita agricultural production during second sub-period over the districts of West Bengal, it is observed that the former is higher than the later for both male and female labourers in all the districts, except Bankura and Darjeeling (and also Howrah and Hooghly only for male labourers). In Darjeeling growth in real wages is, however, insignificant (statistically). In other three districts (Bankura, Howrah and Hooghly), though the increase in real wages is significantly positive and remarkable, it is lower than the agricultural per capita output growth mainly because of relatively higher growth in the later. However, this higher growth of per capita real wages in relation to the growth of per capita crop-output for most of the districts gives an indication that during LFG regime in West Bengal the conditions of the agricultural labourers have improved appreciably with the growth of agriculture. But to get a clear view on this particular issue one should consider the movement of total labour income instead of wage rates. This could not be done in the study because of non-availability of time series data on the volume of agricultural labour employment. However, with the growth of agricultural production along with the limited use of agricultural machineries (refer to Tables 3.3 and 3.4) it can be safely presumed that the agricultural labour employment per labourer has not declined in West Bengal during LFG regime and so higher wages with lower regional and gender disparities may be taken as the indication of general improvement (both in absolute and relative terms) in the economic condition of the agricultural labourers.

7.2 Broad Measures of Rural Equity

Turning to the general measures of rural equity, the pattern in the distribution of land holding in West Bengal has been considered first (land, being the prime asset of an agriculturally dominated society, like rural West Bengal, is the major source of income for the rural people). The Table 7.3 presents the measured Gini coefficients on land distribution in West Bengal over different NSS rounds. Examining the values of Gini coefficient over different NSS rounds [i.e., from 8th round (1953-54) to 50th round (1993)] it is seen that the inequality in the distribution of land holding has not changed over the period (the coefficient of t is not statistically significant). The linear trend equation, fitted to these Gini coefficients gives the following regression results.

$$(\text{Land Gini})_t = 0.6776^{*} + 0.00006t \quad \text{.....................} \quad (7.1)$$
$$(29.501) \quad (0.067)$$

$DW = 3.326^{+}, \bar{R}^2 = -0.3313, F = 0.0045$

[* significant at 1% level, t values in the parentheses, + negative autocorrelation, t = time (1953-54 = 1 and so on). These notations are used for all the equations in this chapter].

From the estimated equation (7.1) it is observed that there is no trend in the Gini coefficients of land distribution. Further the magnitudes of these Gini coefficients (Table 7.3) are relatively high, implying high concentration in the distribution of ownership land holding in rural West Bengal, inspite of the implementation of land reforms under LFG rule. But the Gini coefficient, being a summary measure, should be interpreted cautiously as the inter-group transfers are not reflected here.

The pattern of land distribution signifies only the potential state of rural equity. The actual state of equity, contributed by the growth of agriculture, should be viewed in terms of the flow concept— income, consumption, etc.—rather than the stock concept, i.e., asset holding. In Table 7.4 the Gini coefficients on the distribution of monthly per capita consumption of the rural population in West Bengal are measured. From the lower values of these Gini coefficients (compared with the Gini coefficients on the land distribution), it is observed that the consumption distribution among rural population is more equitable. Further, the inequality in the consumption distribution of rural West Bengal has remarkably declined over the NSS rounds (from 0.327 in the year 1957-58 to 0.250 in 1993-94). This decline is also statistically significant, as observed from the following estimated trend equation (fitted on the measured Gini coefficients on consumption distributions of the years 1953-54 to 1993-94) :

$$(\text{Consumption Gini})_t = \underset{(35.056)}{0.3274^*} - \underset{(-5.441)}{0.00195^* t} \quad \ldots\ldots\ldots\ldots\ldots (7.2)$$

$DW = 2.238^{N}, \bar{R}^2 = 0.7607, F = 29.608^*$

[N implies that the disturbance term is non-autocorrelated].

As an extension of this summary measure, the percentile distributions of the rural consumption in West Bengal over different NSS rounds have been calculated in Table 7.5. From these estimated percentile distributions it is observed that the share of bottom 40 per cent of the rural population in consumption has increased remarkably from 3.29 per cent in 1953-54 to 21.19 per cent in 1993-94. The share of top 10 per cent of the rural population remains more or less constant over the NSS rounds (from 18.25 per cent in 1953-54 to 20.87 per cent in 1993-94) and actually the share of middle 50 per cent of the population has significantly declined (from 78.46 per cent in 1953-54 to 57.94 per cent in 1993-94). Separate linear trend equations have been fitted to these shares over the NSS rounds and the statistical significances of the inter-temporal movements in these shares are tested. The estimated regression equations are given below :

Estimated Trend Equations on the Percentile Distribution of Consumption in Rural West Bengal.

Population Class	Estimated trend equation	$\bar{R}^2$ [F value]	DW
bottom 40%	$SH^c_{Bt} = 5.336^{**} + 0.5701^{*}t$ (2.593) (7.207)	0.8498^{*} [51.936]	1.699^{N}
middle 50%	$SH^c_{Mt} = 73.771^{*} - 0.5935^{*}t$ (30.522) (-6.389)	0.8157^{*} [40.823]	1.594^{N}
top 10%	$SH^c_{Tt} = 20.898^{*} + 0.0228t$ (16.169) (0.459)	-0.0961 [0.211]	1.971^{N}

[**significant at 5% level; SH^c denotes percentage share in consumption for different classes]

From the estimated regression equations presented in the above table it is also observed that the consumption share of bottom 40 per cent of rural population in West Bengal has increased significantly over time at the cost of significant decline in the share of middle class. On the other hand we get no trend movement in the consumption share of rich persons (top 10 per cent of the rural population).

The reduction in the overall value of Gini coefficient associated with the increasing share of bottom 40 per cent of the population in the rural consumption distribution should be readily reflected in the low incidence of rural poverty. From Table 3.11 (section 3.3, Chapter 3) it was observed that the incidence of rural poverty (measured by Head Count Ratio) declined remarkably (from 68.3% in 1977-78 to 40.3% in 1993-94) in West Bengal during LFG regime. This decline was more prominent from the year 1986-87 after which the growth of West Bengal agriculture was also highest (see Table 4.30, sec. 4.4, Chapter 4). To examine whether there is any trend in the incidence of rural poverty in West Bengal, linear trend equation has been estimated. The regression results are as follows:

$$HCR_t = 106.4558^{*} - 1.8219^{*}t \text{ .... } (7.3)$$
$$(8.600) \qquad (-4.637)$$

$DW = 2.317^{N}$, $\bar{R}^2 = 0.7455$, $F = 21.50^{*}$

[HCR = Head Count Ratio and t = time (taking 1977-78 = 21 and so on)].

It is observed from this estimated trend equation (7.3) that there has been significant decline in the incidence of rural poverty in West Bengal during LFG regime. It may be noted in this connection that K.N. Ninan [(1994), Pp. 1544-1551] obtained significantly positive trend [$HCR_t = 51.24 + 1.67^{**}t$] in the incidence of rural poverty in West Bengal during the period 1957-58 to 1968-69.

The general conclusion that can be drawn from the above analysis is that the equity in rural West Bengal is being achieved more and more over time, specially in respect of agricultural wages, rural consumption distribution and incidence of rural poverty. However, no significant decline in the inequality is observed in the case of land distribution. This result may be used to infer that the land reforms in West Bengal has had no effect on the inequality in the land distribution in the State. However, if the positive effect of agricultural growth on inequality in land distribution is taken into consideration, the above inference may not be a valid one. On the contrary, it can be said that the positive effect of agricultural growth and the negative effect of land reforms have jointly contributed to a not statistically significant decline in the inequality in land distribution in the State.

7.3 Impact of Agricultural Growth and Stability on Rural Equity

From the analyses of Chapter 4 and 5 it has been observed that West

Bengal agriculture during LFG regime has experienced higher growth with more stability in its production in comparison to that in the preceding period. Further, from the last two sections of this chapter it has been established that inequalities in the distribution of agricultural wages, rural consumption and incidence of rural poverty have significantly come down during the same period. Now, we are in a position to examine statistically the existence of trickle down theory in West Bengal under LFG rule, i.e., to examine whether the equity in rural sector of West Bengal is directly contributed by its recent agricultural growth and stability or not.

Regressing both money and real wage rates of male agricultural labourer against one period lagged value of trend output (of all crops combined) during the period 1978-79—199-92 for each of the districts (Table 7.6) it is observed that they are significantly positively related in all cases, except for Birbhum and Darjeeling (those too only for real wages). The similar relationship is also observed in the case of agricultural wages of female labourers (not mentioned here). From these estimated significantly positive elasticities one may infer that agricultural growth is a determining factor for agricultural wages in West Bengal. But these positive relations appear to be spurious, rather than real, because both agricultural wages and agricultural trend outputs have significant trend in them, that may contribute to this significant positive relation between them. One period lag relationship between trend output and wage (estimated to measure the effect of bargaining strength of the labourers on wage determination) could not outweigh this spurious relationship. The fluctuations in production may have some effect on the wages of the agricultural labourers. For example, the sharp decline in agricultural production may have dampening effect upon the wages of agriculture labour whereas the sharp rise in it may have positive effect. On the basis of this presumed relationship the agricultural wages have been regressed upon instability in agricultural production. But, however, no significant relationship between agricultural instability and wages (estimated parameters not reported here) could be established for any district. Further, from the cross section study (which is more reliable here to establish the relationship between wages and agricultural performance) the following regression results are obtained.

Real Wages

$$GRRW_j = 3.046^{**} - 0.0354\ GRO_{2j} - 0.8810\ INS_{2j} \ldots\ldots (7.4)$$
$$(1.828)(-0.097) \qquad (-0.057)$$

$\bar{R}^2 = -0.1651$, $F=0.0082$

Money Wages

$$GRMW_j = 10.488^{*} + 0.301\ GRO_{2j} - 4.0985\ INS_{2j} \ldots\ldots (7.5)$$
$$(6.438)\ (0.084) \qquad (-0.273)$$

$\bar{R}^2 = -0.1595$, $F=0.0373$

[GRRW=growth rate in real wages of male labourers (Table 7.2), GRMW=growth rate in money wages of male labourers(Table 7.1), GRO_2=growth rate in output of all crops combined (Table 4.26) during second sub-period, INS_2=level of instability in the production of all crops combined (Table 5.7) during second sub-period and j denotes the districts]

These two estimated equations [(7.4) and (7.5)] also imply that the trend in agricultural wage does not depend on agricultural performance in West Bengal during LFG regime. The similar result is obtained in case of wages of female labourers also. Replacement of GRO_2 by $GRPCO_2$ (growth rate in per capita agricultural output) in the regression equation does not change the result indicating any positive relation between agricultural wage and per capita agricultural growth. The estimated regression equation with $GRPCO_2$ as an explanatory variable is noted below :

$$GRRW_j = 2.965^{**} + 0.159\ GRPCO_{2j} - 3.931\ INS_{2j}$$
$$(2.250)\ (0.199) \qquad (-0.497)$$

Therefore, it may be concluded that growth in agricultural wages in West Bengal during LFG regime is not determined by the market forces, rather determined by the non-market institutional forces (i.e., by the rising bargaining strength of the agricultural labourers with new political power base). If total labour income could be used in place of wages, the different result might be obtained. However, to examine the operation of trickle down theory ultimately inequality in the rural consumption and rural poverty have been used as explained variable instead of total labour income (due to non-availability of data on it).

The regression of consumption Gini on the land Gini may not be advisable, though there may be theoretical arguments in favour of this, since the data on these variables are not synchronous, i.e., they relate to different time points. However, the effect on consumption Gini of the land Gini with lag of two to four years may be expected to be significant. On this ground the consumption Gini was regressed upon the land Gini but no significant relationship could be established.

$$\text{Cons.Gini} = -0.0414 + 0.4827 \text{ Land Gini} \quad \ldots\ldots\ldots\ldots (7.7)$$
$$(-0.104) \quad (0.836)$$
$$\bar{R}^2 = -0.064, F = 0.699, DW = 1.168^{x}$$

Consumption Share of Bottom 40% Population

$$\text{Share} = 91.324 - 110.326 \text{ Land Gini} \ldots\ldots\ldots\ldots (7.7)$$
$$(0.924) \quad (-0.768)$$
$$\bar{R}^2 = -0.0894, F = 0.59, DW = 1.310^{x}$$

[^{x} implies DW in the inconclusive range].

Consumption Gini coefficient has also not been used as explanatory variable in the regression analysis of rural poverty to avoid multicollinearity problem (discussed in section 3.7, Chapter 3) as it is observed from previous section that consumption Gini has a strong trend effect (and so appears to be correlated with the trend output).

Gini coefficient in rural consumption distribution and rural poverty have been finally regressed separately upon two explanatory variables—3 years lagged moving averages of both trend output of all crops combined and its instability, $|X_t|$ (defined in section 3.5), calculated from sub-periods' kinked exponential trend equation. The regression results are presented below.

Consumption Inequality

$$\text{Cons. Gini}_t = 0.7004^{*} - 0.0863^{**} \text{ Trend Output}_t + 0.2795 \,|X_t| \ldots\ldots (7.8)$$
$$(5.813) \quad (-3.680) \qquad\qquad (0.822)$$
$$\bar{R}^2 = 0.7698, F = 11032^{**}, DW^{s} = 2.004^{+}$$

Poverty

$$HCR_t = 273.0651^{*} - 44.5723^{*}\text{Trend Output}_t + 307.1840^{**}\ |X_t| \quad .... (7.9)$$
$$(4.995)\quad(-4.210)\qquad\qquad\qquad(1.708)$$
$$\bar{R}^2 = 0.8063,\ F = 15.564,\ DW^s = 1.948^{+}$$

[*Notes* : ^{S}correcting autocorrelation through "Cochrane-Orcutt two stage procedure", ** significant at 20% level, t denotes time (1977-78 and onwards), $^{+}$ denotes non-autocorrelation].

From these two estimated regression equations [(7.8) and (7.9)] it is observed that agricultural growth (i.e., trend output) has exercised significantly negative effect upon both consumption inequality and poverty in rural West Bengal during the LFG regime. These negative relations may be considered as the effect of trickle down process of agricultural growth. But these negative relations, as pointed out earlier, may simply crop up due to the trend movements of both independent (i.e., trend output) and dependent [i.e., cons. Gini or rural poverty (HCR)] variables. There may be the other causes that might lead to declining inequality/poverty over time along with the growth of agricultural output. Under this circumstances, the dependence of rural equity on the agricultural performance should be determined jointly from the consideration of both the estimated coefficients of growth and instability.

From equation (7.8) it is further observed that higher instability in agricultural production has led to greater inequality in consumption and *vice versa*; but this relation is statistically insignificant. This means that with higher output instability, the people belonging to different sections in the society do not experience any significant change in their consumption distribution. The poor as well as the rich have maintained their respective positions in respect of consumption pattern. However, the poorers are the worst sufferers. This is evident from the significant positive relationship between output instability and rural poverty [equation (7.9)]. Therefore, it may be concluded that agricultural performance in West Bengal is a determining factor of the incidence of rural poverty. And during LFG regime better agricultural performance (in the forms of higher growth and lower instability in agricultural production) may be considered as the cause of lower incidence of rural poverty.

In this connection it is important to note that in the conventional measure of instability, $| X_t |$ [used as explanatory variable in the equations (7.8) and (7.9)], positive and negative deviations from trend are treated at par. But the negative deviations pose greater problems in more disastrous in achieving equity than the positive deviations. To differentiate the relative importance of the negative deviation from that of the positive deviation, it is desirable that the instability be used in its original value, instead of absolute value. Applying this method the following regression results for consumption Gini and rural poverty, measured by HCR, are obtained.

$$\text{Cons. Gini}_t = \underset{(6.624)}{0.7445^*} - \underset{(-4.246)}{0.0944^*} \text{ Trend Output}_t - \underset{(-0.334)}{0.0801} X_t \quad \ldots\ldots(7.8')$$

And,

$$HCR_t = \underset{(4.804)}{304.5587^*} - \underset{(-4.021)}{50.1022^*} \text{ Trend Output}_t - \underset{(-0.468)}{71.1630} X_t \quad \ldots\ldots(7.9')$$

In both these cases, it is seen that the effect of X_t on either consumption inequality or poverty ratio (HCR) is not statistically significant, indicating thereby that X_t, instability in agricultural output, does not have any effect upon either of them. These statistically insignificant effect may be attributed to smaller variance in the instability measures, which, in turn, results from the three-year lagged moving average (smoothing) of the instability measures.

Perhaps more convincing results would have been obtained if separate regression equations (separately for negative and positive deviations of X_t) could have been fitted to the data sets. But as the number of data points is very small, this exercise could not be pursued. Considering all these aspects, the estimated coefficient of $|X_t|$ may be considered to be more reliable than that of X_t.

7.4 Inequality in Land-Holding, Income and Consumption Distributions : Results from Village Level Studies

In the above section (7.2), the extent of inequality in land holding and consumption distribution, as measured from NSS round data, has been discussed. Certain problems are, however, encountered while using the NSS data for this purpose. Only eight NSS rounds supply us information on land holding distribution of the rural households, spanning over a period from 1953-54 to 1993. The trend results obtained by fitting trend lines to these eight observations on Gini coefficient of land holding distribution may not be very convincing. Moreover, no data are available from NSS rounds on the income distribution of the rural households, whereas, these data are very important, particularly in the context of various income generating programmes, including land reforms, launched by the West Bengal Government. To fill in these gaps in the NSS round data and to make up the deficiencies in our analysis, primary data on land holding, consumption and income distribution of the rural households have been collected through a village level study. Three villages have been selected and the sample households selected from these three villages constitute the sources of primary data. Data relate to the land holding, consumption and income distribution.(The sampling design, types of data collected, etc. are all discussed in section 3.3.2 of Chapter 3).

Inequalities in land holding, income and consumption distribution (for the reference year, 1994-95) have been estimated by measuring Gini coefficients. The results are presented in Table 7.7. From Table 7.7 it is seen that the inequality, measured by Gini coefficient, in the distribution of land holding among the sample households in the less developed village (Village III) is high (0.557), while that in the moderately developed village (Village II) is comparatively less (0.531) and that in the most developed village (Village I) is maximum (0.577). Thus, it is seen that with the agricultural development the inequality in the land holding distribution has declined and then it has again increased quite significantly. The similar patterns of inequality in the distributions of income and consumption are also observed over the sample villages. However, as expected, the extent of inequality in income distribution is less severe than that in the land distribution and further this is less severe in consumption distribution than that in the income distribution in all the sample villages.

If these results can be used with certain amount of confidence for generalization of the situations then Kuznet's 'inverted U-shaped hypothesis' regarding the behaviour of inequality with growth does not find empirical support. One possible explanation for this U-shaped behaviour of inequality (inequality initially reduces and then widens with the subsequent stages of development) is that as the development starts with the extension of less expensive, less sophisticated technology of production, more people including small and medium farmers, even marginal farmers too, can take advantage of the process of development with their own means and wherewithals available from the government and/or other sources and their incomes/consumptions increase. On the other hand, the rich farmers do not reap much from this as these facilities have already been available to them. As a result, the inequality among different classes comes down. As further development is achieved with the application of more sophisticated technology, including mechanization of agriculture and such other methods of production, the medium and rich farmers take full advantage of the condition and engage themselves on a bigger scale in various non-agricultural activities with their already high incomes. The marginal and small farmers can hardly take advantage of this development. As a result, inequality in income and consumption distribution goes up and skewed distribution of land holding is observed, both as the cause and consequence of the formers. If appropriate policy measures are not taken, the land reforms alone will not be able to reduce rural inequality in years to come in West Bengal.

In the following paragraphs an attempt is made to measure the contributions of different sources of income, or income components, to the inequality in income distribution among the rural households. Four income sources such as income from farms (y_1), off-farm income from formal sector (y_2), off-farm income from informal sector (y_3) and income from other sources (y_4) which include income from poultry, dairy, cottage industries, community works, etc. are considered here. Table 7.8 presents the values of Gini coefficients in these income sources and other relevant results.

It is seen from Table 7.8 that Gini coefficients for the distribution of each of the income components are much higher than that for the distribution of the total incomes (Gini coefficient is 0.51). It is because the components of incomes are more unevenly earned by different income groups. A particular component income is concentrated among some particular groups in the villages. The distributions of the income components, y_3 and y_4, are more unequal (Gini coefficients are 0.890 and 0.815 respectively). But the inclusion of y_1 (Gini coefficient is 0.631) and y_2 (Gini coefficient is 0.768) has apparently contributed to the decline in inequality in the distribution of total income of the households in the villages. However, these values of G_k's do not help us to know the contribution of each component to the inequality in the total income distribution. The figures in column 5 measure the contributions of different income sources to the inequality in total income distribution while the figures in col. 6 are the measures of the marginal effects on the inequality in total income distribution.

It is observed from the table (col. 5) that the contribution of farm income (y_1) to the overall income inequality is the highest (I_1 = 0.684), since the farming is the major source of income and this farm income is heavily skewed towards the rich. Though the distribution of off-farm income (y_3) and income from other sources (y_4) are extremely unequal, their contributions to the overall income inequality are marginal (I_3 = 0.0057, I_4 = 0.0507). The contribution of off-farm income (y_2) to the total income inequality is quite significant (I_2 = 0.26).

The decomposition analysis of income distribution by I_k's is only an *expost* accounting. Much more important than this, particularly from the view point of policy prescriptions, is to know how the marginal changes in the size of a particular income source affect income inequality. As discussed in Chapter 3, this can be known from the sign and the magnitude of $I_k - S_k$. From Table 7.8 it is seen that the marginal contributions of the farm incomes (y_1) and off-farm incomes from formal sector (y_2) to the total income inequality are positive whereas the marginal contributions of the other income components y_3 and y_4, are negative.

These results indicate that taxing the farm income (y_1) as well as off-farm income from formal sector (y_2) and augmenting the income from non-formal sector as well as from other sources such as poultry, dairy, cottage industries, etc. would result in substantial reduction in inequality in the total income distribution in West Bengal.

Table 7.1 : Estimated Growth Rates and Levels of Money Wage Rates of Agricultural Male and Female Labourers, West Bengal by Districts, 1978-79 — 1991-92.

District	MaleLabourers		Female Labourers	
	Initial level of money wages rate[a] (in Rs.)	Growth rate in money wage rate[b]	Initial level of money wages rate[a] (in Rs.)	Growth rate in money wage rate[b]
Burdwan	7.08	11.158* (24.462)	6.85	10.852* (25.119)
Birbhum	6.59	10.277* (14.725)	5.57	10.846* (18.287)
Bankura	7.49	9.813* (17.477)	6.94	10.057* (16.476)
Midnapore	7.05	10.545* (27.278)	5.92	11.507* (30.056)
Howrah[S]	7.67	9.585* (32.828)	5.68	10.559* (9.012)
Hooghly	7.28	10.131* (27.046)	7.06	10.336* (14.784)
24-Parganas	7.58	9.997* (13.517)	6.01	11.664* (18.113)
Nadia	5.35	11.972* (18.954)	4.73	12.430* (16.398)
Murshidabad	7.30	9.545* (18.847)	5.92	10.399* (17.004)
West Dinajpur	5.25	9.505* (12.800)	4.14	10.104* (25.454)
Malda	4.34	13.491* (18.335)	3.31	15.171* (26.965)
Jalpaiguri	6.85	10.035* (18.252)	6.10	9.431* (14.103)

(Contd.)

Table 7.1: *(Contd.)*

District	Male Labourers		Female Labourers	
	Initial level of money wage rate (in Rs.)[a]	Growth Rate in money wage rate [b]	Initial level of money wage rate (in Rs.) [a]	Growth rate in money wage rate [b]
Darjeeling	7.25	8.176* (10.368)	6.18	7.851* (8.932)
Coochbehar	5.85	9.661* (20.711)	4.60	9.423* (15.886)
Purulia	5.67	9.606* (11.009)	5.40	10.068* (12.030)

Notes : [a] Initial level of money wage rate implies initial three years' (i.e., 1978-79 — 1980-81) average money wage rate per day per labourer (in Rs.).

[b] growth rate, $\hat{\beta}$ x 100, is calculated by estimating equation :
Ln (Money Wage Rate) = $\alpha + \beta t + u$.

* Significant at 1% level,

t values in the parentheses

[s] Number of observations are small for Howrah district owing to non-availability of data.

Table 7.2 : Estimated Growth Rates in Real Wage Rates of Agricultural Male and Female Labourers and in the Production of All Crops Combined, West Bengal by Districts, 1978-79 — 1991-92

District	Growth Rate of real wage rate[a]		Per capita growth rate in output of all crop (combined) [b]
	Male agricultural labourer	Female agricultural labourer	
Burdwan	3.347* (5.718)	3.409* (5.556)	2.43
Birbhum	2.956* (3.408)	3.618* (4.602)	0.48
Bankura	2.462* (3.582)	2.706* (3.482)	3.12
Midnapore	3.221* (5.198)	4.189* (7.285)	2.97
Howrah[S]	2.257* (4.607)	4.202*** (2.250)	2.70
Hooghly	2.824* (5.523)	2.951* (4.128)	2.83
24-Parganas	2.732** (2.769)	4.336* (4.827)	0.06
Nadia	4.847* (5.791)	5.118* (5.164)	1.70
Murshidabad	2.225* (3.236)	3.093* (3.758)	1.02
West Dinajpur	2.178** (2.653)	2.787* (5.949)	0.53
Malda	6.146* (7.622)	7.869* (9.285)	1.16

(Contd.)

Table 7.2 : *(Contd.)*

District	Growth Rate of real wage rate[a]		Per capita growth rate in output of all crop (combined) [b]
	Male agricultural labourer	Female agricultural labourer	
Jalpaiguri	2.637* (3.631)	2.087* (3.104)	-0.70
Darjeeling	0.992 (1.132)	0.527 (0.504)	1.22
Coochbehar	1.557** (2.175)	2.100* (3.758)	0.16
Purulia	2.265** (2.470)	2.859* (3.278)	1.53

Notes :[a] growth rate $= \hat{\beta} \times 100$, estimating equation:

Ln (real wage ráte) $= \alpha + \beta t + u$.

[b] For per capita output growth rates, see Table 6.14, col. 10 and here time period = 1977-78 — 1994-95.

[s] Number of observations small due to non-availability of data.
*, ** & *** imply significant at 1%, 5% and 10% level respectively.
t values in the parentheses.

Table 7.3 : Inequality in the Distribution of Ownership Land Holding, Rural West Bengal, Different NSS Rounds.

NSS Round	Year	Gini Coefficient
8th	1953-54	0.7290
16th	1961-62	0.6944
26th	1971-72	0.6476
37th	1982	0.6966
43rd	1988	0.6721
50th	1993	0.6844

Table 7.4 : Inequality in the Distribution of Monthly Per Capita Consumer Expenditure, Rural West Bengal, Different N.S.S. Rounds.

NSS Round	Year	Gini Coefficient	NSS Round	Year	Gini Coefficient
8th	1953-54	0.342	40th	1985	0.285
13th	1957-58	0.327	42nd	1987	0.252
16th	1961-62	0.298	43rd	1988	0.262
28th	1973-74	0.312	45th	1989-90	0.256
32nd	1977-78	0.299	50th	1993-94	0.250
38th	1983	0.278			

Table 7.5 : Percentile Distributions of Population and Per Capita Consumer Expenditure, Rural West Bengal, Various N.S.S. Rounds.

N.S.S. Round	Year	Percentile distribution of per capita rural consumer expenditure of		
		Lower 40% of Population	Middle 50% of Population	Upper 10% of Population
8th	1953-54	3.29	78.46	18.25
13th	1957-58	5.38	75.43	19.19
16th	1961-62	7.67	71.89	20.44
28th	1973-74	11.36	64.25	24.39
32nd	1977-78	21.49	54.74	23.77
38th	1983	22.89	57.47	19.63
40th	1985	22.48	55.05	22.47
42nd	1987	24.17	55.81	20.02
43rd	1988	24.69	53.25	21.96
45th	1989-90	24.87	53.58	21.55
50th	1993-94	21.19	57.94	20.87

Table 7.6 : Relationship between Wage Rate[1] and Trend Output[2] in Agriculture, West Bengal by Districts, 1978-79 — 1991-92

District	Estimated Regression[3] Results			
	Money wage rate (Male labourers)		Real wage rate (Male labourers)	
	$\bar{R}^2$	$\hat{b}$	$\bar{R}^2$	$\hat{b}$
Burdwan	0.7465	1.902* (5.779)	0.8795	0.695* (9.016)
Birbhum	0.5616	1.332* (3.885)	0.7759	0.588* (6.252)
Bankura	0.6862	0.969* (4.782)	0.6095	0.297* (4.075)
Midnapore	0.7589	1.194* (5.969)	0.7320	0.417* (5.571)
Howrah	0.8605	1.201* (7.096)	0.7308	0.303* (4.766)
Hooghly	0.9050	1.674* (10.739)	0.6958	0.480* (5.333)
24-Parganas	0.5614	1.384* (4.044)	0.4152	0.505* (3.085)
Nadia	0.9384	1.738* (12.983)	0.8029	0.738* (6.767)
Murshidabad	0.7809	1.707* (6.880)	0.5481	0.490* (4.095)
West Dinajpur	0.8122	1.949* (6.969)	0.5450	0.567* (3.765)
Malda	0.8485	2.479* (7.913)	0.7354	1.131* (5.619)
Jalpaiguri	0.5996	3.957* (4.355)	0.3117	1.058* (2.853)

(Contd.)

Table 7.6 : *(Contd.)*

District	Estimated Regression[3] Results			
	Money wage rate (Male labourers)		Real wage rate (Male labourers)	
	$\bar{R}^2$	$\hat{b}$	$\bar{R}^2$	$\hat{b}$
Darjeeling	0.7219	1.901* (5.670)	0.0544	0.288 (1.300)
Coochbehar	0.7227	2.174* (5.906)	0.2327	0.407** (2.223)
Purulia	0.3793	0.763** (2.779)	0.2838	0.251** (2.315)

Notes :[1] wage rate implies money / real wage rate of agricultural male labourers.

[2] Trend output implies predicted value calculated from kinked trend equation fitted on all crops (combined) of the district.

[3] wage rate has been regressed against trend output of one period lag, i.e., $\text{Ln (wage rate)}_t = a + b \text{ Ln (trend output)}_{t-1}$.

t values in the parentheses.

* significant at 1% level,

** significant at 5% level.

Table 7.7 : Measures of Inequality in the Distributions of Land Holding, Income and Consumption of Rural Households in Sample Villages of West Bengal, 1994-95

Sample Village	Number of sample households	Gini Coefficients[+] in the distribution of		
		Land	Income	Consumption
Sample Village I[a]	50	0.5770 (0.5634)	0.4374 (0.4288)	0.3220 (0.3132)
Sample Village II[b]	53	0.5308 (0.5170)	0.4108 (0.4018)	0.3045 (0.2954)
Sample Village III[c]	50	0.5572 (0.5455)	0.5143 (0.5051)	0.4169 (0.4089)
All Villages	153	0.5682 (0.5569)	0.4641 (0.4562)	0.3556 (0.3476)

Notes : [a] agriculturally developed village

[b] agriculturally moderately developed village

[c] agriculturally less developed village

[+] implies true Gini Coefficients (explained in section 3.7 of Chapter 3) Gini lower bounds are in the parentheses.

Table 7.8 : Decomposition[+] of Income Inequality by Income Sources of Rural Households, All Sample Villages Combined, West Bengal, 1994-95

Income sources, y_k	Gini [a] Coefficients, G_k	Share of income sources, S_k	Gini Correlation, R_k	Contribution of income source to income inequality, I_k	Marginal effects of income sources, $(I_k - S_k)$
(1)	(2)	(3)	(4)	(5)	(6)
y_1	0.6307	0.5962	0.8440	0.6838	0.0876
y_2	0.7684	0.2340	0.6706	0.2598	0.0258
y_3	0.8902	0.0242	0.1228	0.0057	-0.0185
y_4	0.8153	0.1456	0.1982	0.0507	-0.0949

Notes :[+] Decomposition is done using the formulae, developed by Lerman and Yitzhaki (1985) [explained in section 3.7, Chapter 3].

Gini Coefficients, as calculated here, consider inequality within and among classes [explained in Section 3.7, Chapter 3].

8

Summary and Conclusions

In this study we have tried to examine the performance of West Bengal agriculture during LFG regime (i.e., 1977-78—1994-95) *vis-à-vis* the pre-LFG regime (i.e., 1957-58—1976-77) in terms of agricultural growth and stability and to assess the contributions of different factors to the overall agricultural growth and stability. An attempt is also made here to study the effect of agricultural growth and stability on rural equity in West Bengal during LFG regime.

From the literature survey in Chapter 2 it has been observed that the earlier studies in this field, both at the National and at the State levels, suffer from various methodological weaknesses in measuring agricultural growth and stability and also in measuring their overall impact on the rural equity. Moreover, it has also been observed that there have been general disagreements among the scholars on the estimated results on agricultural growth, stability and equity in West Bengal, particularly during LFG regime. This study is a modest attempt to improve the methodological weaknesses of the earlier studies and to assess the agricultural performance of the State properly by taking relatively longer time period (1957-58—1994-95) and carrying out disaggregation, whenever possible. This study, however, like other studies in the same area cannot avoid some inherent limitations and restrictions that are imposed by the secondary level data base.

In this last chapter the major findings of the study (derived from the statistical and econometric analysis of the data) have been summarized and conclusions have been drawn.

Findings on Growth

From the whole period's (1957-58—1994-95) estimated growth rates in agricultural production it has been observed that (i) the crops that experienced higher growth rates than others are sesamum, mustard, potatoes, wheat, rice and jute and (ii) the districts having better growth performance for all crop groups—total rice, total foodgrains, total non-foodgrains and all crops combined—are Burdawn, Midnapore, Howrah, Hooghly, Nadia and Murshidabad.

As regards the estimated sub-periods' growth rates it is observed that during the LFG regime (in comparison to the pre-LFG regime) the impressive growth rates of only two crops, namely, aman rice and mustard, lead to significant positive trend breaks in the major crop-groups of West Bengal. The growth rates actually decline in the cases of boro rice, wheat, barley, all pulses, linseed, sesamum, jute, sugercane and tobacco. From a district-wise analysis it is seen that only two districts—Bankura and Midnapore—have recorded steady and significant (in the statistical sense) growth of all the four crop- groups during the second sub-period (compared with the first sub-period). The higher growth rates of total agricultural production in the districts of Burdwan, 24-Parganas, Malda and Purulia during the second sub-period have their origins in the significant growth in the total foodgrains production. On the other hand, the significant growth in the total non-foodgrains production contributed to the higher growth in the total agricultural production during the second sub-period in the districts of Nadia and West Dinajpur. In all other districts, the growth rates in the total agricultural production remained statistically unchanged between two sub-periods.

From the decomposition of output growth rate into growth rates of area and productivity, with crop-wise and district-wise disaggregation, it is observed that productivity growth has been the main source of growth of total agricultural production in West Bengal in the whole period of 1957-58—1994-95 as well as in the sub-periods. However, only in case of non-foodgrains, the higher growth in area was the main contributor to the output growth.

As regards the individual crop's contribution to the overall agricultural output growth in West Bengal, it can be said that the crops which gained increasing significance over time were aman rice, mustard, sesamum and potatoes. The positions of pulses and tobacco remained unchanged and the positions deteriorated for all other individual crops between the two sub-periods. In respect of the districts' shares in overall State's agricultural output growth it is observed that over time the central and northern districts' shares declined while the shares of southern and western districts increased.

Findings on Stability

From the measurement of the nature of instability in crop-output it is observed that, for 52 crops out of total 82 crops and crop-groups, the change of instability over time was statistically insignificant. Again, for 25 crops and crop-groups out of the remaining 30 crops and crop-groups, the instability declined significantly. Only in 5 cases [namely, three crops—gram, sesamum and sugarcane—at the State level and total foodgrains of two districts—Malda and Darjeeling] instability increased significantly over time. So, on the whole, the instability in agricultural production in West Bengal has either declined significantly or remained unchanged during LFG regime, compared with pre-LFG period. However, in respect of most of the major crop-groups in the State and in the production of all crops combined in three districts—Howrah, Hooghly and Nadia—the instability declined significantly.

In the State, levels of instability are seen to be higher in case of individual crops than in the case of crop-groups and within individual crops the levels of instability are seen to be higher in the case of wheat, boro rice, other cereals (i.e., jowar, bajra, etc.) and oil seeds than in other crops. As regards the instability of crop-groups in the districts, the magnitude of instability is seen to be lower for the total agricultural production than for other crop-groups and it is higher for total non-foodgrains production. Further, the levels of instability are generally lower in the northern districts and they are always high for all crop-groups in three districts of Bankura, Howrah and Purulia.

The decomposition of output instability reveals that in the case of individual crops in the State, variability in area plays a significant role to determine output instability during the first sub-period. And during the second sub-period both area and productivity variabilities are significant sources of output instability. But in case of 'all crops' in the districts, all the three components—area and productivity variabilities and their synchronous movements—are equally powerful in explaining output instability for all the periods (except, during second sub-period the contribution of area component was insignificant). Further, over the sub-periods the relative strength of the components in determining the instability in the total agricultural production in the districts remain unchanged.

In order to ascertain the relation between growth and instability, correlation coefficients between growth and instability have been estimated cross sectionally over crops for the State and over districts for each crop-group. These coefficients do not indicate any significant relationship between growth and instability in any case for any period, except for (i) individual crops of the State for the whole period and for the first sub-period and (ii) total non-foodgrains of the districts for the first sub-period [in these exceptional cases higher growth is associated with higher instability]. No definite relationship could also be observed from the cross sectionally measured correlation coefficients between change of growth rates and change of instabilities. But from the tabular exposition it is observed that for most of the crop groups in the State, change of growth rates between two sub-periods is negatively associated with the change of instability over time. But in the case of the districts no such relationship is observed for any of the major crop groups.

Findings on the Factors Determining Growth and Stability

From the time series analysis of the total agricultural production in West Bengal it is observed that the contribution of the composite technological factors to the land productivity during the LFG rule is positive and significant. The cross section study also reveals significant positive joint effect of the technological factors on the land productivity in West Bengal agriculture for all three time points, i.e., 1970-71, 1980-81 and 1990-91. But in case of institutional factors, the results do not indicate any definite pattern. Though the institutional factors jointly affect the land productivity of the districts significantly and positively in the year 1970-71 and 1980-81, their joint effect on the land productivity, as measured for the year 1990, has been statistically insignificant. Further, it is observed that for none of these years there is any significant relationship between farm size and productivity in West Bengal agriculture.

From *apriori* theoretical considerations the variables such as irrigation, crop base, locational base, etc., have been chosen to measure instability in agricultural output. From time series regression analysis it is seen that these factors could sufficiently explain output instability only during the second sub-period. These factors did not have any significant effect upon the instability in West Bengal agriculture during the first sub-period. Thus, it appears that in the years of the first sub-period rainfall was a major factor contributing to instability in agricultural production in the State. From time series as well as the cross sectional regression analyses one can draw the following conclusion:

(i) over time only squeezing crop base and more irrigation from assured sources have exerted stabilizing effect upon agricultural production in West Bengal.

(ii) From the cross sectional analysis it is seen that the crop-base and the holding-size have negative effects on the output stability in the early years (1970-71 and 1980-81). But during the later year (1990-91) these factors have exercised positive effect upon output stability.

(iii) The current fallows, which are normally marginal lands without irrigation facilities, had positive effect upon output instability for the years 1970-71 and 1980-81. But with the expansion of irrigation in the later years, this factor is found to have no significant effect on output instability in later year, 1990-91. But in the time series analysis (for the State as a whole) it, being highly negatively collinear with irrigation, only expresses indirectly the extent of area under irrigation and, because of the presence of this collinearity between irrigation and current fallow, the effect of the current fallow could not be separated out.

Contrary to the earlier findings of insignificant influence of rainfall on the recent agricultural production in West Bengal, it has been observed (applying Durbin's ranking method) that West Bengal agriculture has been significantly dependent on rainfall for all the periods. More specifically, it has been observed that variations in rainfall have led to variations in agricultural production in West Bengal during the first sub-period as well as the second sub-period.

In order to measure the effect of demographic factor on agricultural production, both the agricultural output growth rate and per capita output growth rate of the districts have been regressed upon different lagged values of rural population growth rate. The regression results reveal that the impact of the population growth rate on the output growth rate, both total and per capita, in West Bengal has been felt most strongly after a lag of approximately 25 years. The regression results indicate that population growth in West Bengal has had a significant positive effect upon agricultural growth rate in both total output and per capita during the whole period (1957-58 — 1994-95). This empirical finding supports the induced innovation theory of population in West Bengal agriculture. However, from the period-wise analysis of demographic effect, it is seen that this significant positive effect (supporting the operation of the induced innovation theory of population) is strongly felt in the first sub-period, while in the second sub-period there is actually negative effect of rural population growth on the growth in total agricultural output as well as per capita output in West Bengal. This negative effect of the growth of population on the output growth testifies, on the other hand, to the operation of Neo-Malthusian theory of population in West Bengal agricultural. This means that the West Bengal agriculture during the second sub-period has been hard pressed by the rapid growth in population in the State. This finding hints at the fact that if the population growth rate is checked, the West Bengal agriculture can register higher growth rates than that achieved so far.

Findings on Equity

From the estimated trend growth rates of agricultural wage rates in the districts it is seen that the agricultural wage (both money and real) rates for both male and female in all the districts of West Bengal have increased very significantly during the LFG regime. Further, it is observed that (i) the districts with lower base level of wages are, in general, experiencing higher growth in wages in relation to the districts with higher wage base and (ii) the growth of female labourers' wages with lower base is higher than the wages of their male counterparts. These lend support to the fact that during LFG regime in West Bengal not only agricultural wages have gone up appreciably but also the regional and gender disparities in wages have narrowed down.

Equity in rural West Bengal has also been measured in terms of Gini coefficients in the distributions of land holding and rural consumption and also in terms of the incidence of rural poverty in West Bengal. From the estimated results it has been observed that (i) there is no declining trend in the Gini coefficients of land distribution, (ii) the inequality in rural consumption and the incidence of rural poverty have, however, shown significant declining trends in West Bengal during LFG regime. The inequality in rural consumption in West Bengal has declined owing to the significantly increasing share of the lower 40%, along with the decreasing share of the middle 50% of the rural population. However, the analysis of rural consumption data shows that the share of the top 10% of the rural population in consumption has remained unchanged. Thus, our results show that whatever decline in inequality has been achieved in respect of rural consumption during the Left Front regime in West Bengal, it has been achieved through redistribution of consumption between the lower 40 percent and middle 50 percent of rural population. There has been no change in the position of the top ten percent of rural population.

To enquire into the operation of trickle down process in West Bengal agriculture, the agricultural wages, Gini coefficients of rural consumption distribution and rural poverty have been regressed upon the trend values of agricultural output and the values of instability in agricultural production during LFG regime. From the estimated regression results it is observed that all these dependent variables are significantly affected by the trend value of agricultural output, implying thereby that the higher value of trend output has led to higher agricultural wages, lower inequality in rural consumption and lower incidence of rural poverty in West Bengal during the LFG regime. But as all the dependent variables have strong trend movement, the above relations should not be taken seriously. To determine the actual dependence of these dependent variables on the agricultural performance, the estimated coefficients of instability statistic should be considered. But, as is observed, instability in agricultural production neither affects agricultural wages nor consumption inequality. The dependence of agricultural wages on agricultural performance is also not established from the cross section study. However, in case of rural poverty the results are different. It is observed that better performance in agriculture has contributed to significant decline in rural poverty in West Bengal during LFG regime.

The above study on the extent of inequality in the distribution of land holding and rural consumption is based on the data collected from different NSS rounds. These findings relate to changes in inequality over time. As this problem of inequality or its counterpart equity assumes special significance against the avowed objective of the government to wipe off inequality between rich and poor, a study at the micro level has also been conducted in order to have greater insight into the problem. The primary survey of three villages (at different stages of agricultural development) have been conducted for measuring rural equity. It is found that the inequality in the distribution of land holding is higher than that in income, which is again higher than that in consumption. Further, with the development of agriculture, the inequality in land distribution first decreases and then increases; the similar pattern is also observed in the cases of inequalities in income and consumption distributions. By decomposing incomes inequality it is observed that farm incomes along with the rural incomes from organized sectors are the major contributors to the income inequality, whereas incomes from informal sources and from cottage industries, dairy, poultry, public works, etc. exert negative influence on income inequality.

Conclusions

From the above mentioned findings of the study it can be concluded that during LFG regime (compared with pre-LFG regime) agriculture in West Bengal has recorded higher growth rate (not so high as was established by some of the earlier studies). It is further observed that during this period the instability in the agricultural production has either declined or remained unchanged in quite a number of cases. There has been, however, increase in instability in the agricultural production of some crops and in some districts. Higher equity has also been achieved in respect of agricultural wages, rural consumption and through the reduction in the incidence of rural poverty. But this situation is different in the case of land holding distribution. This higher growth in agricultural production, whether in the State or in the districts, has, however, been achieved by squeezing the crop base and locational base. This means that only a few crops have dominated the field of the farmers by elbowing out other crops, like high protein pulses, wheat, etc.

It is further observed from our study that West Bengal agriculture still depends on rainfall, which many of the earlier studies failed to establish. Good rainfall in successive few years during the Left Front Government rule has significantly contributed to the higher agricultural production in the State. Further, contrary to the popular belief that institutional changes have been responsible for this increase in agricultural growth in the State, it is found that recent agricultural improvement has taken place only because of the use of various technological inputs. Moreover, it has been established that rural population growth has detrimental effect upon the recent growth in agricultural production in West Bengal. Therefore, for sustainable growth in both productivity and production, provisions have to be made for steady supply of non-land inputs and also for widening the agricultural base. To restore stability along with growth, policies have to be framed for balanced expansion of irrigation from assured sources. Along with these, growth in rural population is to be checked and the number of off-farm income sources have to be increased for achieving rural equity with growth and stability. Otherwise, the whole process of recent agricultural development in West Bengal would be thwarted and the stagnation in agriculture may set in.

Bibliography

Ahluwalia, M.S. (1978), 'Rural Poverty and Agricultural Performance in India', *Journal of Development Studies (JDS)*, Vol. 14, No.3.

__________ (1986), 'Rural Poverty, Agricultural Production and Prices : A Reexamination' in J.W. Mellor and G.M. Desai (eds).

Alagh, Y.K. and Sharma, P.S. (1980), 'Growth of Crop Production : 1960-61 to 1978-79—Is It Decelerating?', *Indian Journal of Agricultural Economics (IJAE)*, Vol. 35, April-June.

Bandyopadhyay, A. (1989), 'Growth And Instability In The Production of Main Cereal Crops of West Bengal And Punjab-Haryana, 1950-51 to 1984-85—A Note', *IJAE*, Vol. 44, No. 2.

Bandyopadhyay, N. (1981), 'Operation Barga and Land Reforms Perspective in West Bengal : A Discursive Review', *Economic and Political Weekly (EPW)*, Vol. 16, No. 25-26.

__________ (1988), 'The Story of Land Reforms in Indian Planning', in A.K. Bagchi (ed), *Economy, Society, Polity etc.*, OUP, Calcutta.

Bansil, P.C. (1974), *Agricultural Statistics in India (2nd edn)*, Arnold Heinemann, New Delhi.

Bardhan, P.K. (1973), 'Size, Productivity and Returns to Scale : An Analysis of Farm-level Data in Indian Agriculture', *Journal of Political Economy*, Vol. 81, No. 6.

__________ (1984), *Land, Labour and Rural Poverty : Essays in Development Economics*, CUP, New York.

__________ (1986), 'Poverty and "Trickle-Down" in Rural India: A Quantitative Analysis', in Mellor and Desai (eds), OUP, Delhi.

Barker, R.,Gabler, E.C., and Winkelman, D. (1981), 'Long-term Consequences of Technological Change on Crop Yield Stability', in Valdes, A.(ed), *Food Security for Developing Countries*, Westview Press, Boulder Co.

Bhalla, G. S. and Alagh, Y. K. (1979), *Performance of Indian Agriculture: A District-wise Study*, Sterling, New Delhi.

_________ and Chadha, G.K. (1982), 'Green Revolution and the Small Peasant : A Study of Income Distribution in Punjab Agriculture', *EPW*, Vol. 17, No. 20-21.

_________ and Singh, G. (1997), 'Recent Developments in Indian Agriculture : A State Level Analysis', *EPW*, March 29.

_________ and Tyagi, D.S. (1989), 'Spatial Pattern of Agricultural Development in India', *EPW*, June-24.

Bhattacharya, N., Chattopadhyay, M., and Rudra, A. (1987-88) 'Changes in Level of Living in Rural West Bengal etc.', *EPW*, July 11, Aug. 15, Sept. 5-12, Oct. 31, Nov. 28 and May ('88) 28.

Blyn, G. (1966), *Agricultural Trends in India, 1891-1947 : Output, Availability and Productivity,* University of Pennsylvannia Press, Philadelphia.

Boyce, J.K. (1987), *Agrarian Impasse in Bengal : Institutional Constraints to Technological Change*, OUP, New York.

_________ (1984), 'Agricultural Growth in West Bengal: A Review of the Evidence', *EPW,* Vol.19, No. 13.

Chattopadhyay, M. and Bhattacharya, G. (1986), 'Growth of Indian Agriculture : A Reappraisal', *IJAE*, Vol. 42, No. 1.

Chinn, D.S.L. (1979), 'Rural Poverty and the Structure of Farm Household Income in Developing Countries: Evidence from Taiwan', *Economic Development and Cultural Change (EDCC),* Vol. 27.

Cleaver, H.M. (1972), 'The Contradiction of Green Revolution', *American Economic Review (AER)*,Vol. 72.

Centre for Monitoring Indian Economy (CMIE) (1993), 'Performance of Agriculture in Major States, 1967-68 to 1991-92', Bombay.

Coale, A.J. and Hoover, E.M. (1958), *Population Growth and Economic Development in Low-income Countries : A Case Study of India's Prospects,* Princeton University Press, New Jersey.

Dandekar, V.M. (1980), 'Introduction', *IJAE,* Vol. 35, April-June (No.2).

________ and Venkataramaiah, P. (eds) (1975), *Data Base of Indian Economy, Volume II: Role of Sample Survey*, Statistical Publishing Society, Calcutta.

Dantwala, M.L. (1987), 'Growth and Equity in Agriculture', *IJAE,* Vol. 42, No. 2.

Das, P.S. (1978), 'Growth and Instability in Crop Output in Eastern India', *EPW,* Vol. 13, No. 41.

________ (1981), 'Patterns of Growth in Crop Output in Eastern India', *Unpublished Research Note*, University of Calcutta, Calcutta.

Dasgupta, B. (1995), 'Institutional Reforms and Poverty Alleviation in West Bengal', EPW, Oct. 14 & 21.

Datta, A. (1983), *Growth and Equity : A critique of the Lewis-Kuznets Tradition*, OUP, Calcutta.

Datta, P.K. (1981), 'Statistics of Bargadars and Extent of Barga Cultivation in West Bengal: An Analytical Study', *Directorate of Land Records and Surveys*, G.O.W.B. , Mimeo, Calcutta.

Datta Ray, S. (1994a), 'Agricultural Growth in West Bengal, *EPW*, Vol. 29, No. 29, July.

________ (1994b), 'Growth of Agriculture in West Bengal', *EPW*, November.

Dayal, R. (1966), 'Agricultural Growth Rates and their Components', *IJAE*, Vol. 21, No. 4.

Desai, G.M. and Singh, G. (1973), 'Growth of Fertilizer Use in Districts in India: Performance and Policy Implications', *Indian Institute of Management*, Ahmedabad.

Dey, A.K. (1975), 'Rates of Growth in Agriculture and Industry', *EPW*, Vol. 10, No. 25-26.

Dhawan, B.D. (1983), 'Impact of Irrigation on Instability in Farm Output: A Case Study of Tamil Nadu State', *EPW*, Vol. 18, No. 39.

Dovring, F. (1959), 'The Share of Agriculture in a Growing Population', *Monthly Bulletin of Agricultural Economics and Statistics*, Vol. 8.

Dreze, J. and Sen, A. (1989), *Hunger and Public Action*, OUP, Oxford.

Frankel, F.R. (1971), *India's Green Revolution : Economic Gains and Political Costs*, Princeton University Press, Princeton.

Gastwirth, J.L. (1972), 'The Estimation of the Lorenz Curve and Gini Index', *Review of Economics and Statistics*, Vol. 54.

_______ and Smith, J.T. (1972), 'A New Goodness of Fit Test', Proceedings of the *American Statistical Association*, Pp. 320-22.

Gazder, H. (1992), 'Rural Poverty, Public Policy and Social Change: Some Findings from Surveys of Six Villages', *WIDER Working Paper*, No. 98.

Ghosh, R. and Nagaraj, K. (1978), 'Land Reforms in West Bengal', *Social Scientist*, 66-67, Pp. 50-67.

Grabowski, R. (1979), 'The Implication of an Induced Innovation Model', *EDCC*, Vol. 27.

Griffin, K. (1974), *The Political Economy of Agrarian Change: An essay on the green revolution*, Harvard University Press, Cambridge.

_______ and Ghosh, A.K. (1979), 'Growth and Impoverishment in the Rural Areas of Asia', *World Development*, Vol. 7, No. 4-5.

Harriss, J. (1983), 'Making out on Limited Resources, or What Happened to Semi-Feudalism in West Bengal?' *Center for Regional Ecological and Science Studies in Development Alternatives*, Transactions, Vol. 2.

___________ (1992), 'Does the "Depressor" Still Work? Agrarian Structure and Development in India — A Review of Evidence and Arguments', *Journal of Peasant Studies*, Vol. 19. No. 2.

___________ (1993), 'What is Happening in Rural West Bengal? Agrarian Reform, Growth and Distribution', *EPW*, June 12.

Hayami, Y. (1980), 'Induced Innovation, Green Revolution and Income Distribution: Reply', *EDCC*, Vol. 30.

Hazell, P.B.R. (1982), 'Instability in Indian Foodgrains Production', *International Food Policy Research Institute (IFPRI)*, Research Report No. 30, May.

___________ (1984), 'Sources of Increased Instability in Indian and U.S. Cereal Production', *American Agricultural Economic Association,* March.

Jamal, H. and Zaman, A. (1992), 'Decomposition of Growth Trend in Agriculture: Another Approach', *IJAE*, Vol. 47, No. 4.

Johl, S.S. (1975), 'Gains of the Green Revolution: How they have been shared in Punjab', *JDS*, Vol. 9, No. 2.

___________ (1995), 'Agricultural Sector and New Economic Policy', *IJAE*, Conference Number, Vol. L, No. 3.

Johnston, J. (1972), *Econometric Methods*, 2nd Ed., McGraw Hill Kogakusha Ltd., Tokyo.

Johnston, B.F. and Kilby, P. (1975), *Agricultural and Structural Transformation*, OUP, Oxford.

___________ and Mellor, J.W. (1961), 'The Role of Agriculture in Economic Development', *AER*, Sept.

Jorgenson, D. (1966), 'Testing Alternative Theories of the Development of a Dual Economy', in Adelman, I. and Thorbecke, (eds), *The Theory and Design of Economic Development*, Johns Hopkins Press, Baltimore.

Jose, A.V. (1988), 'Agricultural Wages in India', *EPW*, Review of Agriculture, June 25.

Judge, G. *et al.* (1980), *The Theory and Practice of Econometrics*, John Wiley, New York.

Kada, R. (1983), 'Changing Rural Employment Pattern: Role of Off-Farm Employment for Balanced Rural Development', in Maunder and Ohkawa (eds).

Kakwani, N.C. (1980), *Income Inequality and poverty: Method of Estimation and Policy Applications*, OUP, New York.

Khasnabis, R. (1994), 'Tenurial Conditions in West Bengal: Continuity and Change', *EPW*, Dec.31.

_________ and Chakraborty, J. (1982), 'Tenancy, Credit and Agrarian Backwardness', *EPW*, Vol.17, No.13.

Kmenta, J. (1971), *Elements of Econometrics*, Macmillan Publishing Company, New York.

Kohli, A. (1987), *State of Poverty in India*, Cambridge University Press (CUP), Cambridge.

Koutsoyiannis, A. (1977), *Theory of Econometrics: An Introductory Exposition of Econometric Methods*, ELBS, Macmillan, Hampshire.

Krishnaji, N. (1980), 'Measuring Agricultural Growth', *IJAE*, Vol. 35, No. 2.

Kuznet, S. (1995), 'Economic Growth and Income Inequality', *AER*, March.

__________ (1961), 'Economic Growth and the Contribution of Agriculture: Notes on Measurement', *International Journal of Agrarian Affairs*, Vol. 3.

Leaf, M.J. (1983), 'The Green Revolution in a Punjab Village, 1965-78', *EDCC*, April.

Learman, R.I. and Yitzhaki, S. (1985), 'Income Inequality Effects by Income Sources: A New Approach and Applications to the United States', *The Review of Economics and Statistics*, Vol. LXVII, No. 1.

Lewis, W.A. (1954), 'Economic Development with Unlimited Supplies of Labour', *The Manchester School*, Vol. 22.

Lieten, G.K. (1990), 'Depeasantisation Discontinued: Land Reforms in West Bengal', *EPW*, Vol. 25, No. 40.

__________ (1992), *Continuity and Change in Rural West Bengal*, Sage Publications, Delhi.

Maddala, G.S. (1988), *Econometrics*, 5th Printing, McGraw Hill International Editions, Economic Series, Singapore.

Mahalanobis, P.C. (1946), 'Recent Experiments in Statistical Sampling in the Indian Statistical Institute', *Journal of the Royal Statistical Society*, 109, Part-4.

Mahendradev, S. (1987), 'Growth and Instability in Foodgrains Production: An Inter-State Analysis', *EPW*, Sept. 26.

Maitra, T. and Lahiri, R.K. (1975), 'Land Utilisation and Crop Yield Estimation', in Dandekar and Venkataramaiah (eds).

Makridakis, S. *et al.* (1983), *Forecasting: Methods and Applications*, 2nd Ed., John Wiley, New York.

Mallick, R. (1992), 'Agrarian Reform in West Bengal : The End of an Illusion', *World Development*, Vol. 20, No. 5.

Marten, G.C. (1988), 'Productivity, Stability, Sustainability, Equitability and Autonomy as Properties for Agroecosystem Assessment', *Agricultural Systems*, Elsevier Applied Science Publishers Ltd., England.

Maunder, A. and Ohkawa, K. (eds.), (1983), *Growth and Equity in Agricultural Development*, Gower Publishing Co., Aldershort.

Mehra, S. (1981), 'Instability in Indian Agriculture in the Context of the New Technology', *IFPRI*, Research Report No. 24, July, Washington.

Mellor, J.W. (1966), *The Economics of Agricultural Development*, Cornell University Press, Ithaca.

__________ and Desai, G. M. (eds.) (1986), *Agricultural Change and Rural Poverty : Variations on a Theme by Dharm Narain*, OUP, Delhi.

Minhas, B. S. (1966), 'Rapporteur's Report on Measurement of Agricultural Growth', *IJAE*, Vol. 21, No. 4.

__________ and Vaidyanathan, A. (1965), 'Growth of Crop Output in India, 1951-54 to1958-61: An Analysis by Component Elements', *Journal of the Indian Society of Agricultural Statistics*, Vol.17, No. 2.

Ministry of Food & Agriculture (1966), 'Growth Rates in Agriculture' (1949-50 to 1964-65), *Economic and Statistical Adviser, Directorate of Economics and Statistics*, New Delhi.

Mishra, S.N. (1997), 'Agricultural Liberalisation and Development Strategy in Ninth plan', *EPW*, Vol. 32, No.13.

Morgan, J. (1962), 'The Anatomy of Income Distribution', *Review of Economics and Statistics*, Vol. 44.

Mukherjee, S.N. (1956), *'A Brief Agricultural Geography of West Bengal'*, G.O.W.B., Calcutta.

Mukherjee, B. and Mukhopadhyay, S. (1994), 'Impact of Institutional Change on Productivity in a Small Economy: The Case of Rural West Bengal', Paper Presented at a Seminar on Agricultural Development in Eastern India, *Nabakrushna Choudhury Centre for Development Studies*, Bhubaneshwar, January.

Narain, D. (1977), 'Growth of Productivity in Indian Agriculture', *IJAE*, Vol. 32, No.1, Jan-March.

__________ (1979), 'Unpublished Notes', in Mellor and Desai (eds.), OUP, Delhi, 1986.

Nayyar, R. (1991), *Rural Poverty in India*, OUP, Delhi.

Neale, W.C. (1978), 'Twenty Years of Community Development: Progress or Ripof', *Seventh Wisconsin Conference on South Asia*, Madison, Wisconsin.

Nicholls, W.H. (1963), 'An Agricultural Surplus as a Factor in Economic Development', *Journal of Political Economy*, February.

Ninan, K.N. (1994), 'Poverty and Income Distribution in India', *EPW*, June 18.

Parikh, A. (1966), 'Statewise Growth Rate in Agricultural Output: An Econometric Analysis', *Artha Vijnana*, Vol. 8, No.1.

Parik, K.and Narayana, S. (1995), 'Strategies for Agricultural Liberalisation, Consequences for Growth, Welfare and Distribution', *EPW*, Vol. 30, No. 29.

Parthasarathy, G. (1988), 'Agricultural Growth, Rural Development and Poverty', Selected Writings, Ch. 2 (Part 1), *Dept. of Co-operation and Applied Economics, Andhra University*, Visakhapatnam.

Paul, S. (1989), 'Green Revolution and Income Distribution Among Farm Families in Haryana, 1969-70 to 1982-83', *EPW*, Dec. 23-30.

__________ (1990), 'Green Revolution and Poverty Among Farm Families in Haryana, 1969-70—1982-83', *EPW*, Sept. 29.

Pindyck, R.S. and Rubinfeld, D.L. (1981), *Econometric Models and Economic Forecasts*, International Student Edition, McGraw-Hill, Singapore.

Pray, C.E. (1981), 'The Green Revolution as a case Study in Transfer of Technology', *Annals of the American Academy of Political and Social Science*.

Rajaraman, I. (1975), 'Poverty, Inequality and Economic Growth: Rural Punjab—1960-61—1970-71', *JDS*, July.

Ranis, G. and Fei, J.C. (1961), 'A Theory of Economic Development', *AER*, Vol.51.

Rao, C.H.H. (1975), *Technological Change and the Distribution of Gains in Indian Agriculture*, Macmillan, New Delhi.

__________ (1986), 'Changes in Rural Poverty in India: Implications for Agricultural Growth', Mainstream, January 11.

__________ (1992), 'Agricultural Policy and Performance', in Jalan, B.(ed), *The Indian Economy: Problems and Prospects*, Viking, Delhi.

__________ (1995), 'Liberalisation of Agriculture in India: Some Major Issues', *IJAE*, Vol.L, No.3, Conference Number, July-Sept.

__________, Gupta, D.B. and Sharma, P.S. (1986), 'Infrastructural Development and Rural Poverty in India: A Cross-sectional Analysis', in Mellor and Desai(ed), OUP, Delhi.

__________, Ray, S.K. and Subbarao, K.(1988), *Unstable Agriculture and Droughts: Implications For Policy*, Institute of Economic Growth, Vikas Publishing House Pvt. Ltd., Delhi.

Rao, V.M., Nadkarni, M.V. and Deshpande, R.S. (1980), 'Measurement of Growth and Fluctuations in Crop Output: An Approach Based on the Concept of Non-Systematic Component', *IJAE*, Vol. 35, April - June.

Rath, N. (1980), 'A Note on Agricultural Productivity in India during 1955-78', *IJAE*, Vol. 35 No. 2.

Ray, S.K. (1983), *Growth and Instability in Indian Agriculture, Institute of Economic Growth*, June, Mimeo, Delhi.

Reddy, V.N. (1977), *Statistical Fitting of Growth Curves with Illustrations from Data on Indian Economy*, Indian Institute of Management, Mimeo, Calcutta.

__________ (1978), 'Growth Rates', *EPW*, Vol.13, No.19.

Rogaly, B., Harris-white, B. and Bose, S. (1995), 'Sonar Bangla? Agricultural Growth and Agrarian Change in West Bengal and Bangladesh', *EPW*, Vol. 30, No. 29, July 22.

Rudra, A. (1970), 'The Rate of Growth of the Indian Economy', in Robinson and Kidron (eds), *Economic Development in South Asia*, Macmillan, London.

__________ (1978), 'Class Relations in Indian Agriculture', *EPW*, Vol.13, June 3, 10 and 17.

__________ (1981), 'The Self-contained Village Society II', *Frontier*, Vol.13, No. 20.

__________ (1982), *Indian Agricultural Economics: Myths and Realities*, Allied Publishers, New Delhi.

__________ (1992), *Political Economy of Indian Agriculture*, K.P. Bagchi & Company, Calcutta.

Sagar, Vidya (1977), 'A Component Analysis of the Growth of Agricultural Productivity in Rajasthan : 1956-61 to 1969-74', *IJAE*, Vol. 32, No. 1, Jan-March.

__________ (1980), 'Decomposition of Growth Trends and Certain Related Issues', *IJAE*, Vol. 35, No. 2.

Saha, A and Swaminathan, M. (1994a), 'Agricultural Growth in West Bengal in the 1980s : A Disaggregation by Districts and Crops', *EPW*, March 26, Vol. 29, No. 13.

__________ (1994b), 'Agricultural Growth in West Bengal', *EPW*, Vol. 29, No. 31, July 30.

Sarma, J.S. (1971), 'Estimates of Foodgrains Production: Some Reflections', *Agricultural Situation in India*, Vol. 25, No. 5.

Sawant, S.D. (1983), 'Investigation of the Hypothesis of Deceleration in Indian Agriculture', *IJAE*, Vol. 38, No. 4, Oct-Dec.

__________ and Achuthan, C.V. (1995), 'Agricultural Growth Across Crops and Regions: Emerging Trends and Patterns', *EPW*, Vol. 30, No. 12, March 25.

Sen, A. (1996), 'Economic Reform, Employment and Poverty: Trends and Options', *EPW*, Special Number, September.

Sen, A.K. (1973), *On Economic Inequality*, Radcliffe Lectures, 1972, Clarendon Press, Oxford and OUP, Bombay.

Sen Committee (1983), 'On Agricultural Productivity in Eastern India', Appointed by the *Reserve Bank of India*, Bombay.

Sen, S.R. (1967), 'Growth and Instability in Indian Agriculture', *JISAS*, June; reprinted in Firma K.L.M., 1971, Calcutta.

Sengupta, S. (1981), 'West Bengal Land Reforms and the Agrarian Scene', *EPW*, Vol. 16, No. 25-26.

__________ and Gazder, H. (1996), 'Rural Poverty and Public Policy in West Bengal', in Dreze and Sen (eds), *Indian Economic Development and Social Opportunity*.

Shah, C.H. (1976), 'Growth and Inequality in Agriculture', *IJAE*, Vol. 31, No. 4.

Shand, R.T. (1987), 'Income Distribution in a Dynamic Rural Sector: Some Evidence from Malaysia', *EDCC*, Vol. 39.

Srinivasan, T.N. (1979), 'Trends in Agriculture in India, 1949-50—1977-78', *EPW*, Vol. 14, No. 30-32, Special Number.

__________ (1986), 'Agricultural Production, Relative Prices, Entitlements and Poverty', in Mellor and Desai (eds), OUP, Delhi.

__________ and Bardhan, P.K. (eds) (1974), *Poverty and Income Distribution in India*, Statistical Publishing Society, Calcutta.

__________ and Vaidyanathan, A. (1972), 'Agricultural Statistics', in Rao, C.R. (ed), *Data Base of Indian Economy*, Vol. I, Statistical Publishing Society, Calcutta.

Swaminathan, M. (1990), 'IRDP in Onda Block', Working Paper, *WIDER Project*, Santiniketan.

Todaro, M.P. (1977), *Economic Development in the Third World*, Longman, New York.

Vaidyanathan, A. (1979), 'on Analysing Agricultural Growth', Dr. Rajendra Prasad Memorial Lecture, the 33rd Annual Conference of *Indian Society of Agricultural Statistics*, Trichur, Dec. 21; also Published in *JISAS*, Vol. 32, No. 1, April, 1980.

__________ (1987), 'Agricultural Development in Eastern India', *EPW*, Vol. 22, No. 52, Dec. 26.

Vyas, V.S. (1983), 'Growth and Equity in Asian Agriculture: A Synoptic View', in Maunder and Ohkawa (eds).

__________ (1994), 'Agricultural Policies for the Nineties : Issues and Approaches', *EPW*, Vol. 29, No. 26, June 25.

Webster, N. (1989), 'Agrarian Relations in Burdwan District, West Bengal; from the economics of green revolution to the politics of panchayati raj', Working Paper 89.2, *CRD*, Copenhagen.

__________ (1990), 'Panchayati Raj and the Decentralisation of Development Planning in West Bengal: A case study', *Centre for Development Research*, Research paper 90.7, Copenhagen.

Yntema, D.W. (1933), 'Measures of the Inequality in the Personal Distribution of Wealth or Income', *Journal of the American Statistical Association*, Vol. 28.

INDEX

ANCIENT SETTLEMENT PATTERN IN ORISSA
(With Special Reference to Bhubaneswar)

GYANENDRA NATH SRIVASTAVA

AGAM KALA PRAKASHAN
DELHI

First Published : 2006

ISBN : 81-7320-056-4

Published by :
Dr. Agam Prasad, M.A., Ph.D., Diploma in Museology
for AGAM KALA PRAKASHAN
34, Central Market, Ashok Vihar,
Delhi -110 052
Ph : 27212195, 55688806-7
Fax : 91-11-27212195

Type setting by :
SAI GRAPHICS
Delhi - 110052
Ph.: 27223896, 9811659747

Printed at :
Salasar Imaging System, Delhi-35.

Printed in India

CONTENTS

Foreword

The mainstream of Indian Civilization in diverse stretches had astonishing projections of an urban cast in Kalinga region. It reached a level of substantial growth in the fields of institutional arrangements, ecology and social mobility. For the delineation of this secular phase an obviously more explicit reappraisal methods has to be adopted to dovetail the archaeological, epigraphical, sculptural and literary evidence to substantiate the systematic structural analysis of cities and urban areas.

Many scholars have contributed studies on art and temple architecture of Bhubaneswar and other centres of Orissa. There are other works as well. In all these works no attempt has been made to explore the urban fields of development or give a comprehensive picture of the society and cultural and economic life of the period. Dr G.N. Srivastava, the author of the present work who utilized the opportunity of his posting at Bhubaneswar in the Archaeological Survey of India and was associated with the study of temple architecture and various excavation projects in the area, must be congratulated for his in depth study of the history, life style of people, closer observation and data analysis of evidence from habitational sites and settlements and interpretation for evolution of society from the sixth century to the late-medieval period of history.

A factual analysis of the extensive material has been made by the author by systematizing the knowledge from various sources, more of material content than otherwise, to imbibe a scientific outlook and bring precision to the work. The result of physical and economic geography and ethnogenetics has been of great assistance in the present field of research. The spectrum of society emerges clearly from investigations into the whole area of Kalinga. The trans-development of community from rural economy to ordered urban settlement has such pronounced features in the time of the advent of Aśoka, the great and emergence of Mauryan rule over the region that a very impressive image of the proliferation of material culture emerges. After the consolidation of the conquests, the society achieved greater success in syndicating the management of business that spread throughout the region and beyond. A regular contact with the hub of the Magadha empire paved the way of Kalinga entrepreneurs to travel in far off areas and develop intimate contacts in the fields of polity, trade and religion.

The Chedi rulers who succeeded Aśoka's successors in the region could not control the area to a longer time after Kharavela and it witnessed the overlordship of the Satavahanas and Murundas till the third century CE. This was the time when the Mauryan cities of Tosali and Samapa started loosing their grandeur and Sisupalgarh lost its status of the seat of administration by the fourth century CE. A period of sluggish political environment engulfed the entire region for the next two centuries towards the end of which *āchāryas* of Lakuliśa-Pāśupata cult of Śaivism influenced the area with their propagation of the advance doctrines of Śaiva philosophy and religion which was also attracting the attention of masses throughout the length and breadth of the nation during the period. It replaced Jainism of the time of the Chedis and the region between Krishna and Mahanadi received fresh impact of Saivite traditions. Ekamra-Bhubaneswar started gaining the status of not only a religious settlement but also of a city in the sixth century when gradually it received the attention of the rulers making it military base besides the social mobility due to the importance attached to it because of existence of a *svayambhu lingam* or *lingaraja* where the localities of learned Brahmanas also received for religious and cultural education.

The present work is, thus, a welcome addition for the knowledge and study of urbanization, growth of settlements and society in the hub of Kalinga region and traditional culture continuing till date of the last one and a half millennium passing through various upheavals and process of developments.

Buddha Purnima
23 May 2005

Dr B.R. Mani
Director, Institute of Archaeology
Archaeological Survey of India
New Delhi

Preface

The word settlement is used in different meanings, e.g. in law, a compromise or agreement between litigants to settle the matters in dispute between them in order to dispose of and conclude their litigation. The word settlement is also applied to a disposition of property to be held in trust. In this study the word *settlement* stands for the places of human dwellings and human activities. Thus in archaeological context a settlement remain may be a house, a cattle pen, stable, posthole, fortification, moat, tanks, embankments to rivers, lakes, roads and lanes, gateways, monastery, temple furnace and workshop etc.; the building materials such as timber, clay, mud-bricks and kiln-burnt bricks, dressed stone slabs and laterite blocks as well as stone rubbles, lime and other remains of mortars also are the important settlement remains. The human settlements developed variously; at some places people lived in very backward conditions while at other places people lived in highly advanced and prosperous stage. A settlement may expand in a very large area or it may be of very small shape and size. Various particular characteristics of dwellings, economy, and way of life decide the particular nature of a settlement. On the basis of some basic traits the human dwellings have been classified in settlement-units like village, town or city and country etc.

In the present day census-operations various settlement-units are categorized as hamlet, village, town, market-town, industrial-town, and city. The various factors which can stamp a settlement of past were widely debated by western scholars. V. Gordon Childe, taking into consideration the achievements of the early civilizations in the old world and meso-America laid down the ten abstract criteria to distinguish the earliest city from any older or contemporary village. In the ancient India concepts of various settlement units existed which are referred to in etymological writings as well as by legislators for their interpretation and definition. Thus we come across the-terms like janapada, mahajanapada, rashtra, vishaya or vijit, grama, ghosha, nagara and samvah. In the canonical writings which are called *Silpasatra* or *Vastugrantha* and are various in number, the settlement units of different type are elaborately discussed and defined on the basis of layout-planning, population and profession of inhabitants and other several considerations of theoretical and empirical nature. Thus, we come across the terms for town as : nagara, pura, rajdhani, pattan, durga, kheta, kharvata, sivira, senamukha, cherika, sthaniya, dronamukha, nigam, matha or vihara; and for the villages: manduk, prastara, bahualika, paraka, chaturmukha, purvamukha, mangal, visvakarmaka, devarata, visvesa, kailasa and nityamangala and others.

The land of Orissa is full of enduring archaeological remains, majestic monuments and narrative art panels, ruins and antiquarian remains of civil and religious nature bespeaking about the human life and their settlements. Besides, the surviving cultural trends with very less degree of adulteration also provide a picture of human activities and settlements developed by the man. I had the fortunate opportunity to see the land and the people closely during the beginning of my career in archaeology without any bias or preconceived notion of superiority or inferiority for any place and society. The richness of the material and variety of monuments specially the temples had already attracted the great scholars of India and abroad and the rich yield of research works from many pioneer scholars; e.g. Lt. Kittoe, M.M. Chakravarty, R.L. Mitra, W.W. Hunter, Andrew Sterling, B.C. Mazumdar, M.M. Ganguly, N.K. Bose, R.D. Banerji, S.K. Sarsvati, K.P. Jaiswal, B.M. Barua, P. Acharya, S.N. Rajguru, K.C. Panigrahi, K.N. Mahapatra, N.K Sahu, Debala Mitra, Alice Boner, K.V. Saunderrajan and B.B. Lal and many others are available for understanding the Human-past in Orissa. On the other hand, the recent works also opened new vista of the study of antiquarian remain and monuments of Orissa. I had got an opportunity to go through those preceding works leisurely as well as to see the repositories of past physically during my nine years prolonged stay at Bhubaneswara and wandering throughout Orissa between 1980 to 1989.

While executing village to village survey scheme for antiquarian remains and usual tours to monuments and sites, I noticed various village-settlements located in different terrains i.e. in the coastal plains and hilly tracts and interacted with the people of common folk, and to my utter dismay they were not only aware of the awesome Kalingan war fought by Asoka, the ruling dynasties of the Kesharis, Gangas and Gajpatis and their manifestations but they were also capable of narrating the stories of the decline of Hindu kingdom in the hands of Afgan rulers of

Bengal. Thus the political history had become a part of folk lore which was a mighty symbol of the preserving the past in the simplest form that may perpetuate through generations. The people depending upon the agrarian economy and small scale domestic horticulture, local trade and commerce but the temples and shrines were the greater hub of the activities of the people, and they were associating with those temples and shrines for different kinds of services. Thus a common social set-up and the way of life was vivid to me; and all these casual observation provided an insight to peep into the past of the people through the present. My frequent visits to Sisupalgarh, Jaugada and Malhar, the early historical fortified cities and Buddhist sites viz. Ratnagiri, Lalitagiri and Udayagiri, the rock-cut caves of Khandagiri and the Old-Town of Bubaneswara again inspired me to ponder and visualize about the development of these settlements of varied nature. The village settlements of Ganjam and *Cuttack* districts were far different and as I noticed some villages in the dist. Ganjam were having close resemblance with the prescribed layout of the Architectural Canons. Thus, they have one long and broad road and the houses were standing in rows on the either side of the road. Each house had a large raised platform meant for storage of the paddy underneath. It was though peculiar but a traditional storage system. The whole village was obviously visible at a glance, and the temple was located at the entering end. Such examples oriented my studies towards the history human settlements of the past and patterns developed thereof. Bhubaneswara, the present capital town of Orissa has exhibited the major potential by yielding the material remains and perpetuating human settlements in this area for more than 2500 years. The whole region provides a chronological sequence in a virtual horizontal section for the human activities in the above period. The rock-edict of Dhauli witnessed the event of Kalingan war, the ramparts of Sisupalgarh bespeak of the early historical planned urban settlement and honeycombed rock-cut-caves provides the glimpses of ascetic's abodes and the Old-Town settled around the great Lingaraja temple representing a crystallized society with unbreakable moorings to the Svayambhu Lingam or Lord Lingaraja, the presiding deity of the place and a host of temples and pious tanks making a tirtha or religious settlement. The classifications available in the Indian Vastugranthas have provided the very concept of human settlements and have been used as touch stone while identifying the ancient human settlements in Orissa and particularly at Bhubaneswara. The natural resources, geomorphology, availability of food and water and safe-dwellings are some of the basic requirements for any place to become the habitat of human beings; and Bhubaneswar provided those basic necessities, therefore, a variety of human-settlements sprang there. Thus my studies in this book oriented to an inquiry in the archaeological remains and monuments; the natural settings and the role of human hand behind those manifestations, and the very reasons and circumstances for the emergence and development of the settlements in the region and their social set-ups. The book contains seven chapters, discussing in detail the extent, geomorphology, climate, habitation-sites and nature of settlements, identification and corroboration of the settlements with other evidences. The political and religious background of the region is given in support of evolution of settlements and their decay. The feeding economy of different settlements, the evolution of society and its crystallization in particular forms and other aspects have been dealt with an analytical method.

The present book is an outcome of my studies undertaken for the award of Ph.D. degree in the auspices of the Dept. of History of Dr. B.R. Ambedkar University, Agra under the scholarly and authoritative guidance and supervision of my learned supervisor Professor Agam Prasad Mathur, the former Vice-Chancellor of the Agra University and successfully completed with his spiritual bliss in 1996. The dissertation was highly praised by my both the Examiners; namely; Dr. R.C. Sharma, Professor and Head of Department, St. Jones College, Agra and Profesor S.P. Gupta of Aligarh Muslim University; and Dr. A.K. Singh, the then Head of the History of Dept. of the University. I, therefore express my sincere gratitude to them. In my studies I was frequently helped and guided by my friends and seniors and my acknowledgments are due to them. I am very much beholden to Sri K.K. Muhammad, Dr. Arundhati Banerji, Dr. B.R. Mani, Sri P.B.S. Sengar, Dr. G.C. Chauley, Sri Shankar Nath, Sri K.N. Dikshit of the Archaeological Survey of India for their helps and sincere guidance at various stages of this work. My colleagues Sri O.D. Shukla, Sri S.K. Kulshreshtha and Dr. R.S. Dvivedi helped me in this work and I am highly indebted to them.

My acknowledgments are due to Archaeological Survey of India for kindly permitting the copying of the photographs and maps from its publications. Acknowledgements are also gratefully made to the authors of the books and journals-articles, referred to in this work.

My sincere thanks are due to Dr. Agam Prasad of the Agam Kala Prakasan for kindly agreeing to undertake the publication of work.

My wife Seema and son Akash deserve special thanks for not only their encouraging attitudes but also for their active participation in the checking and arranging of the manuscript. In my studies, I have always been encouraged by my elder brothers Sri Narendra Nath, Surendra Deo and Ashwani Kumar for their unbounded love and affection to me which I cannot adequately express. I am beholden to my parents, Smt. Chandrasakhi and late Banwari Lal whose bountiful blessings always empowered me tackling the onerous tasks. However, the word of gratitude becomes meaningless when the works are done by the almighty; and therefore, I could not afford the words to express adequately to my Master Revered Parthasarathi Rajagopalchari and because of his grace only this became possible.

Gyanendra N. Srivastava

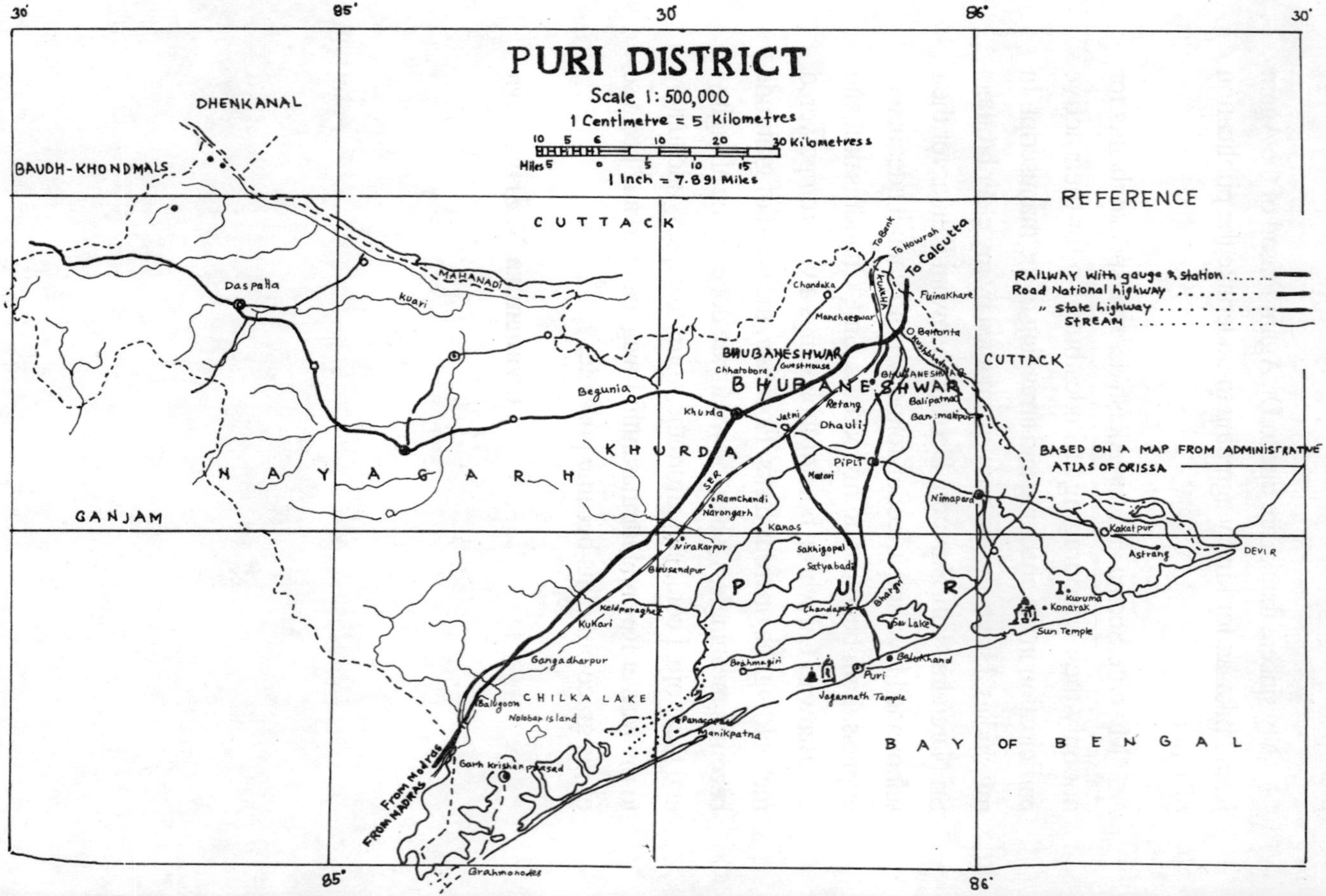
PURI DISTRICT
Scale 1: 500,000
1 Centimetre = 5 Kilometres
1 Inch = 7.891 Miles
REFERENCE
RAILWAY with gauge & station
Road National highway
" State highway
STREAM
BASED ON A MAP FROM ADMINISTRATIVE ATLAS OF ORISSA
DHENKANAL
BAUDH-KHONDMALS
CUTTACK
GANJAM
NAYAGARH
KHURDA
BHUBANESHWAR
PURI
BAY OF BENGAL
CHILKA LAKE
MAHANADI
Daspalla
Begunia
Khurda
Jatni
Dhauli
Pipli
Nimapara
Kakatpur
Astrang
Konarak
Sun Temple
Puri
Jagannath Temple
Brahmagiri
Chandaka
Balugaon
Nolobar Island
Manikpatna
To Calcutta
From Madras

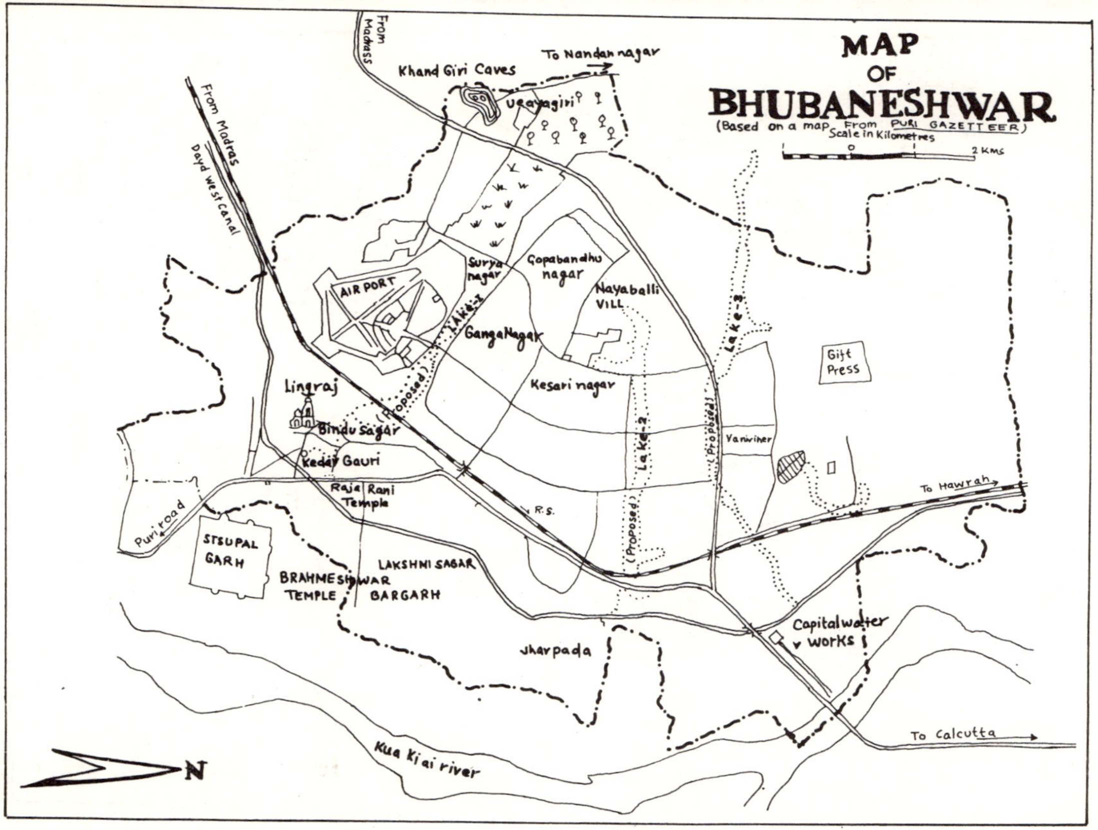
MAP OF BHUBANESHWAR
(Based on a map From PURI GAZETTEER)
Scale in Kilometres
0
2 Kms
From Madrass
To Nandan nagar
Khand Giri Caves
Udayagiri
From Madras
Dayd west Canal
AIR PORT
Surya nagar
Gopabandhu nagar
Nayaballi VILL
Lake-1
Ganga Nagar
Kesari nagar
Lingraj
(Proposed)
Bindu Sagar
Kedar Gauri
Raja Rani Temple
Lake-2
Lake-3
(Proposed)
Vaniviner
Gift Press
To Hawrah
R.S.
Puri road
STSUPAL GARH
BRAHMESHWAR TEMPLE
LAKSHNI SAGAR
BARGARH
Capital water works
Jharpada
Kua Kiai river
To Calcutta
N

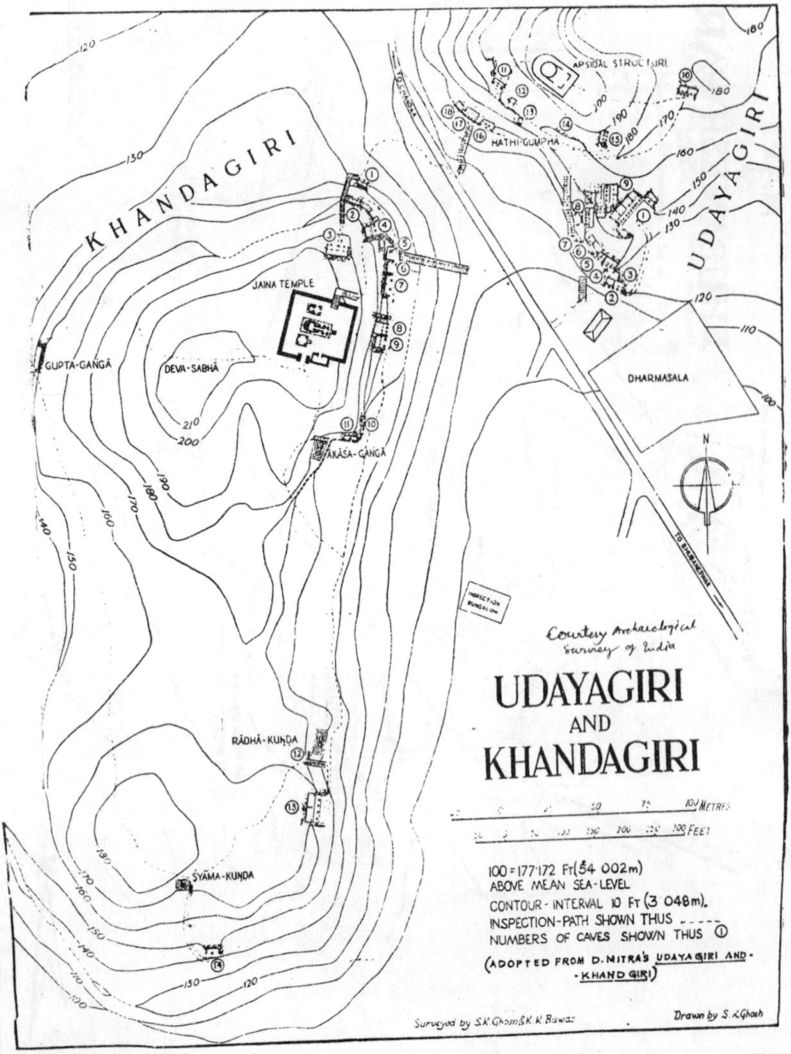
KHANDAGIRI
UDAYAGIRI
JAINA TEMPLE
GUPTA-GANGĀ
DEVA-SABHĀ
ĀKĀSA-GANGĀ
RĀDHĀ-KUNḌA
ŚYAMA-KUNḌA
APSIDAL STRUCTURE
HATHI-GUMPHA
DHARMASALA
INSPECTION BUNGALOW
TO BHUBANESWAR
N
Courtesy Archaeological Survey of India
UDAYAGIRI
AND
KHANDAGIRI
METRES
FEET
100 = 177·172 FT (54·002m) ABOVE MEAN SEA-LEVEL
CONTOUR-INTERVAL 10 FT (3·048m).
INSPECTION-PATH SHOWN THUS
NUMBERS OF CAVES SHOWN THUS ①
(ADOPTED FROM D. MITRA'S UDAYAGIRI AND KHANDGIRI)
Surveyed by S.K. Ghosh & K. K. Biswas
Drawn by S. K. Ghosh

1

Introduction

Bhubaneswar (Lat. 28°15' N, Long. 85°15' E), being the capital city of the state of Orissa, is well known in the political map of India. It is located on the Howrah – Madras main railway line as well as on the National Highway-5. There are other State Highways and district roads connecting the city with the district headquarter Puri and other cities of Orissa and neighbouring states. Besides, Bhubaneswar is also connected with the national capital New Delhi and other important cities, namely : Calcutta, Madras, Bombay and Hyderabad by airways and railways. There is no doubt that Bhubaneswar is one of the most beautiful modern cities of India with well-planned habitation, broad avenues, market-complexes, office-buildings, educational institutions and industries. However, it is better known for its glorious past which is survived by splendid monuments and settlement remains scattered in a vast area. The place has a long-cherished history of human occupation and settlements It has been an established *Śaiva-tīrtha* of farfame since the days of some early *purāṇas* and epics. The traditional sanctity is well attested by a series of majestic temples and sacred lakes and tanks which had been built in various phases and periods. There also existed the leading centres of Jainism and Buddhism in the Early Historical Period. The political importance of this area is established by the Asokan Rock-Edicts, the fortified remains and the inscription of king Khāravela. Further, the traditions, the euologies and dedicatory inscriptions relate to the king Sasanka, the rulers of various dynasties, viz. the Somavaṁsī, the Gaṅga, the Sūryavaṁsī and others with this place.

1. EXTENT

The monuments and the archaeological sites are spread in a vast area in the various localities and mauzas of modern Bhubaneswar city. *"Bhuvanesvara"* is one of the epithets of the *liṅga* now called *Liṅgarāja* enshrined in the biggest temple of the place. In the old revenue records the locality around this is referred to as Bhubaneswar. Thus the place derived its name from the presiding deity of the area. As a revenue mauza Bhubaneswar covers almost the whole area known as Old Town. In the revenue map it is bounded by the villages, namely : Rampur and Bargarh on the north, Mahabhoisasan and Nuagaon on the east, Kakuraghai, Haripurpatna, Kapileswar and Kapilprasad on the south; kapilprasad, Bhimpur and Siripur on the west. But the ancient Bhubaneswar extended in a much wider area as witnessed by the location of Brahmeśwara temple. An inscription originally attached to this temple speaks of the locality of the temple at *Ekāmra*, the other name of Bhubaneswar[1]. The development of modern city as a new State capital was enacted in 1948. The urban local body of Bhubaneswar first constituted as a notified area committee on 1st February 1948 under Bihar and Orissa Municipal Act 1922. Thereafter it become Notified

Area Council on 1st October 1952 under the Orissa Municipal Act 1950[2]. It covers 28 villages or rather mauzas which are revenue units. These mauzas are Purba Bargada, Pachhim Bargad, Bhubaneswar, Kapileswar, Haripurpatna, Lakshmisagar, Lakshmisagarpatna, Bhimpur, Rampur, Bomikhal, Govindprasad, Kalraput, Sunderpada, Kapilprasad, Pokhriput, Barna, Nayapalli, Barmunda, Jagamara, Jharpada, Charbatia, Nuagaon, Garhgopinathprasad, Pandara, Garkan, Chandrasekharpur and Damna[3]. Later it was constituted into a Municipality on 14th April 1979 by Govt. Notification No. SPL. Com. 10/78-11360 dated 29th March 1979 under the same Act. The whole area is now organised and distributed in 42 units with a proper layout. The mauza Bhubaneswar, which has been known as such for many centuries, has given the name to the modern township.

The above extent of Bhubaneswar covers almost all the ancient sites and settlements to be dealt with in this work. But, a few most important sites lay beyond the limits of Municipal area. These ancient sites are located at Basuaghai, Sisupal, Mahabhoisasan, and Dhauli which are contiguous mauzas to the Municipal area. In the rapid growth and expansion of the city these mauzas no more lay outside of Bhubaneswar urban area and notionally those are accepted as part and partial of Bhubaneswar. Further, the new Master Plan of greater Bhubaneswar includes many more villages all around the Municipal area. Though, this extension covers all the sites to be dealt within this dissertation and the extent of Bhubaneswar may be considered as shown in the Master Plan area, but this extent does not bear any permanent landmark and the boundaries of modern Bhubaneswar may further extended with the increase of population, building activities or any new planning of the Government.

On the other hand traditional extent of Bhubaneswar, as quoted in *Purāṇas* and other orthodox Saṅskrit literature, has more certain boundary marks. According to *Brahmapurāṇa* the sacred land of *Ekāmra* extended upto two and half *yojana* on each cardinal direction from the temple of *Svaymbhū-Kṛittivāsas*[5]. The measurement of *yojana* is seldom used in mythical expressions; however, in established Saṅskrit Lexicons like *Śabdakalpadruma, yojana* is referred to as equal to four *kosa*[6]. One *kosa* is equal to two miles. Thus *Ekāmra* extended upto twenty miles from each cardinal direction from Liṅgarāja temple. The *Svarṇādri Mahodaya*[7], an orthodox text of late medieval period, speaks of the extent of Bhubaneswar as three *yojana* on each cardinal direction from the deity Liṅgarāja. The text further intimates the area of sacred and meritorious land and five *kosa* and mentions the landmarks determining the limits on each cardinal direction; thus, the extent of Bhubaneswar is fixed in the area lying between *Khaṅdāchala* (Udayagiri-Khandagiri hill) on the west and Kuṅdaleśvara temple (at village Tonkapani) on the east; and goddess *Varhīm (Bhalha)* on the north and Vahiraṅgeśvara temple (at Dhauli hill) on the south. *Ekāmra Chaṅdrikā* also confirms it[8].

The uncertainty of the old expressions of distance measurements is obvious from the above varying statements; yet the landmarks referred to there are of permanent nature and provide a certain defined boundary of the traditional extent of Bhubaneswar; and this extent is being referred to for the area and extent of Bhubaneswar in this dissertation.

2. GEOMORPHOLOGY

Geomorphologically, the whole area of Bhubaneswar is located in two distinct landforms; vis. *(i)* Laterite tract and *(ii)* Alluvial plains. Besides, there are isolated hill-spurs of Udayagiri-Khandagiri on the west and of Dhauli on the south.

(i) The Laterite Tract

It is higher on the west and slopes towards the east. Gangua,[9] a local streamlet demarcates it from the alluvial plains. It is eroded at many places and has an undulated surface. The local watercourses draining the rain-water of vast laterite stretches towards alluvial plains ultimately reach to river Daya. The agelong water-flow cut deep the land surface and resulted in the formation of large lakes. Due to simultaneous silting and human activities these lakes (water courses) either turned into marshes or paddy fields or levelled for roads and buildings. Out of three watercourses two are shown as proposed lakes in the map of Bhubaneswar published in *Puri District Gazetteer*[10], though in the subsequent town planning map, these are marked as green belts. For our reference we shall call these water courses as Lake-1, Lake-2 and Lake-3.

Lake-1: This is the most important water-course which lay on the southern part of modern Bhubaneswar and flows in the modern Bhubaneswar area in most of its length. Beginning near modern Ganganagar locality in the west, this lake in its original form extended upto the area lying on the north of Liṅgrāja temple and further flowing towards north-east in the shape of a narrow streamlet, it is known as Gangua in its later part; as another longer stream is also called Gangua, therefore we refer to the former as the Shorter Gangua[11]. A careful study and observation of the original contours and course of this lake reveals that the present Bindusagar lake was originally a part of this lake which was separated from this by creating a bundh. This separation would have taken place before 6th or 7th century A.D. as witnessed by the temple of *Uttareśwara* which stands on the bank of this lake; the temple is datable to circa 7th-8th century A.D. We also notice that the area clinging to Liṅgarāja temple is a raised one, and steeply sloping in all directions (only in western portion it is less sloping), but the inner courtyard is in lower level. These contours witness that a extensive earthern ramp was created during the construction of Liṅgarāja temple[12]. A half burried temple, having *ākāśaliṅga* in its pinnacle, in front of the new market building in Old town and another half-burried temple standing near opposite to the main gate of Liṅgarāja temple; and the existence of *Sahasraliṅga* tank in front of Liṅgarāja temple (hidden behind a series of shops mostly in wooden cabins) confirms that the original ground level was much lower than that of the preset one. Therefore, it is probable that once the water of Lake-1 was touching the *Svayambhūliṅga,* the presiding deity[13]. Since, we have references that *Ekāmraliṅga* or *Svayambhūliṅga* was located on the banks of river *Gaṅdhavatī* or *Gaṅdhasiṅdhu*; and the flow of Lake-1 after Kedāra-Gaurī area is also called Gangua a corruption of *Gaṅdhavatī*, therefore, it can be said that Lake-1 was the so-called *Gaṅdhavatī* or *Gaṅdhasindhu* of *Purāṇas* and inscriptions which served a major water source for the religious settlement of *Ekāmra*-Bhubaneswar and is also liable for many perennial springs. It is interesting to note that we get the reference of further channeliztion of *Gandhasindhu* by Choḍagaṅga[14]. It appears that subsequent to the rise of ground level due to the reasons mentioned earlier water of Lake-1 started spreading towards the east where already existed several temples; namely : Kotitīrtheśvara, Svarṇajāleśvara and Tāleśvara etc. This water, creating swamps and floods during rainy seasons, would have disturbed those temples. Therefore, a channel was dug to guide the waterflow properly which also made available vast stretches of dryland for the construction of other temples. At present Lake-1 starts from Ganganagar and runs towards east with mild southward bend along the area lying on the northern periphery of Airport and reaches the railway line where it takes southward turn and widens, then reaches the *Bardaṇḍa* road, here it is tured into almost shallow swamps and called *Asurakiārī* (the field of demons). As, during the Car-festival (*ratha-yātrā*) of Lord Liṅgarāja the wheels of the vehicles used to plunge in the swamps (due to settling down of the earthern road built upon it); and such incidents were taken as the mischief of demons, therefore, this area was believed as the field of demons (the demons are pacified when the car reaches in this area by offerings). After this point the lake runs for a short distance touching a little piece of land an the road which bifurcates from Bindusagar (Bindusagar is linked with this lake by a culvert), the it flows in zigzag way for a little distance passing through the areas lying behind Anantvāsudeva temple and running towards the east; again it creates vast swaps behind O.R.T. Garage (this area is now full of modern constructions). The lake further takes the shape of a stream and enters the alluvial plains after crossing the Puri road. Here, it flows about 4 kilometre parallel to major Gangua stream and ultimately unites with the latter.

Lake-2: It takes its origin near Nuapalli village and runs in eastward direction via Satyanagar where it is called as "Lake". After some distance the water of the lake becomes invisible.

Lake-3: This lake lies on the northernmost fringe of Bhubaneswar. It is though longest but narrow water course in west-east direction which originates in far west near Jokalandi mauza passing via Vanivihar area and terminates behind Capital Water Works.

The soil profile of laterite area is very thin and is totally nil at some places, therefore the land is unfit for crop-cultivation. However, heavy rainfall throughout the year and high humidity in atmosphere help lush forest growth in this area.[15] The porosity of the laterite bed causes the water percolation in greater quantity which raises the subsoil water-level. The raised water-level is also conducive for the growth of *ficus* as well as scrub-plants in laterite area. This underground water gives rise to many perennial springs, the famous one is called Kedāra-Gaurī. The hard and high laterite areas usually lay barren and locally called "*tāngī*". These lands served as good quarries of laterite slabs for building purposes. As a building material, laterite is used since very early period. It is used as main building material in civil and religious edifices of ancient times at Bhubaeswar, e.g. at Sisupalgarh-fortification and in temples and tanks.

(ii) The Alluvial Plains

The alluvial plains of the east are formed by local drainage as well as by major rivers; namely : Kuakhai, Kushbhadra, Daya and Bhargavi. This is a fertile agricultural belt and would have served as an important source of food-supply to the human-settlements in the past which flourished in the neighbouring area. This alluvial land forms the part of well-known long strip of Costal Plains of Orissa.

(iii) The Hill Spurs

The hill-spurs of Udayagiri-Khaṅdagiri lay on the western zone of Bhubaneswar. These are the sedimentary formations of Upper Gondwana phase.[16] The highest contour is 210 feet at peak called *"devasabhā'* located in the western side hill i.e. Khandagiri.[17] These hills served as shelter for Jạina monks and were honeycombed with beautiful rock-excavations under the patronage of king Khāravela. The coarse-grained sand stone of these hills was used in the temples of Bhubaneswar. On the otherhand the Dhauli hills, lying extreme south of Bhubaneswar on the western bank of river Daya, are volcanic formations of Archian's Age.[18] These are composed of upheaved breccia with quartose rock intermixed.[19] The rock is now termed as Khondalite. The hill bears most significant relics of the past i.e. a set of fourteen-rock-edicts of Asoka.

Main Rivers, Distributaries and Tributaries[20]

The principal rivers which flow in the neighbourhood of Bhubaneswar are the Kuakhai, the Bhargavi, the Kushbhadra and the Daya. These rivers also served as the main waterways for inland trade and public transport.

The Kuakhai: The river Kuakhai is an offshoot of the river Kathazori and it takes off near a point located opposite to

Cuttack. After flowing in south-south-easterly direction for 19.2 km. it throwsoff Kushbhadra then travels nearly due south (along the eastern periphery of greater Bhubaneswar area) about 11.2 km. until it reaches village Sardeipur where it divides into two big branches, the Daya to the west and the Bhargavi to the east.

The Kushabhadra: This river leaves the river Kuakhai at Balianta and flows in south-easterly direction for some 64 km. till it enters the bay of Bengal near the shrine of Rāmachandī (9 km. from Konark). Near the mouth of this river traces of a port town of medieval period are discovered in a recent excavation.[21]

The Daya: This river as already stated, takes off from the Kuakhai, runs to the south for about 12.8 km. then makes a sharp turn towards west, after that it continues its course southward for the rest of its length emptying itself into the north-eastern corner of the Chilica lake some 60 km. from its offtake from river Kuakhai.

The Bhargavi: This river is 91 km. in its total length. After take off from river Kuakhai it follows a circuitous course and empties itself into the outfall of the river Daya breaking into numerous branches.

The Gangua: This is a local stream which has two main sources; and both are called Gangua. For our convenience, we have named them as (1) major Gangua and (2) shorter Gangua. The shorter Gangua has already been discussed while describing Lake-1. The major Gangua has its source in the hilly tract to the west of Mancheswar, some 9.6 km. north of Sisupalgarh. From Sisupalgarh it flows further 11.2 km. in south-south-west-south direction and joins the river Daya. The shorter Gangua stream which initially flows in east-south orientation near Kedāra-Gaurī becomes fully south oriented after covering a distance of one kilometer or more, turn almost in right angle towards west and runs the same distance in that direction. Here, it encircles the localities of Bhubāsunī and Kapileswar, then again turns southwards where it meets the major Gangua, here both the courses run almost parallel. It is a noteworthy point that whereas the shorter Gangua inspired the rise of *Ekāmra*-Bhubaneswar, the major Gangua was exploited by the occupants of the fortified settlement of Sisupalgarh mauza.

Rivers as Waterways[22]

For a major part of the year, the Kushbhadra, the Bhargavi and the Daya serve the purpose of navigation. The Kushbhadra dries up in its upper reaches but in its lower part country-boats are to be seen throughout the year. The Daya and the Bhargavi are navigable upto the month of January. These two rivers get empty in Chilika lake (actually it is a part of the sea), in their respective mouths. Upto a distance of 10 miles (16 km.) upstream they are navigable even in hot dry weather. Traders from the Ganjam district with their heavy load of bamboos and other commodities come by way of the Ganjam canal to Chilika and then catch these rivers.

These waterways would have served as the means of trade-traffic in the ancient days also. The existence of port town in the mouth of river Kushbhadra in the medieval period suggests that inland trade centres and cities were linked with maritime settlements by this river. It may be possible that in ancient times the river had a good level of water in its whole length. Similarly, the Bhargavi and the Daya would have linked some ports on Chilika. Some recent archaeological excavations at the village Manikpatna, located on the banks of lake Chilika, revealed the trade links of this earstwhile port town with the Roman-World.[23] The much earlier excavations at Sisupalgarh near Bhubaneswar also yielded similar evidences of maritime trade with the Roman-World. Therefore, it is guessed that the port-town of Manikpatna was connected with the fort-city of Sisupalgarh by the way of river Daya and Bhargavi which are the offshoots of river Kathjori flowing near Sisupalgarh.

3. CLIMATIC CONDITIONS

Bhubaneswar is situated at an altitude of 45 metres above the mean sea level. It has a bracing climate with maximum and minimum temperature of 31.0° C and 16.0°C during winter and 38.0°C and 27.0°C during summer respectively. The average rainfall in a year is 152.4 cm.[24] Apart from the usual rains of mansoon season, Bhubaneswar also receives heavy downpours from pre and post-masoon cyclones. From April to August the prevailing wind is from south and south-west and from September to March it is from North and North-West. The velocity of the wind is maximum in summer. The nearness to sea maintains a high percentage of humidity in the air. The plentyful availability of water and lush-green vegetation are important features of natural environment of Bhubaneswar. The climate of Bhubaneswar is pleasing due to sea breezes coming across the verdant delta from south; on the other hand the forest-cladded hilly surrounding towards extreme north and a beautiful and charming panorama to the city. Here, about 15 km. from Bhubaneswar, an extensive landscape has been developed in zoological park and botanical gardens. This place, now known as Nandanakana, had been the natural abode of various wild animals since remote past. There are big natural lakes at Nandanakanana and Sarangagarh (there is a fort belonging to the period of the rulers of the Gaṅga dynasty). A nearby forest known as Chhandaka is the natural habitat of wild elephants; the State Government of Orissa has developed a sanctuary for elephants here. Besides, ruins of a fort of the Gaṅga period have been also noticed here amidst the forest-clad hills. Unlike the Nandankanana area this forest is dry and devoid of any natural water-reservoir. The scrub-forest and bamboo-bushes in this area are highly susceptible for wild conflagrations during summers due to particular topography and the climate of the area.

The original topography of Bhubaneswar area is rapidly changing due to fast urbanization, growth in population, various kinds of land use in industrial development and other allied factors. However, there has been little change in the climatic conditions of this locality during the last two thousand years.[25]

4. CONCEPTS AND SCOPE OF STUDY

In the above extent of Bhubaneswar, we get many well identified relics of human habitation which take the history of human settlements in this land to a hoary past. The world settlement is used in different meanings, e.g. in law, a compromise or agreement between litigants to settle the matters in dispute between them in order to dispose of and conclude their litigation.

The word settlement is also applied to a disposition of property to be held in trust. In this study the word *settlements* stands for the places of human dwellings and human activities. Thus in archaeological context a settlement remain may be a house, a cattle-pen, stable, posthole, fortification, moat, tanks, embankments to rivers and lakes, roads and lanes, gateways, monastery, temple, furnace and workshop etc.; the building materials such as fiber, clay, mud-bricks and kiln-burnt-bricks, dressed stone slabs and laterite blocks as well as stone rubbles, lime and other remains of mortars also are the important settlement remains. The human settlements developed variously, at some places people lived in very backward conditions while at other places people lived in highly advanced and prosperous stage. A settlement may expand in a very large area or it may be of very small shape and size. Various particular characteristics of dwellings, economy, and way of life decide the particular nature of a settlement. On the basis of some basic traits the human dwellings have been classified in settlement-units like village, town or city and country etc.

In the broad classification of settlement-units the term urban and rural are used. In the present day census-operations the various settlement-units are categorized as below:[26]

1. *Hamlet:* No definition has been adopted by the census, but by this term we mean a small cluster of homesteads in rural territory.

2. *Village:* Village is recognized as the basic area of habitation. In the rural areas the smallest unit of habitation, viz. the village, generally follows the limits of a revenue village that

is recognized by the normal district administration. The revenue village need not necessarily be a single agglomeration of habitations. But the revenue village has a definite boundary and each village has a separate administrative unit with separate village accounts. It may have one or more hamlets. The entire revenue village is one unit. There may be unsurveyed villages within forests, etc., where the locally recognised boundaries of each habitation area are followed within the larger unit of, say, the forest's range-officer's jurisdiction.

3. *Town:* To qualify for an urban area, a place should have either a municipal corporation or municipal area, or be under a town committee or cantonment board. In addition, each census has adopted a number of census-towns, which do not enjoy any statutory label of administration. This has been considered desirable in order to obtain a truer measure of urbanization, as it is usual for an administrative label to fall some way behind actual achievement. The census-town were determined on the basis of a number of empirical tests :

(a) with a density of not less than 1,000 per square kilometre;

(b) with a minimum population of 5,000;

(c) with three-fourths of the occupations of the male population being outside of agriculture; are

(d) with according to the Superintendent of State, a few pronounced urban characteristics, the definition of which, although leaving room for vagueness and discretion, is yet meant to cover newly-founded industrial areas, large housing settlements, or places of tourist importance which have been recently served with civic amenities.

4. *Market-town and industrial-town:* The functional character of the town is determined by the percentage of workers under one occupation to the total workers in the town. If workers under one of the occupations from 40 p.c. or ore of the total workers, it is a monofunctional town, e.g. industrial town or service town etc.

5. *City:* Any town having a population of 100,000 or more is termed a city.

In the above classification if we put aside the modern terminology and administrative practices, as surveyed villages, revenue-units, municipal area, town-committee etc, the following broad points have been taken into consideration for the above classification:

(1) The village is the basic area of habitation, though the existence of smaller units, hamlets, is not precluded.

(2) A higher density of population and a larger population, largely non-agricultural, would distinguish a town from a village.

(3) The modern criterion of the city having a population of an arbitrarily-fixed figure only means that it is larger than the town, there being no functional difference between the two.

The various factors which can stamp a settlement of past were widely debated by western scholars. In this connection

V. Gordon Child's ten criterion are worthmentioning. Taking into consideration the achievements of the early civilizations in the Old World and meso-America he laid down the following ten abstract criteria to distinguish the earliest city from any older or contemporary village:[27]

(1) In point of size the first cities must have been more extensive and more densely populated than any other previous settlements, although considerably smaller than many villages of today.

(2) While there might be peasants cultivating lands outside the city, all cities must have accommodated in addition non-food-producing classes — full-time specialist-craftman, transport worker, merchants, officials and priests supported by the surplus produced by peasants.

(3) Each primary producer paid over his tiny surplus to a deity or divine king who thus concentrated the surplus.

(4) Truly monumental buildings not only distinguish each known city from the village but also symbolize the concentration of social surplus.

(5) All those not engaged in food-production were supported in the first instance by the surplus concentrated in temples or royal graneries and were dependent on the temple or court. But naturally, priest, civil and military leaders and officials absorbed a major share of the concentrated surplus and thus formed the 'ruling class', which did not confer substantial benefits

upon their subjects in the way of planning and organisation.

(6) The ruling class was compelled to invent systems of recording and exact, but practically useful, sciences. The mere administration of the vast revenues of a Sumerian temple or an Egyptian pharao by the perpetual corporation of priests and officials obliged its member to device conventional methods of recording that should be intelligible to all their colleagues and successors, that is, to invent writing and numeral notation. Writing is thus a significant, as well as a common trait in Egypt, Mesopotamia, the Indus valley and central America.

(7) The invention of writing enabled the leisured clerks to proceed to the elaboration of exact and predictive sciences-arithmatic, geometry and astronomy.

(8) Other specialists gave a new direction to artistic expressions. Artist-craftman-full-time sculptors, painters or seal-engravers began to carve, model or draw according to conceptualized and sophisticated styles.

(9) A further part of the concentrated surplus was used to pay for the importation of raw materials needed for industry or cult and not available locally. 'Regular foreign trade' over quite long distance was a feature of all early civilizations.

(10) So in the city specialist-craftsmen were provided with raw materials needed for the employment of their skill

> and also guaranteed security in a state organization based on residence rather than kingship. Itinerancy was no longer obligatory. The city was a community to which a craftsman could belong politically as well as economically.

In ancient India concepts of various settlement units existed which are referred to in etymological writings as well as by legislators for their interpretation and definition. In the early historical period the composite unit of settlement was the *janapada*, like the English word 'country', had the dual meaning of territory and rural area. We also come across the term *mahājanapada*, which is very clear from its prefix that it was greater than the others, and therefore would have came into being by engulfing the weaker ones, a process which is clearly discernible in the time of Buddha and Mahāvīra. The *janapada* gradually ceased to mean a territorial unit and came to denote only an ill-defined region or countryside. Territorial units then were to be known as *rāsthra, vishya* or *vijita*.

Within the *janapada*, there were different types of settlement. According to one enumeration, that of Patañjali, the types were *grāma, ghosha, nagara* and *saṁvāha*.[28]

The primary unit was the *grāma* (village), which even now contains the bulk of the Indian population. According to Kautilya, the population of a village should be mostly composed of *Śūdra* cultivators of soil - an attitude which is different from that of the law-givers, who would like pious *brāhmans* to live in the village; but it is more realistic, as who else but the cultivators could feed the non-food-producing population of the village itself and of the city. There were specialized villages also is clear from numerous references to villages called after craftsman, traders and professionals and even castes like the *brāhmaṇa* and outcastes like the *Chāṇḍāla*. Villages were of different sizes. A village should consists of one hundred to five hundred families, says Kautilya, but the *Jātakas* refer to villages with thirty to one thousand families. In the *Vinayapiṭaka*, village is referred to as consisting of a single *kuṭī* or two, three or four *kuṭīs*, perhaps indicate about the existence of small hamlets.[29]

Then, as now, the predominant occupation of the villagers was agriculture, which alongwith a lesser density of population, is one of the features distinguishing the rural areas from the urban ones all over the world and throughout the ages after the growth of urbanization. Pastoral settlements were also known and were called *ghosha*.

Nagara was an urban settlement and there grew up different, types of urban settlement, is clear from literature. The *Saṅskrit* lexicons give the following words as synonymous, all standing for town or city : *pura, purī, nagarī, pattana, putabhedana, sthāniya* and *nigama*. In early *Vedic* literature *pura* meant 'rampart', 'fort', 'stronghold', and did not bear any urban significance. In the later times the meaning of rampart or fortification assumed lesser importance, and the derivative word *pura* became identical in meaning with *nagara* or *nagarī*.[30] The origin of the word *nagara* is uncertain. However, '*Nakar*', the *Tamil* word for the city, is used variously in ancient *Tamil* literature to mean a house, a temple, a palace or a castle, and a city.[31]

In the canonical writings which are called *Śilpaśāstra* or *Vāstugraṅtha* and are various in number, the settlement units of different types are elaborately discussed and defined on the basis of layout-planning, population and profession of inhabitants and other several considerations of theoretical and empirical nature. In their classification we also met with many references of earlier period. Each text enumerates a variety of settlement types of its own including some common types also. In summary a list of ancient important settlement-types with brief description is given below for an understanding of concepts of settlement in Indian contexts:[32]

Nagara: It is already referred to for its various meanings. A *pura* and *nagara* belong to same category. It is a fortified town. This town is inhabited by all classes of people and is a place of commercial activities as well as the seat of administration.

Rājadhānī: Māyāmata gives a description of the royal capital with its imperial castle. Śukrāchārya says that it must have the *sabhā* or council hall in the centre. It has a belt of walls surrounded by ditches or a moat. Military encapments are located outside the town. The town has lofty entrance gates with ornamental wreaths carved in stone. Gardens are essential feature of a *rājadhānī* town.

Pattana: It is a commercial town situated on the bank of a river or sea.

Durga: It is a fort. Kautilya laid down elaborate provisions in a layout of Durga. It was much more than a fort : it was a full-fledged city with the king's palace and his offices in the centre, with roads and temples and with *Brahmanas*, merchants, workmen etc., as its inhabitants - infact everything that would constitute a capital. Weapons, food and fuel should be stocked there to the extent that would be necessary for years and should be replenished from time to time. *Durgas* might be constructed in the midst or rivers, on plain land, on hills, in deserts and in forests and, failing all of them in the midst of *janapada*.

Kheta: This is a small town situated on the river bank or in a forest by the side of hillocks. It is surrounded by mud walls.

Kharvāṭa: This town is similar to the *Kheṭa* but is defended by an inferior type of wall.

Śivira: As the name denotes *Śivira* is a military camp of a king. According to *Śilparatna* it is a camp of soldiers close to the scene of a prospective battle. Another variety of this type is *Senāmukha*. It is a military base at some distance away from the battle field. According to *Māyāmata* it may also be a suburban town linked to a city to guard the latter's main approach. It therefore occupies a strategic position. It may be compared to a modern military cantonment. In the description of king Harsha's palace the space in front of royal palace complex is referred to as *Śibira* which obviously was meant for guarding armies.

Sakndhāvāra: This and its suburb *Cherika* fall under the above category. It is protected by the rivers, mountains and woods. It is a military encampment inhabited by the common people alongwith soldiers.

Sthāniya: According to Kautilya it is a fortress-town situated amidst 800 villages, along a river bank or at the foot of hill. This town is a district headquarter alongwith police force. According to *Śilparatna* it is the headquarter of the dignatory who is incharge of defending the frontier of that part of the country.

Dronamukha: This is a market town situated on the bank of a river, sea-shore or at their confluence. This is a smaller trading place than a *pattan* and is situated in the midst of four hundred villages.

Kotamolaka: This town is situated in the hilly or wild terrain.

Nigama: This is market town mostly inhabited by the artisans. Nigam means a trade route. Hence it is a small market place for traders and caravans.

Maṭha or Vihāra: It is a university town and is also known as *Vidyāsthāna*, meaning the place of learning. *Maṭha* and *Vihāra* literally mean residence of students or colleges. This was therefore a residential university where religious itinerants (*parvrājakas*) or monks reside alongwith students. The monasteries are found located amidst the town or city or in isolated places as hills or forests in the neighbourhood of cities or towns. Nalanda and Takshasila are the noted examples of monastery-town in ancient India.

Likewise, the village settlements were also classified variously. *Viśvakarmā-Vāstuśāstra* describes 12 types of villages according to their characteristics, they are (1) *Maṇḍūka*, (2) *Prastara*, (3) *Bahulīka*, (4) *Paraka*, (5) *Chaturmukha*, (6) *Pūrva Mukha*, (7) *Maṅgala*, (8) *Viśvakarmaka*, (9) *Devarāt*, (10) *Viśveśa*, (11) *Kailāsa* and (12) *Nitya-Maṅgal*. There is no hard and fast rule about the fortification of first three types. The next four are surrounded by ramparts and may be deemed as special habitations. The Viśvakarmak is a very large village. The ninth village is a temple village and the tenth is a commercial centre. The eleventh variety is situated on the sea-coast or on the slopes of hill and protected by ramparts. The Nitya-Maṅgala is a village having a population of about 6000 people.

Mānsāra classifies a villages into eight types according to their shape, the method of street planning and the method of temple and folk planning. These categories are : *Daṇḍaka, Sarvatobhadra, Naṅdyāvarta, Padmaka, Svāstika, Prastara, Karmuka* and *Chaturmukha*.

It is evident from the above narrative that settlements were classified on various grounds and termed accordingly. It is noteworthy that certain concepts of canonical classification are applicable in respect of the settlement existed in Bhubaneswar area. However, handling with the past through some material remains, literary references and some documentary evidences, is though logical but speculative in nature. Therefore, all the canonical characteristics cannot be observed in the settlements to be found in the prescribed extent of this dissertation. Further, settlement remains helps to guess about the contemporary people or the society; similarly various references of the peoples' way of life or the society affirms the nature and status of a settlement. Thus a settlement unit and its society both are inter-related; therefore, social patterns may also define the nature of a settlement. We may quote here the writings of A. Ghosh, who, while defining the rural and urban characteristics, has given the following broad features : (1) The population of a city is denser than in rural area; (2) only a limited part of its area, if any at all, is utilized for agricultural purposes, (3) its population is predominantly non-agricultural in occupation and is dependent on rural areas for the supply of food and raw materials; and (4) this would involve the presence of merchants to supply the citizens with the necessities of a life. A greater diversity of occupation is not necessarily a feature of the city, for there may be cities with specialized functions, where this cannot be a feature. Thus, a city may be only a political and administrative centre, another a centre of trade and commerce and a third one a religious centre. But all would share the above features, which would distinguish them from the rural areas, where (1) land is less restricted (2) food need not be brought from outside, and (3) comparative self-sufficiency makes the need of a permanent population of merchants more limited.

The above criteria are also applicable in this study while ascertaining the nature and type of settlements.

The settlement-remains apart from the living monuments and localities at Bhubaneswar are scattered in a far and wide area; and are varied in nature. While at one habitation site and settlement appears of a citadel type at the other site it is of monastic nature. The distribution of numerous temples and tanks in vast laterite tract, portray a vivid picture of the greater religious sanctity and thus we notice the existence of a holy city i.e. a *tīrtha* or *kshetra* (sacred place), the glory of which is sung in orthodox literature as well as in inscriptions. The rural settlements, that may be the villages of peasants or *Brāhmanas* or hamlets of crafts-professionals or out-castes, also existed in this area;

some of these are referred to in the inscriptions while some are known to us by their physical presence. Several scholarly treatises have already been published on the monumental architecture, art and on paleographic finds. Yet an integrated study on various ancient settlements and the contemporary society thereto is lacking. Therefore, I have undertaken a systematic study of the settlement remains of past lying in Bhubaneswar area with a view to trace their origin and evolution and to search their historical identity as well as their place in the classification of settlement units in Indian context. In the human habitation human beings are bound together for self maintenance and self perpetuation and sharing their own institution and culture. This human behaviour is very part of human life which makes him social. Thus, a society is an integral part of all the human settlements. Therefore, in this dissertation an attempt is also made to reconstruct the picture of contemporary society from the available sources.

The most conspicuous material remain of human settlement in this area is a set of rock-edicts of Asoka which provides at least one certain data of a defined human settlement and identity of the land. Thus, we come to know that this land was a part of Kaliṅga country under the dominions of Asoka who annexed it in his eighth year of coronation. We also come to know that a nagaram (city-settlement) named Tosali existed here at that time. Therefore, Asoka's conquest of Kaliṅga has been taken as the datum line to begin this study. Further, Bhubaneswar enjoyed an important political and strategic status apart from its religious sanctity during the period of last great independent Hindu dynasty of Orissa i.e. the Sūryavaṁśīs, and very soon after the fall of this dynasty, Orissa was conquered by the Afgan sultan of Bengal in A.D. 1568-9 which marks the end of Hindu rule in Orissa. This event would have resulted the decline of the political and administrative functions of Bhubaneswar and also proved a major setback to the religious institution. Therefore, this event has been taken to limit the period of this study.

This is a place oriented study within a given period and extent. In this study the archaeological and epigraphical materials, literary sources and cultural traditions have been enquired and employed with proper analysis. The modern publications relevant to the subject have been also consulted with analytical approach. Further, my prolong association with archaeological field work has enabled me to interpret the things on impirical grounds also.

NOTES AND REFERENCES

1. S.N. Rajaguru, *Inscriptions of Orissa*, Vol. IV, Bhubaneswar, 1966, pp. 244-52; see, the title "Brahmeśvara Stone Inscription of the Time of Uddyotakesrī (R.Y. 18)", this inscription informs about the construction of Brahmeśvara temple by Kolāvatī, the queen-mother of Uddyotakesarī in the latter's regnal year 18. The verse-12 refers to the name of this temple as Brahmeśvara and the site of its location has been called Siddhartīrtha-Ekāmra herein:

   श्रीमद् ब्रहमेश्वरस्य प्रणतमलहृत: स्पशर्तोमुक्ति
   दस्य प्रसादो भ्रड़कषाग्रस्तरणिरथयव्या हृतिस्थाणुकल्प:।।
   एकाम्रे सिद्धतीर्थे चतुरमरकुली चारूशाला
   समेत: कोलावत्या तयैष: क्षितिमुकुटनिभ: कारित: कीर्तिराज:।।

2. S.N. Sarangi, *Proposed Master Plan of Bhubanswar*, Bhubaneswar, 1988.

3. *Puri District Gazetter*, (e.d.) N. Senapati, Cuttak, 1977, p. 708.

4. ibid pp. 714-15 "The Draft Master Plan for Bhubaneswar was Published under the notification No. 2630-PL-2/68-TP, dated 30th July 1968 under the Orissa Town Planning and Improvement Trust Act 1956. Under the said act, the Town Planner, Orissa, Bhubaneswar has been declared the Special Town Planning Authority for Bhubaneswar, but a proposal has been submitted by the Town Planning Organization to the Government for the creation of a Development Authority for Bhubaneswar. So the Master Plan has not yet been finalised in view of the fact that once the Master Plan is approved, it will be very difficult for the Development Authority to effect the change subsequently."

   However, now the Bhubaneswar Development Authority is in function. Recently (1988) Mr. S.N. Sarangi, a retired I.A.S. Officer of the Orissa Secretariat, published a Master Plan of Bhubaneswar with the title "Proposed Master Plan of Bhubaneswar" in which he included 65 villages in addition to the Municipal area.

5. *Brahmapurāṇam (Devanāgari)*, (ed. with Hindi translation) Taranish Jha, Hindi Sahitya Sammelana, Prayag (Allahabad), Chapter 41: Ekāmra Kshetra Māhātmya Kathanam, verse-75

   देवस्य पश्चिमे पूर्वे दक्षिणे चोत्तरे तथा ।
   योजन द्वितीयं साध क्षेत्र तद्भुक्ति मुक्तिदाम् ।।

6. *Śabdakalpadrumah (Devanagari),* Vol. 4, (ed.) Raja Radhakantadev, 1961, p. 62 : Yojanah-chatuh krosyarthe.

7. *Svarnādri Mahodaya*, (pub.) Ratnakar Gargabatu, Calcutta, 1938, Ch.I, verses : 28-29. C.f. S. Tripathy, "Ekāmra-Bhubaneśvara", Orissa Historical Research Journal, Vol. XXXIII, No. 1 & 2, p. 152.

देवस्य दक्षिणे पूर्वे पश्चिमें चोत्तरे तथा,
योजनं तृतीयं यावच्छेत्रं मुक्तिप्रदागोत्तमम् ।
पंचकोशं प्रवक्ष्यामि श्रृणध्वं भे द्विजोत्तमः,
खण्डाचलं समासाद्य यत्रास्ते कुण्डेश्वरः ॥
असाद्य वरहीम् देवीं वहिरंगेश्वरावधि, ...

8. Cf. M.M. Mahapatra, *Traditional Structure And Changes In An Orisssan Temple,* Calcutta, 1981, p. 8.

9. Refer to the heading "Gangua".

10. See the map attached to the appendix of Bhubaneswar in the *Puri District Gazetteer*, op. cit.

11. See the map in p.no. 65 of *Ancient India* No. 5, A.S.I. New Delhi, 1949. In this map both the streams are shown as Gangua.

12. In a temple-construction spiral and inclined ramps were being raised to reach conveniently in greater height. See, V. Dehijia, *Early Stone Temples of Orissa*, New Delhi, 1979, figs. on p. 29, illustrating the spiral and inclined ramps for construction of temples. The biggest inclined ramp would have been made on the southern side of Liṅgarāja temple. The present land-surface beginning from Liṅgarāja temple upto Keutasahi seems to be the residue of that incined ramp.

13. The land-surface in front of the main gate of Liṅgarāja temple is much higher than the level of courtyard of the temple. If the level of outer land is decreased to the lowermost level of courtyard the water of Lake-1 may reach to this temple and further flow towards east and southward direction. Thus, it is probable that in very earlier days when the embankments to the Bindusagara were not raised the Lake-1 or shorter Gangua alias river Gandhavati was flowing to very close of the Liṅga inshrined in the temple of Liṅgarāja.

14. S.N. Rajaguru, *Inscriptions of Orissa*, Vol. V, pt. i, Bhubaneswar, 1975, pp. 211-20: Ananta Vāsudeva Temple at Bhubaneswar, see verses 9 & 10.

    तत्र च क्षेत्रमेकाम्रकस्ताश्रृतम् एकदेवकुल देवकुलोराकुम्दमुतम् ।।
    यत्र गिरिजापतिर्वसति गंधसिन्धोर्मिस
    निधाय निकटे धुनिममर सार्थ सार्थप्रपा
    यद्वैयक कृपाभारन्यधिति कृत्तिवासः श्रियम्
    महेन्द्र पदवीत्वरीं सुभट चोड़गंगनेताम् ।।

15. The neighbouring villages, viz: Rampur, Bharatpur and Bhimpur were full of forests and cleaned for the new capital. See M.M. Mahapatra, *Traditional Structure and Changes In An Orissan Temple*, op. cit., pp. 9-10.

16. *Puri District Gazetteer,* op. cit., p. 20-21.

17. D. Mitra, *Udayagiri and Khandagiri*, A.S.I. New Delhi, (reprint) 1975, p. xix.

18. *Puri District Gazetter*, op. cit., p. 21

19. E. Hultzch, *Corpus Inscriptionum Indicarum,* Vol. I (Inscriptions of Asoka), Varanasi, 1969, p. xiii.

20. The description of major river-courses is based on *Puri District Gazetter,* op. cit. pp. 12-13.

21. *Indian Archaeology* 1984-85 - A Review, A.S.I. New Delhi, p. 59, "Excavation at Khalkatapatna, dist. Puri".

22. *Puri District Gazetteer*, op. cit., p. 13.

23. Excavations were conducted by Orissa State Museum, Bhubaneswar. In a published leaflet the finds of Amphorae and Roulleted ware were reported. Roulleted ware is also reported from the excavations at Sisupalgarh; see, B.B. Pal, "Sisupalgarh 1948 : An Early Historical Fort In Eastern India", *Ancient India*, No. 5, New Delhi, 1949, p. 71 & pl. XLII. The practice of roulleting is of Mediterrainian origin. The

roulleted ware has been found mostly in coastal areas, viz : Salihundam, Kaveripattanam, Tamluk etc. See, A. Ghosh (ed). *An Encyclopaedia of Indian Archaeology*, Vol. I, New Delhi, 1989, p. 258.

24. Data obtained from *Puri District Gazetter,* op. cit., p. 708.

25. K.A Chaudhury and S.S. Ghosh, "Wood Remains from Sisupalgarh", *Ancient India*, No. 8, A.S.I. New Delhi, 1952, pp. 28-32.

26. Cf. A. Ghosh, *The City in Early Historical India,* Simla, 1973, pp. 39-41.

27. V. Gordon Childe, "The Urban Revolution" *The Town Planning Journal,* Liverpoole, 1 April, 1950, pp. 3-17, Cf. A. Ghosh, *The City in Early Historical India,* op. cit., pp. 23-28. See also, V. Gordon Childe, *What Happened In History*, Penguin Books, (reprint), 1978; and *Manava Pragati Ki Kahani* (Hindi version) of Gordon Childe's *Man Makes Himself* by Kripa Shankar Mathur, Lucknow, 1971, pp. 126-160.

28. Cf. A. Ghosh, *The City in Early Historical India*, op. cit. pp. 36-7.

29. ibid.

30. ibid., p. 45.

31. B.B. Dutt, *Town Planning In Ancient India,* Delhi, 1977, p. 29.

32. The Description is adopted from P. V. Begde, *Ancient and Medieval Town Planning in India*, New Delhi, 1978, and from A. Ghosh, *The City in Early Historical India.*

2

Habitation Sites and Nature of Settlements

Human occupations at Bhubaneswar area can be traced back to the pre-historic periods. The stone age equipments such as handaxes, polished stonecelts and microliths have been frequently reported from different spots in this area.[1] However, it was in the Early Historical Period that we find prolific evidence of human habitation in different kinds of settlements. The epigraphical records as well as the archaeological remains available in this area reveal that in the Early Historical Period there were urban settlements with the seat of Government as well as the monastic settlements. Further, the cluster of ancient temples and a number of tanks containing auspicious water, are the living testimony of the celebrated Hindū-*tīrtha*. This *tīrtha* variously called as *Ekāmrakshetra, Kṛttivāsakshetra* and Bhubaneswar survived even in the most adverse conditions, when Hindu kingdoms were in very lamenting days.

Thus, the history of human dwellings in Bhubaneswar area is varied. The human occupation in this area perpetuated from a hoary past to the modern times; and throughout this long period Bhubaneswar has witnessed many ups and downs, shifting of habitations from one site to another which were guided by the decline of a particular settlement and emergence of another new settlement of different economic set ups and cultural forms.

In the Bhubaneswar area, there are four prominent sites; namely; (1) Dhauli Hills, (2) Udayagiri and Khandagiri Hills, (3) Sisupalgarh and (4) *Ekāmra*-Bhubaneswar; which had been inhabited by greater masses or occasioned frequent public gatherings. Each one of these sites lay in diverse physiographical position. Besides, a few stray caves, cult-images and archaeological remains bear the testimony of human habitation in those places. Now, we shall discuss about these habitation-sites and the nature of settlements thereto:

1. DHAULI HILLS

These hills, crowned by a modern Buddhist *stūpa*[2] and an old *Śiva* temple, are discernible from a very long distance. Lying about 8 km. southward from Bhubaneswar town this site can be reached by a metallic road that bifurcates from Puri road near a bridge on river Daya. The south-eastern part of this hill range is a low ridge extending lengthwise while in the north-western portion the contours are highly increased. The total hill range lay in the north-west to south-east orientation. The highest peak has a steep decline towards north while on the other side the contours are gradual. The river Daya flows on the north-western periphery of this hill range. A large waterpool called Kosallagaṅgā is located near the tip of lower ridge. The waterpool is lengthwise in shape and extends in north-south orientations. Apart from a little succession of laterite formation, the hills are surrounded by agricultural lands. The river has also formed fertile alluvium belts on either banks. There are a few present village settlements around these hills; namely; Dhauli, Kosala Ganga, Kosalaypur, Uttarsasn and Purbsasan.

The following monuments and archaeological remains have been found at Dhauli hill area :

(i) Rock Edicts of Ashoka: Engraved on a rock-boulder in the lower-ridge these rock-edicts consist of eleven out of Fourteen Rock-Edicts of Asoka, and two Sperate Rock-Edicts (also called as Kalinga-Edicts)[3]. At the top of the same boulder there is sculptured forepart of an elephant cut out in the rock which is also the execution of the same period, though it lacks the typical Mauryan polish.[4]

(ii) Ruins of a temple and a wall, ascribed to circa A.D. 7th century have been discovered by Department of Museum and Archaeology, Govt. of Orissa at the foot of hill.[5]

(iii) A Small Rock-Cut Cave bearing an inscription, dated in regnal year 93 of Śāntikara (a king of the Bhauma dynasty). The said regnal year is corroborated to A.D. 829-30.[6]

(iv) The Temple known as Bahiraṅgeśvara located on the top of hill datable to circa A.D. 9th century.[7]

(v) The Temple called as Dhavaleśvara located in the foot of hill and dated to circa A.D. 9th century.[8]

(vi) Another Temple in the foot of hill dated to the period of the Gaṅga dynasty.[9]

Besides, a few structural and antiquarian remains were noticed during an excavation conducted by the Archaeological

Survey of India of which any published report is not available, but the finds are referred by late Dr. K.C. Panigrahi who was also participant of the excavation. According to Dr. Panigrahi a trench laid in close vicinity of the Asokan Inscription exposed a thick wall made of rubble and mud-mortar. This wall was similar to the cyclopian walls of the New Rājagriha at Rajgir. In the above excavation some terracotta snakes and multi-spouted vessels were found resembling to those prototypes recovered from the excavation in the Maniar Matha at Rajgir.[10] The early occupation of the site is attested by cultural assemblage consisting of early pottery types and terracotta objects noticed in this area by researchers.[11]

In addition to the above, mention may also be made of Kosalla Ganga the large waterpool. The tank contains the remains of a palace. According to the *Mādlāpāñjī*, a chronicle of Lord Jagannatha Temple Puri, the tank was dug by Gaṅgesvara i.e. Anaṅtavarman Chodagaṅga of the Gaṅga dynasty.[12]

Since no extensive excavations have been carried out in Dhauli area, we have very meagre source-material for the detailed study of the nature of settlements of this area. Even though the epigraphical records of Asoka at this site furnish a substantial clue regarding the existence of a city settlement named Tosalī in the neighbourhood of this hill range. The Separate Rock Edicts of Asoka found at Dhauli and Jaugada (in Distt. Ganjam of Orissa) inform us about the existence of two cities: namely: Tosalī and Samāpā respectively. The first and second Separate Rock-Edicts of Dhauli inscription have been addressed to the judicial officers (*nagalaviyohālaka*), the Prince-Governor (*kumarāmātya*) and the *mahāmāta* stationed at *Tosalī*[13]; the same at Jaugada are addressed to the *mahāmāta* stationed at Samāpā.[14] Therefore, it is implied that the above mentioned cities were located somewhere near the site of respective rock-edicts. However, the availability of settlement remains at Dhauli area as well as at Sisupalgarh of same period made a vexed problem for identification of the proper site of the city of Tosali. The fortified remains of Sisupalgarh are not very far from the site of Asokan Edicts at Dhauli Scholars like Panigrahi[15] and others were very much tempted to identify this contemporary fortification with the city of Tosalī. The site of Sisupalgarh was excavated by Sri B.B. Lal of Archaeological Survey of India. The excavations revealed the occupational deposits from circa 3rd century B.C. to circa A.D. 4th century.[16] Though the settlement appeared contemporaneous to Asoka, however, Sri Lal was not much inclined to corroborate it with the Tosalī of Asokan Edicts, as evident from his writings, "At Tosali opposite the granite boulder bearing the Asokan Edicts there is an open stretch of land where one can find traces of ancient habitation in the form of early pottery and other antiquities. Similar remains are also obtainable along the southern periphery of the hills. A systematic excavation of these areas may reveal evidences regarding the identity of the Mauryan or pre-Mauryan town elsewhere."[17] Sri Lal had observed the principles of *Arthaśāstra* in the layout of Sisupalgarh fort,[18] but, since in his chronology the Asokan period coincided with the pre-defence phase and probably for that reason he had not given much stress to associate with the Tosalī of Asokan Edicts. However, while reporting the finds of next season's excavation at the same site, he has reviewed the chronology in the light of advance and uptodate researches done in the field of Black and Red Ware which was an important dating factor for the earlier occupational levels at Sisupalgarh.[19] The revised dating put the erection of defences in circa 3rd century B.C. which was the period of Asoka's conquest of Kaliṅga. In his article Sri Lal has further stressed the principles of *Arthaśāstra* responsible in Planning of this fort. His theory is confirmed by the discovery of lanes which were laid in grid pattern.[20] Relying upon the above notices we have reasons to accept that the fort of Sisupalgarh was founded by Asoka after his conquest over Kaliṅga. But, the city of Tosali was existing in Pre-Mauryan period also. We have literary references about the existence of Toaslī even in 6th century B.C. Mahāvīra is said to have visited Kalinga and preached at Tosalī. According to *Vyavahāra Bhāshya* there was Jina-image at Tosali guarded by king Tosalīka.[21] It appears that Kalinga and Tosalī were combined political unit in the Early Historical Period as were the *Visālā* and *Mithilā* in North India;[22] and during the Asoka's invasion over Kaliṅga the city of Tosalī was the capital of that country and for this reason Asoka made it the provincial capital. The might of Kalinga can be judged from the deadly resistance rendered by Kaliṅga people in the war with Asoka which is evident from the XIII rock-edicts of Asoka itself. Therefore, the capital city of this country would not have been of ordinary status. The poor-relics represented by pre-defence phase at Sisupalgarh ruled out the possibility of its being the Tosalī of pre-Mauryan period.

While searching the site of ancient city Tosalī, J.W. Mc Crindle suggested its location near the margin of pool called

KosallaGanga.[23] H.P. Sastri thought that the ruins along the road leading to Dhauli-Rock (Rock-edict Boulder) might represent Tosalī.[24] According to P.C. Bagchi and N.K. Bose Tosalī was situated between the river Daya on the west, KosallaGanga on the east and Dhauli on the north or north-east.[25] In this connection mention may be made of a Buddhist work in Sanskrit called "*Gaṅdvyūha*" or *Avataṅsaka-Sūtra* which is regarded by scholars to have been composed during the 3rd or 4th century A.D.[26] This work mentions a city named Tosali in the country of *Amita Tosalī* and indicates that on the north or east of that city was located a hill called *Surabha Parvata*. If we take the above landmark as the guiding factor for the location of the city of Tosalī in Dhauli area we find that the site suggested by Bagchi and Bose appears more correct. We have already noticed that the Dhauli hills lay in south-east to north-west orientation; therefore any settlement towards south of this hill will have to see that hill on its east and north directions. This was favourable location for the city of Tosalī as it can afford water from river Daya as well as from KosalaGanga.[27] Further, the contours of hill on that side i.e. on the south-western side are gradual and it is located on the foot of hills that the site of gradual slope is preferred instead of steep slope. It is, therefore, contended that the original city of Tosali was located on the north-west of Dhauli hills. The landmarks mentioned in "*Gandavyuha*" about the location of Tosalī thus disapprove the claims of Sisupalgarh for the original city of Tosalī though it was a contemporaneous settlement. It appears that the composer of this work was much inclined to the original city of Tosalī which would have been associated with the traditional sanctity of Buddhism.

The Separate Rock-Edicts also mention that the message of Asoka is to be read before the people of Tosali by those officers on several occasions.[28] It is implied in the contents of these edicts that Asoka wanted the publicity of his code of conduct and to reconcile the agrieved public of the city. The location of Rock-Edicts site at Dhauli hills is therefore indicative of the fact that greater masses of people were residing near the Rock-Edicts site and they had a convenient approach to this place. On the other hand the fort of Sisupalgarh was a royal camp lying detached from the original city of Tosalī which was founded by Asoka with elaborate security arrangements in view of the uncertain political atmosphere of the newly conquered country. Though, we do not find clear mention of city of Tosalī in Hāthīgamphā inscription of Khāravela; yet the activities of a much erlier ruler Naṅdarāja are frequently referred therein who is identified with Asoka by Dr. Barua and Dr. Panigrahi.[29] Dr. Barua has also taken the word "*Tanasuliya*" of Hāthīgumphā inscription as a corruption of Tosalī.[30] This reading and interpretation of the inscription not only approves the location of Tosali near the *Kaliṅganagara*[31] or *Nagara* of the Hāthīgumphā inscription but also asserts its co-existence.

The city of Tosalī existed upto A.D. 3rd or 4th century, which is attested by its mention in "*Gaṅdavyūha*" as well as in a *Prākṛit* inscription of Vīrapurushadatta, a ruler of Ikshvāku lineage who ruled in the 3rd century A.D.[32] The inscription is found at Nagarjunikonda. In this inscription Tosalī is listed in many independent kingdoms whose inhabitants received happiness through the Cylonese *Theris*. It appears that by the A.D. 6th century the city of Tosalī near Dhauli hill was abandoned and forgotten, as evident from the various copper plate grants of that period which refer to Tosalī as a kingdom but not as a city in this time; the kingdom of Tosalī now found divided, as references are available for such terms, viz. "*Uttara* Tosalī" and "*Dakshina* Tosalī". It is also evident from the texts of those copper plate grants that the capital headquarter of any of these Tosalī kingdoms was not located at Bhubaneswar area but their headquarters were situated in other places much away from Bhubaneswar.[33]

To know the layout of the ancient city of Toṣalī we have to wait for further discoverise by extensive horizontal excavations; the description of "*Gaṅdavyūha*" enables us to have a glimpse of the glory of this ancient city. According to the text, the city was thickly populated and had the lanes for pedestrians and vehicular paths for the movements of the chariots. The lanes and roads intersected the other lanes and roads, thus formed the carrafours and these carrafours had platforms.[34]

Most of the extant monuments at Dhauli suggest the religious sanctity of this place. It is already stated that the forepart of the elephant at the rock-boulder of Dhauli Edicts is a representation of Buddha. At the end of 6th edict at Dhauli occurs a word *seto* which means the White one. This word is employed for the said figure of elephant representing Buddha.[35] At Girnar Rock a similar citation is found below the 13th edict (at right side) which reads: *sarva sveto hasti sarvaloka sukhahāro nāma*.[36] At Kalsi rock a complete figure of elephant is incised with a label *gajatame*.[37] Therefore, it appears that the elephant figures at Dhauli was in worship under Hīnayāna Buddhism.[38]

The Dhauli hills were also celebrated for the Nāga-worship, a popular cult of ancient India, as attested by the finds referred by Dr. Panigrahi (mentioned earlier). Dr. Panigrahi proposed the co-relation of terracotta-objects from Dhauli area with those found inside Maniarmaṭh at Rajgir.[39] At Rajgir the association of the said objects is established with the Maṇināga cult which was very much popular among the people of Rājagriha.[40] In Orissa the hills and deities bearing the name of Maṇināgeśvara are still found.[41] The Kanas copper Plate of Lokavigraha dated in Gupta year 280 (A.D. 599-600) and of Srī Bhānudatta, year 5 (datable in 6th or 7th century A.D.), record a dedication to a deity named Maṇināgeśvara. The location of this Maṇināgeśvar is not clear; however, the readings of this copper plate by Dr. D.C. Sircar suggest the probability of its location at *Ekāmra* i.e. Bhubaneswar.[42] Earlier we have discussed the boundaries of *Ekāmrakshetra* which extended up to the Dhauli hills. Taking these facts into considerations we can co-relate the site referred by Panigrahi with the Maṇināgeśvara deity of the above mentioned copper plates. Both the copper plates refer to the *Brahman* students inhabiting at this place suggesting that the religious institution of *Maṇināgeśvara Bhaṭṭaraka* was also endowed with a monastery which was a centre of *Vedic* learnings. The natural and rock-cut caves as well as some thatched huts would have been the shelters for the students and sages residing at Dhauli hills. The religious sanctity which was attached to Dhauli hills and the existence of monastic settlements therein are further attested by the small rock-cut cave containing an inscription on its wall. The inscription records that Bhaṭṭa Loyomaka son of the physician Nannaṭa and grandson of Bhīmaṭa and born of

Ijyā and who was an inhabitant of *Virajā* caused to be made this *maṭha* of Āghyakā Varāṭi as a gift in the year 93 during the reign of the illustrious Śāṅtikara. The inscription was edited by Dr. R. D. Banerji in Epigraphica Indica Vl. XIX. In his notes and translation, he has taken maṭha as a temple and Āghyakā Varāṭi as the deity.[43] But, in the Saṅskrit laxicans "*maṭha*" is explained as the dwelling house or shelter for students and monks.[44] The inscription implies that the *maṭha* caused to be made (*kāritaṁ* means executed, see text), was the cave itself. Panigrahi also suggested that the word "*maṭha*" stands here for the auspicious lodge which was excavated here in the rock-boulder.[45] Further, the word Āghyakā Varāṭi does not seem to be the name of any deity, rather it appears the name of some teacher (*āchārya*) or scholar. Here the word Varātī appears to be the real name of the person while the term Āghyakā seems to indicate his origin relating to parents or native place. The architecture of the cave also suggests that it was inclined for some monk. At the Dhauli hills there had been a number of natural and artificial caves in the western extremity of main ridge.[46] The very existence of these caves and the evidences from the above inscription speak of the monastic settlements in the area. The existence of the temples of the Bhauma and the Ganga periods at Dhauli and the pilgrimage of Bhaṭṭa Loyomaka testify that religion sanctity was still attached with Dhauli hills though the city settlement of this area was abondoned by this period.

2. THE JAINA CENTRE OF UDAYAGIRI AND KHANDAGIRI HILLS

The hill-spurs lying on the west of the temple-town of Bhubaneswar have been associated with Jaina monks since the time of Khāravela the Chedi king of Kaliṅga[47] and even earlier to him. The hills located amidst the vast and high lateritic stretches, are about 9 km. from the Old Town of Bhubaneswar. The whole range is locally called as Khandagiri (the broken hill). A narrow defile divides this hill range into two main hills and therefore the above term coined for the name of this hill range; however, the north-western hill bears a separate name Udayagiri (the hill of Sun-rise). In the orthodox Sanskrit literature these hills were collectively called Khaṅdāchala.[48] In the inscriptions available in these caves at these hills the Udayagiri hill is called *Kumārī Parvata*[49] and the other one i.e. Khandagiri is referred to as *Kumāra Parvata*.[50] The hills come under revenue mauza Jagamara which is now a part of Bhubaneswar Municipality. The National High Way-5 passes by the east of these hills and a road coming from Bhubaneswar town passes through the narrow defile and leads towards Chandaka and Nandana Kanana.

Though, these twin hills were the main habitational sites of a particular class of people; however, another much isolated and low spur lies towards east which is called Naligiri. Though this spur is almost secluded from the Khandagiri, yet, a few ordinary caves can be noticed here also which would have been the recluses for the monks and ascetics of Jaina faith.

The hills bear ample testimony of human habitation and the nature of settlement flourished here in the form of rock-cut caves, dedicatory inscriptions, structural remains, cult-images and symbols.

(i) The Rock-Cut Monuments

The twin hills were honeycombed with a number of big and small caves for the rainy retreats of the Jaina-ascetics. These are called "*layana*" or "*lena*" in the inscriptions. The caves were excavated at different heights as per the availability of solid rocky mass. Charls Fabri counted 44 rock-excavations at Udayagiri and 19 numbers at Khandagiri.[51] On the other hand Dr. (Mrs.) D. Mitra has ignored most dilapidated and hence insignificant caves in her numbering. She has also left some unnumbered tiny oblong excavations locally called *Pavanāri-gumphā*. Thus she recorded and described 18 caves at Udayagiri and 15 caves at Khandagiri which cover all the extant rock-excavations having some significance.[52]

The design and plan of the caves were contingent upon the shape and size of rocks where a particular cave was excavated. Yet, a fully developed cave monastery consisted of one or more cells fronted with pillared verandah. The rear wall of verandah is found benched in many cases. The cells may be ranged in one, two or three sides of verandah. Some important cave-complexes, viz: Cave-1, 9 and 10 of Udayagiri and Cave-3 of Khandagiri have levelled ground for courtyard in front. Cave-9, Cave-1 are double-storey and have two and three wings respectively. The other types of caves are open in front and deprived of any verandah. In most of the caves a horizontal chase i.e. projected eave runs over the openings obviously designed as rain-shed to protect from windy showers.

The doorways and ceilings of the cells are low and hardly allow a man to erect. The cells may have two or more openings.

These openings were also provided with movable shutters. The floors of cells sloping from rear end towards front indicate that these cells were essentially meant as dormeteries. One cell was tenanted by many monks.

The excavator-architect of these caves had the conceptions of the structural houses in their mind and imitated also those constructional devices which were not required in such rock-excavations, thus we can notice the arched or concave ceilings, pillars supporting non-functional architraves, struts supporting non-functional (relieved) *chhajjā*-balconies. The interior of the cells is austerely plain though facade and brackets are profusely decorated with carvings which consists of narrative friezes, religious and traditional motifs and landscape etc. furnishing important information on contemporary society and faiths. It is noteworthy that this decoration in the walls of the monuments which were intended for the shelters of monks appears contrary to the life style of monks. However, the artisans working there did not consider that aspect and following the ornamental traditions of the age gave full expression to their creativity. In total the architects simulated the concurrent dwelling-houses which are still being survived by *kuchcha* village houses in the country.

(ii) Structural Monuments

Among the structural monuments mention may be made of an apsidal temple witnessed by the remains on the crest of Udayagiri hill, the available portion of which is built with laterite blocks. This is dated contemporaneously to the time of Khāravela and being corroborated with the structure mentioned in the

Hāthīgumphā inscription which is said to have been built by Khāravela on the *prāgabhāra* of *Kumārī Parvata* i.e. top of Udayagiri.[53] It is being corroborated that the cult object called as *Kaliṅga-Jina* which has been taken away by Nandarāja was brought by Khāravela and installed in the said structure.

Another structural edifice and *Jaina* Temple, identified through its collapsed fragments, was built on Khandagiri in medieval times. Its architectural members suggest that it was built in contemporary Orissan temple style.[54]

(iii) Inscriptions

There are a number of inscriptions engraved in the walls of caves recording the dedications of the donors. Among these the most important is that of king Khāravela inscribed on the brow of Hāthīgumphā cave.[55] This inscription consists of 17 lines and records the details of civil, military and religious achievements of the king. This also mentions that the rock-cut caves were caused to be excavated by him for the rainy retreats of Jaina ascetics (*arhat*). Among the other inscriptions,[56] the one at cave-10 of Udayagiri and other at cave-11 (Lalāteṅdu-guṁphā) of Khandagiri are important in view of chronolgy as they mention the rulers of the Bhauma and the Somavaṁśī dynasties respectively, the latter is also important for the evidences of cult-worship.

(iv) Cult-Images and Symbols

The early cult-symbols are depicted in relief-art of the rock-cut caves, viz: *naṅdipada, śrīvatsa* and *svāstika*; besides, scenes of tree-worship also are shown. However, during the medieval period when many dormitory caves of Khandagiri were converted into sanctuaries, viz: cave-7 (navamunigumphā), cave-8 (bārābhjīgumphā), cave-9, cave-11, we find the images of *tīrthaṅkaras* and *sāsanadevīs* carved on the walls of the caves. At the peak called devasabhā there are a large number of miniature monolithic shrines containing the representations of *tīrthaṅkaras* on their faces.

The exploration and excavation at the site also yielded a variety of antiquarian remains, worthmentioning among those in view of chronology are the Paleolithic tools and the microliths.

The existing step-wells and natural caverns are the evidences of the water sources in immediate vicinity and were important parts of this settlement which had been repaired from time to time.

The available source material at these hills have been variously exploited by the scholars for their particular themes. The chronology and the date of the rock-cut caves and relief-art as well as of the Hāthīgumphā inscription have been subjected for endless controversies by the scholars due to much defaced condition of inscription and lack of other corroborative records of that period in Orissa. But for the habitational history, the following facts seem to be well established; that the hills were occupied by the human beings from a hoary past. A large number of microliths have been found from excavation from this site. Besides, some stry paleolithic tools are also reported from these hills. Therefore, it is evident that this site was a resort of the pre-Historic people. The coarse-grained sedimentry rock is highly susceptible for the caverns still exist at many places in these hills.

These caves would have been the easily and readily available shelters for the pre-Historic man.

However, coming to the Historical Period we find that these hills were occupied by *Jaina* ascetics and monks. The earlier existing caverns would have served as the safe shelter to the wandering ascetics for their pennance. Therefore, the site was preferably inhabited by the monks. We have already mentioned that Mahāvīra himself visited Kaliṅga and praised Tosalī. Tosalī had a Jaina image guarded by King Tosalīka. We have already noticed that the ancient city of Tosali was situated at the foot of Dhauli hill. Taking to that we may corroborate that the said Jaina image might have been placed at Udayagiri-Khandagiri which was the centre of Jainas since very ancient times and therefore this hill might have been the place where Māhavīra had preached during his visit to Kalinga. It also suggests that Kaliṅga and Tosalī were inseparable in that period. Thus in view of the above references the Jaina occupation of these hills appears to be the time of Māhavīra himself.

The Hāthīgumphā inscription of Khāravela mentions that Kaliṅga-Jina (cult object) was taken away by Naṅdarāja long back which was brought back by Khāravela and reinstalled here (at the Udayagiri-Khandagiri).[57] Thus, it can be derived that during Asoka's time the Jaina centre was fully established at this site. However, the settlement received real court patronage during the reign of Khāravela whose date is now fixed in the middle of 2nd century B.C.[58] The king Khāravela caused many rock-excavations with a planning to provide protective resorts for the resident-monks and erected edifice for their religious congregation. In this act of charity the chief queen of Khāravela also contributed to the rock excavations by her name which is now called as Svargapurīgumpha.[59] The other distinguished donors were Mahārāja Kudepasiri and Kumāra Vaḍukha.[60] They seems to have been the successors of Khāravela. Besides, we have a number of epigraphical records on the walls which inform us about the names of the individual donors.[61] In many cases the portfolio held by the donor in royal service is also mentioned alongwith the name. Therefore, we can say that during the rule of Chedi dynasty this settlement of Jains was patronised by the rulers and the people in full zeal.

After the end of Chedi rule we do not find documentary evidence for the occupation of these hills till the end of the Śailodbhava rule; but for this reason it will be unwise to think that this place was abandoned by the *Jainas*. It is known from a copper plate grant of Śailodbhava ruler Māḍhavarāja that Jainism was existing in his territory and it was patronised by him;[62] therefore it may be suggested that the *Jaina* centre of Udayagiri-Khandagiri would have also been under occupation. However, coming to the Bhauma period we get the documentary evidences for the celebrity of this place. An inscription engraved in the Gaṇeśa gumphā referring to the reigns of Śāntikara a ruler of Bhauma dynasty records the piligrim visit of a physian to this place who was the resident of Virajā (modern Jaipur).[63] This record is an evidence of the eminence of this place during the reign of Bhaumakaras. During the rule of Somavaṁśī dynasty Jainism again received court patronage in these hills. This time the western hill got more importance and probably some structural temples were also constructed. In the Khandagiri hill a few old cells were also constructed. In the Khandagiri hill a few old cells were converted into sanctuaries by carving the reliefs of *Tīrthaṅkaras* and *Sāsanadevīs* on the walls. An inscription dated in the regnal year 5 of Uddyotakesarī refers to the restoration of the ruined well etc and installation of 24 *Tīrthaṅkaras* (i.e. carving of those reliefs).[64] The Khandagiri hill remained a living and active centre of the Digambara Jaina cult for a long period that continued upto 15th-16th century, during the rule of Gaṅgas as attested by some crude carvings in the few caves of Khandagiri.[65]

At present, though these caves are not occupied by monks, still the temple on the top of Khandagiri hill is a living temple and is being visited by Jaina pilgrims from all over the country; and *Jaina* or *Parvāra* merchants of Cuttack particularly assemble here once every year. The temple, though a neat modern construction in stone, existed at the time of Sterling's visit to this place in A.D 1925.[66]

From the preceding discussions it is evident that a fulfledged monastic settlement developed at these hills which was fostered by public and royal charity. It is also notable fact that some *Jaina*-cult objects of high veneration were being worshipped here since very ancient times. We do not have any evidence in record or legend for association of these hills with the birth place or the land of pennance of any of the *Tīrthaṅkaras* which would have been the main reason behind the great *Jaina* sanctity of this place. However, we get mention of reputed cult objects in literary tradition as well as in epigraphical record. Whether the *Jaina* image mentioned in *Vyavahāra Bhāshya* and the *Kaliṅga-Jina* of Hathigumpha were the same, is difficult to say in want of any certain corroborative fact. Yet, the greater celebrity of both tends to suggest that the same *Jina* image would have perpetuated to the time of Naṅdarāja and taken away by him which was again brought back by Khāravela. It is very plausible to take on various grounds that the hills of Udayagiri-Khandagiri became important *Jainna* centres due to existence of that celebrated cult object (*Jina* image or *Kaliṅga-Jina*).

Another noteworthy fact is that the caves in these hills were intended for the stay of monks only for a particular period in a year i.e. the rainy season (*varshā-vāsa*) which implies that in the remaining period of the year these caves were lying vacant; but this was not the case, we noticed that the hills were associated with the worship of a cult-object which necessiated a regular establishment and for this reason the hills would have been regularly inhabited by the priestly people and lay devotees. Further, as per the tradition of the monastic life an annual tour is essential for the monks and ascetics, therefore, it would have happened that monks from the other monasteries might have visited this place while the inhabitant monks visited other holy places. Thus the hills were always populated by moving people.

The medieval inscription in Navamunigumpha refers to the name of a Jaina teacher (*āchārya*) and a student (*chhātra*).[67] Some letters written with paint in the wall of a cell i.e.. Tatvagumpha suggest that a student was practising the alphabets of early *Brāhmī* script.[68] These evidences prove that the monastic settlement was not a mere shelter of wandering ascetics but it was a regular institution of religious learning guided by rules and regulations laid in religious scriptures and by presiding monks since the Early Historical Period upto the Medieval period. The place was associated with great religious congregations and the religious discourses.[69] The learning of scripts by the monks suggests that it was the part of their integrated curriculum.

Inhabited by monks, this settlement would have resembled with a village situated in the narrow valley of hills. Though, the caves were excavated in various contours as per availability of solid rocky mass located in different parts of hill, however their maximum concentration can be seen on the either side of the narrow defile. This defile resulted the formation of terraces on the either side, vertical faces of rock fronted by some horizontal space provided by the terraces proved the ideal place for the cave excavations. This narrow pass also served as a common path for the residents of the hills. For easy access the hills were connected with the main approach by steps cut in or inclined ramp. The present artificial ramp leading to the Udayagiri hill was exposed during the debris clearance in 1961-62.[70] Such a location of caves though incidental due to natural reasons, provided visual and natural neighbourhood between the residents of both the hills and facilitated communication between them. We already noticed the great influence of the style and tradition of village houses in the architecture and embellishment of these caves, and it is a noteworthy point that village factor is also present in the settlement pattern as evident from the distribution and location of residential caves. A further notable feature is that the dwelling caves were at low and harmonious contours. They are running horizontally and are found in clusters, whereas the topmost space in the peaks was reserved for the deities and their temples. This is also a usual practice that can be noticed in the settlement pattern of the villages.

Thus, we see that the apsidal structure[71] which was meant for worship place lay on the top of the Udayagiri hill. Similarly, the Khandagiri is crested with a temple and at the topmost terrace also contains a sanctuary now called as Devasabhā having more than a hundred miniature monolithic shrines most of them have been relieved with the Tīrthankara figures.

Thus, the settlement planning and architecture features simulate a close copy of traditional village habitation. The main objective of these rock-excavation was to provide the shelters with minimum amenities to the monks; however, the profuse ornamentation with various motifs and narrative friezes on the door-openings, brackets and exterior wall surface may be quoted in contrast to the very austere nature of the settlement. In this connection it may be noted that this was a usual practice in ancient times and we get the Hindu temples and the Buddhist stupas and other monuments full of delicately carved sculptures and intricate designs.

Water is an essential requirement for any human settlement. Natural lakes full of water existed in the laterite tract, but no such water was available at these hills except for a few hallows in the rocks containing insufficient water for the year. Therefore, to meetout this basic necessity a few reservoirs and cisterns were cut out to hold rain water. Such reservoirs exist on both of the hills. On the Udayagiri, there are two of this kind located near Gaṇenśagumphā (cave-10) one called *Hāthī-Nisūnī* and the other *Lalita Kuṇḍa.* On the Khandagiri, there are four such reservoirs; namely: *Ākāśagaṅgā, Rādhā Kuṇḍa, Śyāma Kuṇḍa* and *Guptagaṅgā*. Of these *Śyāma Kuṇḍa* and *Guptagaṅgā* are natural caverns but retouched by human hands.

The monastic settlement basically depends on charity. The settlement flourished during the royal patronage, however, the main source of subsistence of the ascetics was the alms. The ascetics would have wandered in nearby village habitations and city settlements. As witnessed by antiquarian remains we are able to say that there would have been contemporary village habitations at the site of the present village of Dumduma, Jagmara, Sundarpada and Kapil prasad apart from the urban settlements at Sisupalgarh and Dhauli hills; and *Krittivāsa-Kaṭaka* of Bhubaneswar, though a later contemporary, was another rich urban settlement in the neighbourhood of Udayagiri. Aropada[72] and Lakshmisagarpatna[73] were also the popular villages in Medieval period. Besides, there may have been many other rich villages in the fertile alluvial land of south and east as well as in the western and northern highlands. Thus, the Hills of Udayagiri-Khandagiri served as ideal places for monastic establishments by providing a secluded location to the monks for the meditation and *sādhanā* while its nearness from other settlements favoured to subsist the monks on the charity of the neighbouring people.

It would be worthy of mention that the cult worship and monastic settlement at Udayagiri-Khandagiri hills did not turn into priestly settlement and remained for its total life span the place of ascetics, teachers (*āchāryas* of religious learnings) and students (*chhātra*) learning the religious ethics, practices and associated curriculum, belonging to the particular sects of the Jaina faith.

3. ANCIENT FORTIFIED SITE OF SISUPALGARH AND ENVIRONS

A planned human settlement sprang in the fertile alluvial tract during 3rd century B.C. as stated earlier which is witnessed by the remains lying in the close vicinity of the Old town of Bhubaneswar. This fortification lay in the present revenue mauza Sisupalgarh. The huge defences, rising some 6 to 9 mt. high from the surrounding plains, are clearly visible from the road leading to Puri. Apart from the quadrangular area of the fortification the ancient habitation also extended in Mahābhoisasana and Brahmeśvara temple area towards north and upto Bhuāsunī temple towards west. Among these sites, the area of Sisupalgarh is under cultivation with a small village habitation inside the fortified area. The area of Mahaboisasana is also under cultivation and a little land of it is occupied by modern buildings near Brahmeśvara temple. The Bhuasuni temple area is covered with modern construction. All these sites lay along the Puri road.

Though the fortified site and its environs contain extensive remains of ancient habitation; yet, we lack documentary evidences from the site to identify it or corroborate it with any of the ancient cities of Kalinga or Tosali. The original name of this fortification is forgotten off and even none of the local traditions bear any record about these remains. However, the different occupational strata were identified in an excavation by Archaeological Survey of India under the direction of Sri B.B. Lal in the year 1948 and since that time the same excavation report served as the sole source of information for the cultural sequence and chronology of this site.[74] Sri B.B. Lal has given a total occupational period of about seven hundred years for the fortified settlement and the earliest of the site is dated to circa 3rd century B.C. Sri Lal mentions, "The excavation revealed the site was occupied from the beginning of the 3rd century B.C. to the middle of 4th century A.D. Through this occupation there was one integral culture, although no doubt there were gradual changes in some of the industries."[75] The chronology settled for this total occupation is as below:

I (Early Period) c. 300-200 B.C.

II A (Early Middle Period) c. 200 B.C. – A.D. 100

II B (Late Middle Period) c. A.D. 100 – 200

III (Late Period) c. A.D. 200 – 350

This chronology was based mainly upon a few type-finds; such as: the rouletted ware, the black and red ware, a coin of Huvishka, a gold coin imitating the Kushāna type and Puri-Kushāna coins. The early period represents the formative stage of Sisupalgarh and "defences" did not occur in this phase. In the Early Middle Period (IIA) the Sisupalgarh culture was in full bloom. The "defences" were erected in this period at about the beginning of 2nd century B.C. and they remained in use till about the abandonment of the site, although towards the later part they had structurally degenerated.[76]

But, the excavator, in his latest writings[77] revised the dates of first occupation of Sisupalgarh and subsequently the erection of "defences", on the basis of recent discoveries in respect of Black and Red Ware, which was the main dating factor for the earliest occupational stratum of Sisupalgarh. In his article Sri Lal writes, "since at that time (1948) the chronology setout by Sir Mortimer Wheeler, after his excavation at Arikamedu and Brahmagiri, was followed in toto, the earliest occurrences of the rouletted ware and the black and red ware dated respectively to circa A.D. 50 and 200 B.C. With these two landmarks the total habitation at Sisupalgarh was dated from circa 300 B.C. to A.D. 350. However, in the light of discoveries, particularly in respect of Black and Red Ware made in various part of the country since 1948, it appears that we were rather conservative in dealing the lower levels of Sisupalgarh. The beginning of the site may well go back to around the middle of the 1st millennium B.C. and the earliest occurrences of the Black and Red Ware to circa 4th century B.C.[78]

"The vertical trench laid across the fortification wall showed that the construction of defences did not take place at beginning of occupation of the site, but coincided with the first appearance of the Black and Red Ware. With the revised dating in the preceding paragraph, the defence would appear to go back to circa 4th century B.C."[79]

The pre-Mauryan occupation of the Sisupalgarh site was also suggested by Sri. A. Das basing upon the exploratory finds of punchmarked silver coins from Sisupalgarh.[80] In his inferences Sir Das suggested that Sisupalgarh was in flourishing condition during the Mauryan rule. Thus, the revised dating for period I and IIA of Sisupalgarh by Sri Lal is also confirmed in the light of subsequent discovery from the site itself. However, the total abondonment of Sisupalgarh fort after 4th century A.D. is questioned by Das who is inclined to prove that the Sisupalgarh fort was under regular occupation upto 18th century A.D.[81] He refers to some sculptural finds from Sisupalgarh village and also quotes tradition of *Mādalā Pañjī* in support of his view. Sri Das has given much stress upon an image of a goddess kept in the village for worship. The image also contain an inscription datable to circa 11th-12th century A.D. on its back slab. In this inscription occurs the name of Gangesvara who was the famous king Chodaganga of the Gaṅga dynasty.[82] Another archaeological evidence produced by Sri Das in a Hero Stone containing an inscription in Oriya. This stone was dugout by late K.N. Mahapatra from a paddy field at Sisupalgarh.[83] The historical data furnished by Sri Das is from *Mādalā Pañjī*, the chronicle of Lord Jagannath at Puri. In this record it is said that the fort of (at) Sisupal and Dhauli were destroyed by Muhammad Taqui the Naib Nazim of Cuttack.[84]

These notices of Sri Das are superficial and cannot be considered in right achaeological perspective. The image of goddess containing inscription of Gaṅgeśvara is kept under a shed, its findspot as well as locus (in archaeological terms stratified layer/pit etc.) is not known. We are quite aware of the fact that Ekāmra-Bhubaneswar was a great religious centre during the period of the Somavaṁśīs and the Gaṅgas. The religious activities of Choḍagaṅga at Bhubaneswar are well known. Bhubaneswar, full of such antiqurian remains, is not far off from Sisupalgarh village. Therefore, if any such image exist at Sisupalgarh, it is not very unlikely. It is most probable that the said image would have been brought from the ruins of old temples and monasteries of Bhubaneswar. Ṭhe same thing can be said for the herostone recovered by Mahapatra.[85] Sri Lal's excavation was a scientific work. He was aware of the literary tradition of *Mādāla Pañjī* as for he has mentioned it in a footnote of his excavation report;[86] but, he based only the material that he encountered in the well stratified deposites. As we see the fortified area still contains a small village-habitation and the early habitational remains are buried under the present habitation and agricultural fields. It is also true that while digging the paddy fields here and there villagemen encountered structural remains built in well dressed laterite slabs and these slabs were/are quarried by villagers for use in their own houses. During the archaeological excavation in 1948 trenches were laid out at four different parts of vast fortified area, viz: SP-I, SP-II and SP-IV. Of these the SP-I, which yielded the remains of residential structure and therefore termed as Habitation Area by the excavator, was under cultivation by present day dwellers. Therefore, it is obvious that the present habitation is not a continuation of ancient settlement; rather it may have been a *zamindari* settlement of late Mediaeval Period[87] which has been mentioned in *Mādalā Pañjī*. Thus, these late references of Sisupalgarh adds only that the site was sporadically occupied for ordinary village habitations headed by some local *zamindars*.

The available layout of the ancient settlement of Sisupalgarh clearly shows that this was a fortified town with high defence wall and gateways. The tentative plan of the fortified complex as restored by Sri Lal in his excavation report is as follows: "It formed a square plan whose each side, oriented approximately along the cardinal directions, measured about 1.2 km. length. The fortification consisted of a basic mud-rampart on the top of which were added two walls of kiln-fired bricks with rammed earth in between. Each side of fortification wall was pierced by two gateways placed in a manner so as to divide it into three almost equal parts. For each one of these gateways there ran thoroughfares right upto the corresponding gateways on the opposite side, thus, producing a grid pattern and dividing the whole settlement in nine major blocks. Of these blocks, the one in the centre appears to have been occupied by the ruling chief and nobles etc. The gateways were elaborately laid out with tight security arrangements. There were watch-towers also at the corners."[88]

"Thoroughfares were paved with stones and on their sides stood well-planned houses made basically of stones often supplemented with brick-work. Occasionally wells were also provided to supply the potable waters.[89]

Around the fortification wall a moat was dug which was linked with a perennial streamlet of Gangua to feed it water regularly.[90]

At present the whole site lay under crops-cultivation and village habitation. The excavation conducted here were confined to the Defence wall and Gateway. Only a small portion was dug inside the fortified area. Therefore, we lack the detailed pictures of habitation inside. However, it can be opined in modern terms that there existed an ordinary village like settlement prior to the erection of the Defences. In this habitation people lived in the mud-houses made of some perishable material probably that of timber (as no structural remains encountered).[91] But, this ordinary village habitation got impetus from some royal house during the 3rd-4th century B.C. and an extensive area was laid out in the plan of well defended fort. This planning obviously speaks that it was not an evolution of an ordinary village habitation into an urban settlement in due course; but, here was some political force behind the random emergences of such advanced settlement which is a rare example of planned fortification in the whole of India in the Early Historical Period as reported till now.[92]

The most and the high defence wall all around were intended for security but intricate security arrangements were made for this settlement, a fact which has come to light from the excavation of one of the gateways. This gateway is termed as Western gateway. Sri Lal described the position revealed in excavation as follows.[93] "Here the two ends of the main defensive wall turned outwards and each joined heel of massive L-shaped gateway flank built of laterite blocks of an average size 6' × 2' × 1'. The longar arm of the flank measured 160 feet and the shorter one 63 feet, while the width was roughly uniform, being 27-28 feet. The passage flanked by two longer arms took the form of a slow rising ramp, about 25 feet wide. Though it provided a straight access to the interior of fort, the entire system was well organised from a defensive point of view."

"There were two gates, one near the entrance and another further back at a distance of over 100 feet. Though no remains of door-leaves were obtained (for presumably they were made of wood), door-sockets were duly identified, the width of each gate being 13 feet. Immediately behind the first gate, and built into the southern flank, was a room (?) 9-feet wide. Sections cut across the passage way revealed several successive streets which were made up either by ramming hard mud or laterite gravel or as noticed in two or three cases by paving with brick-bats. Cart-tracks were observed at various levels. Adjacent to the inner gate, and pierced through the northern flank, was an ancillary passage 5 feet and 3 inches wide. It has a stone floor and was approached by three steps from the contemporary street level. It appears that the entire passage way was open to sky except at the two gates where the length of the portals was indicated by the rectangular blocks projecting from the longer arms of L-shaped flanks."

"The entire system at the gateway may be imagined to have functioned as follows. At a certain fixed hours in the night, the inner gateway was closed, stopping all vehicular traffic beyond this point. Pedestrians could, however, get in or out through the ancillary passage. A little later, the outer or main gate was closed, while the guards at its back probably still remained on duty."

Such intricate arrangements for checked entry and exit at the gateways focus the advance stage of town-planning. Inside the fort, though not excavated much, the habitation consisted of houses built in well chiselled laterite blocks. During 1948 excavation was carried out only in 60 square feet area to trace the habitation pattern. This digging revealed houseplans for two residential houses belonging to Period-III, though the remains of laterite structures were also encountered in Period-II A and II B but those could not be traced in plan as the digging in lower levels was restricted to small vertical pits. The two houses built in laterite blocks are adjoining and lay north and south with a two feet narrow space between them. Sri Lal suggested[94] that this narrow space cannot be taken to represent a lane; it only shows that the owners of two houses did not like to have a common partition wall. It may be recalled here that such a practice was recorded by Kautilya in his *Arthaśāstra* and similar instances are also found in the ruins of Sirkap (Taxila) and elsewhere. The northern house consisted of two rooms measuring 10 × 8 feet and 10 × 9 feet in front.[95] This house continued to be used till the late levels of the period. Towards the fag end of the period were built some baked-brick houses in one which can be seen a doorsill.[96]

Some structural remains available at the surface suggest that there was a big building complex at the centre of this fort.[97] These remains are 16 monolithic pillars of laterite some of which are still standing intact, while of the others no more than the stumps remain. The excavation if conducted at this complex may reveal some more pillars and the factual layout of this complex. Eight of the pillars, forming roughly an east-west row, mark the northern edge of the entire area, approximately 90 feet × 80 feet, covered by the group.[98] On the north-east the semblance of a chamber (9 × 8½ feet) is produced by the location of four pillars at the corner points.[99] The remaining pillars lie to the south on a slightly raised ground but no plan can be made out with out an excavation.[100]

The average height of the pillars above the ground is 14-15 feet at bottom and also of 2-3 feet at the top is 2½ square in section while remaining portion is facetted (except two pillars which are circular in section) the faces from 8 to 16.[101]

These pillars certainly belong to the Early Historical Period and not to the Late Mediaeval Period and some of them contain the medallions of the type common at Bharhut and Sanchi and also in a few examples in the Udayagiri-Khandagiri caves.[102]

As already stated the habitation of this settlement extended beyond the limits of fortification. Towards the western side pot-sherds can be picked up to the Bhuasuni temple. On the western side there is a tank with stone embankment alongwith huge stone debris suggesting the area under occupation. On the northern side plot-sherds are scattered upto Brahmeśvara temple and the contemporary antiquarian remains have been recovered even farther that is Bhāshkareśvara temple area.[103] In this side a mound called Mahabhoihuda in mauza Mahabhoisasana is noteworthy for contemporary deposits. At Mahabhoisasan mauza, there lay very big tank adjoining to the northern defence wall of Sisupalgarh. This tank has a very strong and high embankment built with huge sized laterite blocks. The embankment is arranged in two terraces, from the top terrace to the whole depth of tank lead the stepped courses. The dimension of this tank is amazing. Its one arm may be approximately 500 mt. Mahabhoihuda, the tank and the other remains found at Bhāshkareśvara temple suggest that the greater concentration of the habitation was on the northern side. Towards this side some stupa mounds are also noticed in the vicinity of Brahmeśvara temple.

The layout of the citadel and of the extensive habitation remains at near Sisupalgarh reflect its greater political importance during the Early Historical Period. However, its identification with any of the known ancient cities is not yet established. The present name of the fort has been derived from the small village Sisupal located within the boundaries of the fortification itself. Its association with the king Śiśupāl of *Mahābhārat* or Śiśupāl Kesarī of the Somavaṁśī-Kesarī dynasty is ruled out by Sri Lal for unfitting chronology.[104]

The identification of this settlement is proposed either with the Tosalī of Asokan Rock-Edicts or with the Kalinganagar of Hāthīgumphā inscription or with the both. It is to remind that a pre-Mauryan date to the original city of Tosali is already discussed suggesting a distinct site i.e. Dhauli; it is also contended that fortification of Sisupalgarh also developed in a date contemporaneous to Asoka, therefore, it would certainly have been associated in some way with the ancient city of Tosalī. Further, this settlement also continued in the time of Khāravela, therefore, it had also been associated with him and its corroboration with Kalinganagara can be proposed well. But, as the Hāthīgumphā inscription did not say anything about the distance and direction of the city of Kalinga from the Udayagiri-Khandagiri hill, hence left a probability of its being located anywhere far or near irrespective of the location of the inscription. Availing this state of uncertainty some early historians tried to hurriedly identify this city with those places bearing similar names and known from other sources. Thus, a claim was made for the Kaliṅganagara located on the banks of river Vaṁsadhara near Mukhaliṅgam in Srikakulam district of Andhra Pradesh which had been the capital of the rulers of the Eastern Ganga dynasty and frequently referred in their copper plate grants, to be identified with that of Hāthīgumphā inscription. Similarly, following the description of Kālidāsa in the epics *Raghuvaṁśa*, the capital city of Kalinga was located on the sea-shore and a port town named Kalingapatanam was proposed for identification to the Khāravela's Kaliṅganagara.[105]

Though, these corroborations diverted much attentions of scholars but ultimately were discarded in view of unfitting chronology and for want of settlement-remains respectively. Inspite of prevailing controversies the date of Khāravela is put in between 2nd century B.C.[106] and 1st century B.C.[107] On the other hand the Eastern Gaṅgas (Gaṅgas of Kaliṅga) had evidently a post-Gupta origin and their capital at Kaliṅganagara was established in circa 8th century A.D.[108] The excavation at Kalingapatanam do speak of a contemporanious settlement i.e. from 3rd century B.C. to 4th century A.D. But, except for a brick-built stupa (dia. 26 mt.)[109] no such settlement remains, which can be corroborated with the city-discription of Hāthīguṁhā inscription, have been yet reported.

However, in case of Sisupalgarh the documentary evidence (though indirectly) as well as settlement remains both come forth in support of its being erstwhile Kalinganagara. Scholars like S.N. Rajaguru,[110] Dr. Balasubrahmanyam[111] and Dr. B.M. Barua[112] agreed that the Bhubaneswar region was the main political centre during the reign of Asoka as well as of Khāravela and Kaliṅganagara of Hāthīguṁphā was located somewhere near present Bhubaneswar city. Dr. Balasubrahmanyam mentions that it should not be confused with that Mukhalingam in present Andha Pradesh. He also provides a clue from Hāthīguṁphā inscription that, the word *idha* means "here" or at this place which indicates the location of Kalinganagara somewhere near the inscription.

While editing the Hāthīguṁphā inscription Dr. Barua[114] has pointed out more confirmative clues about the location of the Kalinganagara. The line-6 of the inscription speaks, "in the fifth year his majesty caused the canal open out by the king Nanda, 300 or 103 years back to be brought into the capital from the *tanasuliya* road." The identification of Nandāraja and the interpretation of notation of *ti-vasa-sata* had remain highly controversial. Yet, the latter interpretation as 103 years has proved more logical than 300 years and it is accepted in majority of modern works and even K.P. Jayaswal and R.D. Banarji, in their joint work have accepted the reading of 103 instead of 300 and interpreted as the year one hundred and three of king Nanda; thus opined the existence of some Nanda-era.[115] For the identification of Naṅdarāja referred in the insciption, Dr. Barua has proposed the name of Asoka who is more relevant in the case of Kaliṅga than that of the Naṅda dynasty of Magdha. The identification proposed by Dr. Baua is supported by Dr. Panigrahi[116] on substantial grounds which is accepted in this work considering the logical and relevant corroboration. One more word of the above of Hāthīguṁphā inscription i.e. *tanasuliya* remained inigmatic. Dr. Barua has suggested its identification with Tosalī; as said in his own words "*Tanasuliya* road from which the canal opened out by king Naṅda was extended upto the capital of Kaliṅga. I still think that the *tanasuliya* road was no other than a road connecting the Tosali of Asoka with the Kaliṅganagara of Khāravela."[117] Thus, above interpretation leads that Kaliṅganagara was located somewhere near the city of Tosali. We have already suggested the area between Kosallaganga and Dhauli as the probable site of erstwhile Tosalī city, therefore, in all probability the site of Sisupalgarh stands for Kaliṅganagara.

Another aspect related to the settlement pattern also support the same corroboration as we can compare the description of the capital city in the Hathigumpha inscription with the fortification remains of Sisupalgarh. The *pākāra* (Sanskrit : *prākāra* i.e. fotification wall), the *gopura* i.e. portal, the *tadāga* i.e. tank or lake of the inscription[118] can be tallied with the "defences", the "portals" on gateways (and corner-towers now indicated by mounds upon defence wall), and the "big tanks" (one at Mahabhoisasana Mauza and another on the western side of defences) all still present in the form of ruins at Sisupalgarh area. In the same line of the inscription a word is read as *khibīra* by Dr. Jayaswal and he interpreted it as the name of a sage (*rishi*).[119] On the other hand Dr. Barua suggested this word as the name of the city of Kalinga.[120] But, it seems that *khibīra* occurred here as a corrupt form of Sanskrit word *śivira*. The tendency of corruption of *śa* in *ksha* and *kha* is noticed often in usage e.g. *śudraka-kshudraka-khuddaka*; and in Orissa *ksha* is always replaced by *kha*. Further, the letter *va* is always replaced by *ba* in Eastern India. Therefore, in all probability the word *khibīra* stands for *śivira*. In Silpasaatras *śivira* is treated as a settlement type and referred to as military encampment, which may be of an army on war expedition or may be an advance encampment at the outside of the capital city to guard against any attack.[121] It is stated earlier that the habitation of Sisupalgarh extended beyond the fortification towards northern side with a vast tank. The embankment of tank has wide terrace in two stages and provided with spacious steps. The slope of embankment yields a convenient approach for the quadrupeds like elephants and horses. It appears that this tank might have served as the main water source for the armies camping outside the fortification which was an advance camp of *śivira*. Such advance camp of military near the capital city was in practice in ancient times for immediate check of any danger. Therefore, we can say that the tank site was for military encampment.[122]

One more statement of Hāthīgumphā inscription is supported by a circumstantial evidence yielded by excavation at Sisupalgarh. Sri Lal has traced a phase of collapse and subsequent repairs of the gateway flank of fortification[123] and this can be corroborated with the statement of king Khāravela in line-2 of Hāthīgumphā inscription i.e. about the repairs of fortification.

Besides, a seal (surface find) bearing an inscription in early *Brahmī* is found from Sisupalgarh.[124] The inscription reads *amata sapasanka* which means the *amātya* (minister) named *Sapasanaka*. The presence of an *amātya* at Sisupalgarh is a testimony of its greater political importance.

Therefore, the preceding discussions tend to suggest that the fortification at Sisupalgarh were erected by emperor Asoka after his conquest of Kalinga. The *Śilpa-śāstras* invariably emphasized that each new dynasty should establish its own capital. A king residing in a city laid by his enemy or by any other will meat death in no time.[125] In case of Asoka's rule in Kaliṅga a such hostility in between the people of Tosali and Magadhan administration is well reflected in the Separate Kaliṅga Rock-Edict at Dhalui. This hostility and suspicious attitude of the inhabitants towards a new Government which came into power by conquering them was a natural outcome of the political affairs between two countries. However, Asoka assured the people for administration of justice and forbade his officers from committing any prejudiced judgement against any person. However, the safety and security of the Govt. Officers was also important point which necessitated the erection of a new citadel with proper security arrangements. The history bears a greater testimony that the most of the conquerors established their new capital-cities; and in the same analogy it is also true that Asoka would have founded the new seat of his provincial government near the old city separately. The location of ancient Tosali is already pointed out in Dhalui area; thus, the fortification at Sisupalgarh were an extension of the old city, and for this reason we do not find any separate mention of this fort, rather the old place name i.e. Tosalī, is referred to frequently.

However, it seems that this citadel was locally called as *nagara* and by the time of Khāravela this term became so much popular that the city of royal court was referred by that name only. In Hāthīgumphā inscription we come across a word *Kaliṅganagarim* indicating the capital of Khāravela. The same is referred to as *nagarīm* in line-4 and *nagaram* in line-6 of the above inscription. The inhabitants of that city were referred to as *paura* while the remaining population of the country is called *jānapadhān*.

While classifying the different types of the human settlement, the *śilpaśāstras* designated a fortified town as *nagara* or *pura.*[126] This type of town is a place of district headquarters and commercial activities. The word *nagara* also signifies permanence and strength of fortification. *Nagara* and *pura* as a suffix to the names of cities, towns and villages is very common since the very ancient times and still continuing. However, the term *nagara* is also found singularly as the name of localities, villages and towns. At present in India, there are atleast 34 villages and two towns having the name of *nagara* or *nagarā*, 20 village as *nagaram*. 5 as *nagarī* and 13 as *nagariā*. There are villages having names like *nagarabātī*, *nagarabārī* and *nagaraparakara*.[127] Due to increase of population or for other reasons new colonies or hamlet villages always develop in the outskirts of the old cities or villages. This tradition is very old. In western Uttar Pradesh, Hariyana and Punjab, one may frequently come across a term *nagalā* (corrupt from *nagara*) for a village settlement which is originally a hamlet village. Another survey of village/town settlements discloses that a particular locality or a big village or town, containing the houses or palace or a *zamindar* or *raja* is called *nagara*. The tradition still survives, i.e. a village towards north of Fatehpur-Sikari in district Agra (U.P.) is named as *nagara* as there resided a big *zamindar*, the *havelī* (palace) of the *zamindar* still exists. Another example of such tradition is from Orissa itself. The royal abode of ex-*rājā* of Puri is still called *naara* (नअर) and the famous Rānīguṁphā cave of Udayagiri hill was called *rānī-naura* (house of the queen) by local people and both are corrupt forms of word *nagara*. The tendency of naming the places or settlements on this old tradition still survives in the minds of people and we can see that new city of Bhubaneswar where the headquarter of state capital is located, is commonly called the "Capital" by local people and New Delhi is referred to often as *Rājadhānī* in newspapers.

Thus, we can conclude that the fort was founded by Asok for the safe seat of his Provincial Government, which lay detached from the old city. The fort contained the offices, the residences of the officers and the staffs of the Government and the Security Forces as well as the important traders and other persons of elite class. The fort was laid out in classical pattern which resembles the principles stated in *arthaśāsta* of Kautilya for the layout of fortifications (*durgavidhānam*). The fort was provided with complete security arrangements, i.e. the defence wall with bastions, the well-guarded gateways and the water-moat surrounding the whole fort. The grid patterned road and the water-supply arrangements from the perennial stream and reserve stock in tanks also give an idea of sensible planning.

The fort continued as a place of royal court during the rule of king Khāravela and later also. By this time a considerable civil population settled in the neighbourhood of this fort. There was also an advance military camp near the fortification. The popular term for this fortified settlement was *nagara* while the old city was known by its original name Tosalī.

4. EKĀMRA-BHUBANESWAR: THE ŚAIVITE TĪRTHA

This is the pilgrim centre with the seat of Śaivism and became popular by 5th-6th century A.D. The sacred area extended far and wide, yet, the concentration zone lay in the southern portion of laterite tract which is watered by Lake-I. This area is now known as Oldtown. As per the distribution of the temples in this land, it appears that the proper temple-town consisted of the whole revenue mauza Bhubaneswar apart from the portions of Baragada, Basuaghai, Bhimapur and Kapileswar.

The Lingarāja temple (circa A.D. 11th century) occupies the central position and is also the centre of all social and religious activities of the place. This is a stupendous monument enclosed within a vast compound wall (520 feet north-south and 465 feet east-west enclosing an area of 4.595 acres).[128] The height of its main tower (*śikhara*) is 45.11 mt. while the succeeding hall of pyramidal type is 28.96 mt. high.[129] Inside the vast compound, there stand more than 100 temples of small and big size apart from other structures, viz: platforms, wells and kitchen which were added through out the past centuries.

There were about 500 temples at Bhubaneswar according to an old survey. However, due to agelong negligence and the rapid urbanization in this century the number of temples is in decreasing order. The temples other than the Liṅgarāja are located in vast extent of laterite tract either in cluster or in isolated position. Most of the temples were erected either on the banks of Bindusāgara or in the vicinities of Liṅgarāja temple. The temples which were lying away from natural water-sources were provided with the stepped tanks.

The following extant temples are important either for chronology or for architecture or both:

Śatrughneśvara group of temples (6th century A.D.), Paraśurāmeśvara, Svarnajāleśvara, Uttarreśvara and Mohinī temple (circa 7th century A.D.), Vaitāl, Śiśireśvara, Mārkanḍeśvara, (circa 8th century A.D.), Siddheśvara, Kedāreśvara, Rājārānī, Ḍakrā-Bhīmeśvara, Brahmeśvara (circa 10th-11th century A.D.), Megheśvara (cica 12th century A.D.), Ananta-Vāsudeva (circa 13th century A.D.), Chitrakārinī, Sāri-Deul, Bhāskareśvara and Yameśvara (circa 13th century A.D.).

Besides, mention may be made of the well known Mitreśvara, Varuṇerśvara, Makareśvara, Koṭitīrtheśvara, Rāmeśvara and Nāvakeśvara which are the late specimen of the degenerate phase. There are other so many nameless temples. The Kapileśvara temple, though a late construction of degenerate phase, yet, have important role in the celebrations of festivals and religious rituals at Bhubaneswar.

Besides the temples, the monasteries (*maṭha*) were other important institutions of this settlement. R.L. Mitra, who visited in 19th century, mentions the five monasteries in function.[130] The present Bhāratī *maṭha*, Kāpālinī *maṭha* and Śivatīrtha maṭha are of ancient times. We also come to know about a monastery of 13th century A.D. named Siddheśvara maṭha referred in an inscription found in the neighbourhood of Kedāra-gaurī temple.[131] Another inscription originally attached to Megheśvara temple (circa 12th century A.D.) informs us about the existence of several monasteries and cloisters for the studies of *Veda*.[132] Svapneśvaradeva, the founder of that temple also established a village for the pious *Brāhmans* that was referred to as *Brahmapura*.[133] This village was superintended by the Śaiva teacher Vishṇu who had consecrated the above temple.

The priestly class and other caste people resided in the vicinities of temples and monasteries. Each class would have established its own hamlets as evident from the present settlement terms e.g. *Baḍusāhī, Paṅdāsāhī, Miśrasāhī, Gudiāsāhī, Haḍīsāhī, Keutsāhī, Kumaharasāhī, Baurisāhī, Kammābārī,* and *Baḍheibānko*.[134] These are the old and traditional settlements of local people. The dwelling houses in these *sāhī* (localities) were ordinary constructions made of mud-wall (occasionally of laterite blocks) and thatched roofs.[135] However, some of the houses were of stone masonry covered by flat roof bordered with low parapet walls. A depiction of such house is available in the carvings of Paraśurāmeśvara temple where a lady is shown in awaiting pose holding one door-leaf. The parapet wall over the roof is designed in the form of battlemented merlons. The same design is discernible in some present dwelling houses located on the *Barādāṇḍa* road; and one such prominent building is that of *Baḍu-Nijoga*.

The traditional market still exist to the south-side of Lingaraja temple and called *Hāto* in local terms. The main ancient road is the *Barādāṇḍa* (wide path) on which passes the chariot of Lord Liṅgarāja during the Car-festival celebration.

The religious settlement was decorated with a number of tanks of big and small size which gives a pleasing appearance to this place. The *Bindusāgara* is the biggest tank. In the chapter I it is stated that this tank or lake was originally the part of big Lake-I. This tank is referred in *Brahmapurāṇa*[136] as well as in inscriptions[137] also where it is called *Bindusara*. The other noteworthy tanks are *Kedārakuṅda, Gaurīkuṅda, Pāpanāsinī, Kotitīrtha, Champāpukharī,* and *Kapileśvara* tank. One *Lakshmīsāgara* was also a very big tank which is now covered with modern constructions.

The religious settlement of Bhubaneswar evolved mainly due to greater sanctity of the *liṅga* enshrined in the Liṅgarāja temple. This *liṅga* has the various epithets, and derivative of these epithets were used as the synonyms for the name of this settlement. The popular epithets are as follows :

(i) Svayaṁbhū

The word means self-existant. The term is popularly used for *Śiva.* The orthodox text of Bhubaneswar called *Svayaṁbhū* to the *linga* enshrined in this great temple. This *linga* is a natural stone and not fashioned by human hand like the common *linga*. In the *Mahābhārata* this *linga* is referred to as *Svayaṁbhū* and the place is mentioned as *Svayaṁbhū-vana* (the forest of self-existant).[138]

(ii) Ekāmra

According to *Brahmapurāṇa* this place of *Śiva* was originally having a single mango tree and therefore, known as *Ekāmaraka*.[139] We get the frequent mention of this word in literature and inscriptions[140] either singularly or with suffix such as *vana*, *kanana* and *kshetra*. This term is still very popular as place name as well as the deity.

(iii) Kṛittivāsa

In the votive inscriptions[141] of the Gaṅga and the Sūryavaṁsī period this epithet is frequently referred to and accordingly the place was also mentioned as *Krittivāsa-kshetra* and *Krittivāsa-kataka*.

(iv) Bhubaneswar

An inscription of Sūryavamsī period refers to this epithet.[142] Another earlier inscription datable to 12th century A.D. mentions *Tribhuvanādhīśvara*.[143]

The term Bhubaneswar is frequently referred for this temple as well as the deity in *Mādalā Pañji*.[144] The *Ekāmra Purāṇa* and other orthodox text referred to the epithets of *Tribhuvaneśvara* and *Bhuvaneśvara*. According to *Ekāmra purāṇa* the *Gauda*-king Śaśāṅka build the shrine of

Tribhuvaneśvara.[145] Thus, this epithet appears to have been current during circa 7th century A.D.

Bhubaneswar, the constellation of eight holy centres (tīrth), was initially a favourable natural resort for the religious and spiritual people who had engaged themselves in pennance for salvation from the cycle of birth and death; and searching peace being disappointed from temporal world as evidenced by its various designations; such as *Ekāmra-kānana*, *Śāmbhava-vana* and *Siddha-kshetra* (the land of the people having divine powers). On the other hand presence of a cult-object in the form of *liṅga* and the large water sources which themselves were the symbols of holiness, increased the religious importance of the site. The cult-object and the availability of water combined with the inhabitant of ascetics and medicants having supernatural and miraculous powers and performing amazing practices attracted the common folks who were worshipping the cult-objects on their traditional faith and consulting the medicants to get rid of their worldly troubles and fulfillment of their desires. *Ekāmra-kshetra* established among the great *tīrthas* as soon as it was associated with the various religious rituals to be performed among the Hindus for the welfare of the living generation as well as for the salvation of the dead ancestors. It was being advised by the orthodox scriptures that *snana* (ablutions), *śrāddha* (ritualistic sacrifice for the forefathers), *tarpaṅa* (offerings of water) and *dāna* (donation) at *Ekāmra* is meritorious.

The religious sacredity of *Ekāmra* led the temple building activities at this place; and we can say that in the next phase *Ekāmra* assumed the status of a temple-village with the construction of the shrine of Lord *Tribhuvaneśvara* by king Śaśānka in the 7th century A.D. Though, there would have been some earlier temples at this place but the substantial royal patronage brought it into focus and led a follow on action in subsequent periods resulting in the construction of a cluster of temples in the whole area. Further, the building of stupendous Liṅgarāja temple or shrine of *Kṛittivāsas* at the same site of *Thribhuvaneśvara* (whose supremacy was established since the time of king Śaśāṅka) caused the concentric growth of temples and the affluent public settlements. Now this temple became the nucleus of public life. Being a seat of religious learning a few monasteries had already grown at this place, however, the temples alongwith their religious importance also functioned as miniature universities and a number of monasteries were associated with them where subjects like Astronomy, Literature, Religion, Philosophy, Grammer, *Arthaśāstra* (politics) etc. were taught in great detail. Further, various auspicious days, religious festivals and fairs were being celebrated at and around the temple precincts which attracted public gatherings in large-scale. All the above conditions and various job-opportunities at construction sites brought a radical change in the demographic map of *Ekāmra* and yielded the scope for prospering markets. Thus, a fulfledged settlement of religious moorings supported by multifarious economic set-up emerged around the temple. As the temple was the center of all the public activities and the prime factor for the growth of the settlement; and it also governed the way of life of the people, therefore, we can designate it a temple-town/temple-city or *Devarāṭa* type settlement of the classification of Viśvakarmā's *Vāstuśāstra.*[146]

As we have noticed that royal patronage played an important role in the development of *Ekāmra* as a temple-city; however, this patronage was not offered only by issuing a charter of donatory grant from the headquarter of the kings, rather the development of the temple-city was caused by the visit, stay and keen interest in the particular type of architectural design of various kings and their family members and officers, as evident from various inscriptions and traditions. The frequent visits of the kings, the royal family and the important officers of the state required a lot of security measures and separate residential wings within the convenient distance from the temple of the Lord *Krittivāsa*. This necessity of the royal house was liable for the foundation of a well furnished palacial complex guarded by defence walls and sufficient military encampments, in the southern periphery of Lingaraja temple complex that became popular as *Krittivāsa-kaṭaka* during the rule of the Gangas and the Suryavamsis. Though, it would have been founded much earlier probably in the time of the Somavamsis for their halt when they were constructing the great temple of Lord *Krittivāsa* (Liṅgarāja); however, it became a highly sensitive base of royal politics during the last days of the Gangas, and according to the *Madala Pañji*, Kapilendra, the founder of the Suryavamsi dynasty, coronated himself on the throne at Krittivāsakaṭak after abdicating the last ruler of the Ganga dynasty; he issued his edicts towards his feudatories and dissidents with due sanctions of *rājaguru* (royal spiritual preceptor) and the priest of the temples.[147] The seriousness of this event apparently speaks that how much military strength was in store at Krittivāsakaṭaka that enabled a successful coup de etat by their commander-in-chief. Thus we can say that at Bhubaneswar alongwith temples a *kaṭaka*[148] i.e. the royal military establishment also came into existence with full vigour; and played decisive roles in changing of the ruling dynasties.

Though at present no such structural remains are traceable on the southern side of Liṅgarāja temple which can match with any fortification or large palacial complex. However ruins of two extensive palaces had been seen by Andrew Sterling in the first quarter of ninteenth century A.D. at Bhubaneswar. He writes,[149] "The ruins of two extensive palaces belonging to the Rajas of Kesarī line, are seen at or near Bhubaneswar. There is likewise a very superb tank lying north of the temple...." Though the location of these palaces in the account of Sterling is not pointed out exactly, yet this account indicates that the palacial ruins were near the Lingaaja temple which is mentioned Bhubaneswar. From the existing landuse around the Liṅgarāja temple, it is obvious that there was no space for the extensive palaces on northern, eastern and western sides; however, towards the south we find an extensive space for bi-weekly market, newly constructed houses, the building of B.M. High School and its wast playground behind it; and one can still see some ruins stone-alignments in this playground; thus it was the site of royal settlement at Bhubaneswar. It is also confirmed from the local traditions in the Old Town area that the old royal palace was located at the fields B.M. High School (B.M. *paḍhiā* (field) *ru thilā puraṇā rājabāṭī* (old royal palace). From Hatosahi and B.M. High School upto Mehtab road towards south lay an extensive area. In this area either we notice the modern habitation or the open stretches of land and no ancient religious edifice is existing in this area rather traces of some large wall alignments are noticeable near the playground of B.M. High School. Therefore, it appears that the royal settlement extended here in a very large area. The ruins of this royal settlement totally disappeared and the modern occupation took place in the empty land. As happened with most of the ancient sites from where the building materials have been robbed for the new constructions and other various public use, the deserted ruins of the royal settlement at Bhubaneswar also suffered from such public vanndalism. The speedy urban growth in the Old Town area would have expedited the robbing of building material for use in private constructions as well as by government contractors. Since the ruins of this royal settlement were of purely secular nature and there is no such moral fear of any imprications in robbing off or destruction of the ruins of secular nature unlike the edifices of religious nature which are protected against any kind of public vandalism by fear of imprications and religious sentiments. Thus all the physical remains of krittivasa-kataka, the royal settlement at Bhubaneswar (the two extensive palaces seen by Sterling) vanished with ravages of time leaving no evidence of its erstwhile glorious existence.

On the other hand the settlements exclusively of learned brahmins; namely Brahmpur, recalls us the *Maṅgala* type of village of the classification of *Māyāmata* and *Mānasāra*[149] and the *sāsana* villages (though now not available in their oldest form) may be compared with the *Agrahāram* i.e. the village of noble brahmans (*Viprendra*).[150]

The ancient villages; viz: Lakshmisāgarpatnā and Aropada (present Adapada) were the common peasantry settlements inhabited by different class of peoples and were neighbouring the Temple-city. The inhabitants of these villages served as trustees in execution of some dedicatory grants made by generous donors.[151]

5. STRAY CAVES CULT-IMAGES AND ARCHAEOLOGICAL REMAINS

Mention may be made of a few stray caves, cult-images and archaeological remains, which indicate some other habitations in the laterite tract, some rock-cut caves are still to be found in the close vicinity of Bhāskareśvara temple, mostly to the west of it. These caves have been excavated on the edge of immense laterite quarry which seems to have supplied stones for the buildings and gateways of Sisupalgarh, situated about a mile south-east of this place.[152] The laterite quarry usually form a stepped tank; however, cave formation is also noticed in many present quarries, therefore, there is no doubt that these rock-cut caves would have been created due to quarry of laterite slabs but subsequently would have been occupied by *śaiva* ascetics. The location of these caves falls in the area called *Siddhakshetra or Siddhāraṇya* (mentioned earlier). Therefore, it is reasonable to think that these caves would have been the shelters for the *Siddhas* and *Munis*. This fact can be further attested by some verses of the Śailodbhava inscriptions which mentions the ascetics residing in the caves.[153] The most remarkable group of these caves known as *Pañcha-Pāndava-Gumpha* consists of three big chambers and a spacious verandah. This rock-cut cave appears a planned excavation. According to Dr. Panigrahi the proximity of these caves with the Bhāskareśvara temple, indicate that they were intended for the abodes of the *Śaiva* ascetics.[154]

These caves, though, would have been inhabited by medicants and ascetics of any sect, however, it appears that there was no such organized monastic settlement like that of Jainas at Udayagiri-Khandagiri hills, rather, the ascetics of these caves were the stray occupants as usually seen in the *tīrthas* or religious places in present time also.

The images of *Yakshs* and *Nāgas* are reported from Bhubaneswar area. Dr. Panigrahi had discovered four images of *Yakshs,* there from a field in village Dumduma near Jagmara not far from Khandagiri and one from a site near the Brahmesvara temple.[155] Two of the Dumduma statues are in complete form, each measuring 5 feet 7 inches in height and each having a socket on the head. The third specimen, discovered from Dumduma is a torso which has been spilt in the middle dividing it into two half, front and back portions. The fourth specimen discovered from the Brahmesvara area, is a little smaller in dimensions and has holes below the arms. Another *Yaksha* of Dumduma type is being worshiped as a village deity in a small shrine situated near a tank in the village Badagada[156] (a revenue mauza of Bhubanesar Municipality). A sixth specimen is reported to be in the village Panchgan about 8 km. to the west from Oldtown.[157] All these *Yaksha* images are of Early Historical Period and Dr. Panigrahi placed them in the post-Kharavela period or in 1st century B.C.[158]

Apart from the *yaksha* image statues of *Nāgarāja* were discovered By Dr. Panigrahi,[159] one from village Sunderpada and another from Brahmesvara temple area (now both under Bhubaneswar Municipality). Further, three ancient *nāga*-images were discovered by Sri Suryanarayan Das from the same

Sundarpada village.[160] Dr. Panigrahi dated the *nāga*-image discovered by him in the post-Khāravela period while S.C. De dated the three naga images discovered by Sri Das, in the 1st or 2nd century A.D.[161] These images are accepted as independent cult-images of the place by the above scholars.[162] However, the *yaksha* images discussed earlier are being considered as the architectural members due to the certain features. Their posture typical to *Bhārapūtaka* figures e.g. bent under the weight of a structure which they raise, the uplifted hand and socket on the head, led Dr. Panigrahi to consider these *yaksha* figures as architectural members of some *stūpa* like at Sanchi.[163] In this context a noteworthy point is that all these *yaksha* images are of life size and carved in round. Dr. Panigrahi himself mentioned that these are the free-standing figures,[164] but associated with architectural forms due to above reasons. However, as I feel, these *yaksha* images are more akin to those free standing *yaksha* images reported from different parts of India, viz : Parkham, Badoda (both in district Mathura, U.P.), Patna, Pawaya Rajghat and Other places.[165] The Rajghat *Yaksha* is also in the similar pose with the bent knees and uplifted hands and being suppressed by the weight, but, it is a free-standing image having three faces.[166] Therefore, it appears that there were certain conventional poses and postures of the images of *yakshas* which could be followed in both the usage whether it may be an architectural member or an independent cult-deity. Further, the socket on the head of those images do not prove that this was necessarily meant to fix the successive architectural member of an edifice, I have come across several such modern representations modelled in the traditional style in the villages of Orissa as cult-object, some of these images were depicted carrying pitcher or bucket like object. Therefore, in the same analogy it can be said that there would have been also additional picther or bucket like object in those ancient *yaksha* images. Whatever may be in both the case they witness some settlements at or near their findspots.

We have already noticed that two *yaksha* images are still under worship, therefore, it would be more reasonable to suggest that the remaining *yaksha* images were also under worship. Dr. V.S. Agrawal[167] has proved the antiquity and the tradition of *yaksha* and *nāga* worship, which goes back to the Mauryan period and also referred to in Vedic and Buddhist literature and was widespread in all parts of India. He has rightly stated the tradition of having a local god by each village though a temple of Śiva may be commonly worshipped by many villages.[168] This tradition is still alive in Orissa. One can see *Mā-Maṅgalā* as village goddess located in the outskirts of villages. Similarly, there are installed the male statues in the outside of villages or near a path, these statues are called *Mahāpurusha*, but, these statues do not represent any named individual of the village which obviously suggests that the installation of *mahāpurusha* is a follow on action of ancient tradition in which *yaksh* images were installed in each village.

Thus, taking these *yaksha* and *nāga* images as independent ward-deities it is opined that there were ancient village settlements at Dumduma, Jagmara, Brahmeśvara, Bdgada, Sundarpada and Panchgan in the Early Historical Period.

A few archaeological remains of Early Historical Period found in the laterite rock are discussed by Dr. Panigrahi in his writings. Worthmentioning are a bell-capital of Asokan type and few railing pillars and sculpture of a lion found in Bhāskarśvara area. The *liṅga* inshrined in the Bhaskaresvara temple is also identified as Asokan pillar by Dr. Panigraphi.

The archaeological potentiality of the area near Bhāskareśvra temple and Brahmeśvra temple is well established and we have already considered this area as suburban extension of Sisupalgarh fort. Panigrahi has discussed in detail about the linga of Bhāshkareśvara and his arguments are substantial.[169] I would like to add here only that Asokan pillars were mistaken for *Śivaliṅga* by people in other places also, viz: *Lauria* Nandangarh and Lauria Araraj. The very word "lauriā" is a derivative of the word "laurā" (phallus) in Bihar.[170] Therefore, there is nothing surprising that at Bhubaneswar also an Asokan Pillar was worshipped as *Śivaliṅga* by the common people as well as staunch Śaivites; however, it is certain that this pillar was conceived as *liṅga* before fifth century A.D. as evident from its mention in *Brahmapurāṇa*.[171]

The find of bell-capital at Asoka-jhara and the association of traditional festival i.e. *Asokāshtamī* with place suggest the importance of this place in the early settlement history of Bhubaneswar area.

NOTES AND REFERENCES

1. *Indian Archaeology 1961-62 – A Review,* A.S.I. New Delhi, pp.36–7 and 103; 1963-64, p.90 and 1969-70, p.60.
2. Śāṅti-stūpa (peace pagoda) is constructed between 1971 and 1972 by the Kaliṅga Nippon Buddha Sangha founded by

Japanese Indian devotees to commemorate the conversion of Asoka to Buddhism after the bloody conquest of Kaliṅga.

3. E. Hyltzsch, *Corpus Inscriptionum Indicarum*, Vol. I, op.cit., see for the details of rock-edicts of Asoka.

4. S.K. Saraswati, *A Survey of Indian Sculpture,* New Delhi, (reprint) 1975, pp. 30-31. See, also D. Mitra, Bhubaneswar, A.S.I. New Delhi, (reprint) 1978, p.5.

5. *Indian Archaeology 1963-64 – A Review,* p.17.

6. R.D. Banerji, "*Dhauli Cave Inscription of Śāntikara;* The (Gaṅga) Year 93", Epigraphica Indica, Vol. XIX (1927-28), Calcutta, 1929, pp. 263-4. R.D. Banerji corroborated the regnal year referred in the inscription with A.D. 731. But beginning of the Bhauma-*Saṁvat* is now accepted as A.D. 736, therefore, the date of the inscription comes as A.D. 829-30, see, K.C. Panigrahi, *History of Orissa*, Cuttak, 1981, pp. 74 and 383.

7. K.C. Panigrahi, *History of Orissa*, op.cit., pp. 382-3.

8. See the list of the protected monuments of the State Government of Orissa. The original Dhavalesvara temple does not exist any more now, as a new temple was constructed at this site in 1972.

9. K.C. Panigrahi, *History of Orissa*, op.cit., p. 430.

10. Ibid., p. 17.

11. B.B. Lal, "Sisupalgarh 1948: An Early Historical Fort", *Ancient India* No. 5, op.cit., p. 66. The potentiality of this area is also examined by other several scholars. Recently, K.S. Behera of Utkal University has carried out some trial excavations; and Mr. and Mrs. Brandtner have been working in the Early History of Orissa. They noticed some terracotta objects, pot-sherds and rock-cut caves of early period as well as of medieval period. Noteworthy finds are Black and Red Ware, Red Ware

and Lids which are comparable to the pottery of Sisupalgarh. For details, see, Martin Brandtner, "Archaeological Remains at Dhauli", H.C. Das and others (ed.), *Kṛṣṇa Pratibhā,* Delhi, 1994, pp. 29-37.

12. *Puri District Gazetteer,* op. cit., p. 737.

13. E. Hultzsch, *Corpus Inscritionum Indicarum,* Vol. I, op.cit., pp. 92-107. See, first Separate Rock-Edict of Dhuli: "At the word of *Devanāṁpriya* the *Mahāmātras* at Tosali (who are) the judicial officers of the city have to be told (thus)."

    (देवानांपियसि वचनेन तोसलियं महामातं नगलवियोहालका वतविय अं किछि दखामि हकं तं इछामि)

    See, second Separate Rock-Edict of Dhauli: "Ac the word of Devanampriya, the Prince (governor) and the *Mahāmāras* at Tosalī have to be told (thus).

    (देवानं पियसि वचनेन तोसलियं कुमाले महामाता च वतविय अं किछि दखामि हकं तं इ....)

14. ibid, pp. 111-8. See, first Separate, Rock-Edict at Jaugada: "The *Mahāmāras* at Samāpā (who are) the judicial officers of the city, have to be told (thus)

    (देवानं पिये हेवं आहा समापायं महामाता नगलवियोहालक हेव वतविय)

    See, second separate Rock-Edict at Jaugada : "Devanmpriya speaks thus. The *Mahāmātras* at Samāpā have to be told (this) at the word of king. (देवानं पिये हेवं आह समापायं महामाता .....)

15. K.C. Panigrahi, *Historical of Orissa*, op. cit. pp. 15-16.

16. B.B. Lal, "Sisupalgarh 1948 : An Early Historical Fort", *Ancient India* No. 5, op. cit. 70-75.

17. Ibid., p. 66.

18. Ibid., p. 74.

19. B.B. Lal, "Sisupalgarh: Some Reflections on Its Layout", C.P. Sinha (ed.), *Archaeology and Art,* Vol. II, Delhi, 1990, p. 590.
20. ibid., p. 591.
21. Cf. D. Mitra, "Glimpses of Jaina Relics In Orissa", *Orissa Review* (Monument Special), Bhubaneswar, 1976, p. 15.
22. H.C. Raichaudhary, *Political History of Ancient India,* Calcutta, 1953 (sixth ed.), pp. 118-21.
23. J.W. Mc Crindle, "Ptolemy's Geography of India And Southern Asia", *Indian Antiquary,* Vo. XIII, 1884, p. 382.
24. D. Das, *The Early History of Kaliṅga*, Calcutta, 1977, p. 11.
25. ibid.
26. Cf. S. Tripathy, "Ekāmra-Bhubaneśvara", Orissa Historical Research Journal, Vol. XXXIII, No. 1 & 2, Bhubaneswar, p. 135.
27. Though, this waterpool is said to have been excavated by king Choḍagaṇga, however, it is obvious that this is a natural, waterpool like those lakes at Chudangagarh and Bhubaneswar, and existed at this place in the early period also. It is probable that Choḍagaṅga would have cleaned and renovated his water-source.
28. E. Hultzsch, *Corpus Inscriptionum Indicarum,* Vol. I, op.cit. p. 96: And this edict must be listened to (by all) on (every day of) the constellation Tishya. And it may be listened to even by a single (person) also on frequent (other) occasions between (the days of) Tishya. See, p.100 also for the similar versions.
29. K.C. Panigrahi, *History of Orissa,* op.cit., p. 28.
30. B.M. Barua, "Hahigumpha Inscription of Khāravela", *Indian Historical Quarterly,* Vol. XIV, Delhi, (reprint) 1985, p. 474.

31. For the identification of Kaliṅganagara, see, the discussions under the heading : Sisupalgarh.

32. J.Ph. Vogel, "Prakrita Inscriptions From A Buddhist Site At Nagar junikonda", *Epigrahica Indica,* Vol. XX (1929-30), New Delhi, (reprint) 1983, pp. 22-3.

33. S.N. Rajaguru, *Inscriptions of Orissa,* Vol. I, pt. ii, Bhubaneswar, 1958, p. 117.

    उत्तरतोसल्यां वर्तमान भविष्यन्महासामन्त ....., p. 121 (No. 23) : तोसल्यायामष्टादशाधिराज्यया परमदैवताधिदेवत श्रीलोकविग्रहभट्टमहासामन्तो दक्षिणतोसल्या सटिदविषयात् ततिच्छापत्तने ......, p. 125 (No. 24) परममाहेश्वर श्री शम्भूयय्यने (शम्भूयसस्) शासतिं दक्षिणतोसल्यां वोर्त्तनोकात् परम दैवताधिदेवत श्री परमभट्टारक चरणकमलामल क्षौणि हारधिगम प्रतिहत: कलियुगागत दुरितनिचय: (यो) महाराज शिवराज कुशली । p. 138-39 (No. 29) ॐ स्वस्ति ।। जयस्कन्धावारादाग्रतक्षकवासकात् परमदैवताधिदेवत श्री परमभट्टारक पादानुध्यातो महाबलाधिकृतान्तरम महासान्धिवग्रंहिक सोमदत्त: कुशली ।। उडविषये उत्तरतोसल्यां सरेफहार विषे वर्तमानभविष्यन्महासामन्तमहाराज राजपुत्र कुमारमात्योपरिक विष्यपति तदायुक्तक दाण्डवासिक स्थानान्तरिकानन्याश्च चाटभट वल्लभजातियान् महामहत्तर कूटकोलस पुस्तपालाद्यधिंकरणांश्च यथाहं पूजयत्यवगमयति च विदितमस्तु भवतां यथास्माभिरेतद्विषये सम्बद्ध अडयार ग्रामो ....

    And see the notes on pp. 151-2, in the same volume, where Rajaguru suggested the Virja (modern Jajpur) as the probable capital of the kingdom of Tosalī in sixth century A.D.

34. Cf. ibid., p. 150.

35. E. Hultzsch, *Corpus Inscriptionum Indicarum,* Vol. I, op.cit., p. 26.

36. ibid., p. 26.

37. ibid., p. 50 and plate 25.

38. On the crest of the southern hill (Asokan Edict area) remains of brick-building (possibilly stupa) could be seen a century

ago, when Beglar visited the site, see, J.D. Beglar, "Reports of Tours In The South Eastern Province In 1874-75 and 1875-76, Calcutta, 1882, (*Archaeological Survey Reports,* ed. A. Cunningham, Vol. XIII, p. 96).

39. K.C. Panigrahi, *History of Orissa*, op. cit. pp. 17, 347-8.

40. M.H. Kuraishi, (revised by A. Ghosh), *Rajgir,* A.S.I., New Delhi (reprint) 1975, pp. 20-26.

41. D.C. Sircar, "Two Plates From Kanas", *Epigraphica Indica*, Vol. XXVIII (1949-50), Delhi, 1958, pp. 329-31.

42. ibid., pp. 329-31.

(I) सीमालिंगानि संस्थाप्य (चे)क (ा) म्ब (की) यं श्री म(णि) नागेश(व)र भट्टारक म(ठी)य

बलि चरू सत्र प्रर्वतनाय नानागोत्र मैत्राययणीयं ब्राम्हणानां (स्ति) तये ताम्रपट्टीकृत प्रतिपादित:

(लोकविग्रह का ताम्रपट्ट)

(II) चैकाम्ब मणिनाग भट्टारकाय च छात्र मठ ब्राह्मणानाम् पतिपादितस?

(भानुदत्त का ताम्रपट्ट)

43. R.D. Banerji, "Dhauli Cave Inscription of Śānitikara; The (Gaṅga) Year 93", *Epigraphica Indica,* vol. XIX, op.cit., pp. 263-4

श्री शान्तिकरदेव राज्यसम्वत् 90 3
इज्यागर्भजेन विरजोवास्तव्य वैद्यनन्नटपुत्र
भीमट पौत्र भट्टलोयोमकेन आध्यकावराटि
मठोऽयं देयेभ्य कारित .... बारक

44. *Śabdakalpadrumah (Devanagri)*, op.cit. Vol. 3, p. 569.

छात्रादिनिलय: मठन्ति वसन्ति छात्रादयोऽत्र मठ: आदिना परिव्राजक क्षपणोदीनां गृह:

45. K.C. Panigrahi, *History of Orissa,* op. cit., p. 133.

46. *Puri District Gazetteer*, op. cit., p. 740. A large number of rock-cut caves are encountered on all three hills of Dhauli. Brandtner has also suggested that these caves were inhabited by ascetics and monks, some of these caves are described by him, see, Martin Brandtner, "Archaeological Remains At Dhauli", *Kṛṣṇa Pratibhā*, op.cit., pp. 32-3.

47. B.M. Barua, "Hathigumpha Inscription of Khāravela", *Indian Historical Quarterly*, Vol. XIV, op.cit., pp. 459-85.

48. *Svarṇādri Mahodaya*, op.cit., verse 29.

    खण्डाचल समासाद्य ......

49. B.M. Barua, "Hathigumpha Inscription of Khāravela", *Indian Historical Quarterly*, op. cit., p. 468, L-14 of Inscription. *Terasame cha vase supavta-vijaya-chake Kumārīpavate arhate.*

50. R.D. Banerji, "Inscriptions In the Udayagiri and Khandagiri Caves", *Epigraphica Indica,* Vol. XIII (1915-16), Delhi, (reprint) 1982, p. 166.

    ॐ श्री उद्योतकेसरी विजय राज्य सम्वत 5 श्री कुमार पर्व्वत स्थाने ....

51. Charles Fabri, *History of the Art of Orissa,* Bhubaneswar, 1974, p. 17.

52. D. Mitra, *Udayagiri and Khandagiri,* op.cit. The numbering of D. Mitra is followed in this work.

53. ibid., pp. 43-4.

54. ibid., p. 13.

55. This inscription is edited and interpreted by many scholars. The readings and interpretations rendered by K.P. Jayaswal jointly with R.D. Banerji in *Epigraphica Indica* Vol. XX, by B.M. Barua in *Indian Historical Quarterlly* Vol. XIV and by D.C. Sircar in his *Select Inscriptions*, are more popular and more accurate.

56. R.D. Banerji "Inscriptions in he Udayagiri and Khandagiri Caves" *Epigraphica* Indica, Vol. XIII, op.cit., pp. 159-67; and B.M. Barua "Minor Old Brahmi Inscriptions In The Udayagiri-Khandagiri Caves", *Indian Historical Quarterly*, Vol. XIV, pp. 158-66.

57. Line-12 of the Hathigumpha inscription refers to the Khāravela's military expeditions against Magadha and recovery of Kaliṅga-Jina, Jayaswal and Banerji edited this line as following in *E.I. Vol. XX*, pp.80 and 85.

    Māgadhānaṁ cha vipulam bhayam janeto hathi Sugāṁgīyam pāyayati Māgadhaṁ cha rājānaṁ Bahasatimiaṁ pade vaṁdāpayati. Naṅda-rāja-nītaṁ cha Kaliṁga-jinaṁ saṁnivesa.....

    Kaliṅga jina is identified with Śītalanātha who was born at Bhadalapura, which is the same as Bhadrapura or Bhadrāchalam in the Godavari district of the Madras Presidency. ibid. p. 85, note 38. Though, Barua's reading differs in respect of *Kaliṅgajina,* however, former's view appears more appropriate and can be corroborated with the literary sources which speaks about the existence of *Jina* image in Kaliṅga.

58. K.C. Panigrahi, *History of Orissa*, op.cit., pp. 27-32.

59. R.D. Banerji, "Inscriptions In the Udayagiri and Khandagiri Caves", *Epigraphica Indica*, vol. XIII, op.cit. pp. 159-60.

60. ibid., pp. 160-1. D.C. Sircar suggested Vakradeva in place of Kudepasiri; see *Select Inscriptions*, Calcutta, 1965, p. 222.

61. ibid. pp. 162-4.

62. Kalyāṇadevī, the queen of king Dharmarājadeva of the Śailodbhava dynasty granted the land to Darhadāchārya Nāsichaṅdra and his disciple Ekaśāṭa Prabuddhachaṅdra. The donees obviously appear to be some Jaina saints. See,

S.N. Rajaguru, *Inscriptions of Orissa,* Vol. I, pt. ii, op.cit. pp. 223-8.

63. R.D. Banarji, "Inscriptions In the Udayagiri and Khandagiri Caves", *Epigraphica Indica* vol.XIII, op.cit. p. 167 Probably the same person engraved an inscription at small cave Dhauli, but, due to some omissions in engraving or drafting, the name and genealogy of the donor became confusing in Ganesagumpha inscription of Udayagiri hill.

64. ibid., p. 166.

    श्री उद्योतकेसरी विजय राज्य सम्वत् 5
    श्री कुमार पर्व्वत स्थाने जिन्न वापी जिन्न ईषण
    उद्योतित तस्मिन् थाने चतुविस्थ्त्तीर्थकर
    स्थापित प्रतीष्ठ काले हरि ओपजसनन्दिक
    न्नदति प्रथा (?) श्री पारस्यनाथस्य कर्म्मखय:

65. D. Mitra, *Udayagiri and Khandagiri,* op.cit. pp. 6–7.

66. N.K. Sahu (ed.), *History of Orissa by W.W. Hunter, Sterling, John Beams and N.K. Sahu,* Calcutta, 1966, vol. II, p. 271.

67. R.D. Banarji, "Inscriptions In the Udayagiri and Khandagiri Caves", *Epigraphica Indica* Vol.XIII, op.cit. pp. 165-6.

68. ibid., p. 165.

69. K.P. Jayaswal and R.D. Banarji, "Hathigumpha Inscription of Khāravela", *Epigraphica Indica* Vol. XX, op.cit., pp. 80-9. L-15 refers to the sages of various order who came from all directions or hundred directions.

    "*Sukata-samaṇa-suvihitānaṁ cha satadisānaṁ ñāninam tapasi-isinam samghayanam Arahata-nisidiyā samīpe* ....."

    "bringing about a Council of the wise ascetics and sages, from hundred (i.e., all) quarters, the monks (samaṇas) of good deeds ...."

Further, an inscription in the Navamuni cave dated in the year 18 of the reign of Uddyotakesarī refers to the teacher (āchārya), student (śishya) and the particular order or school of monastery. See R.D. Banarji, Inscriptions In the Udayagiri and Khandagiri Caves, *Epigraphica Indica,* op.cit., pp. 165-6: "*Om Srimad Uddyotakesarī devaya pravarddhamānevi jaya rājye saṁvat 18 Sri Āryasaṅgha-pratibaddha grahakula vinrgata-desīgaṇa āchārya Śrīkulachandra bhaṭṭārkasya tasya sishya Subhachaṅdrsya.*"

70. *Indian Archaeology 1961-62 – A Review*, pp. 36-7.

71. *Indian Archaeology 1958-59 – A Review,* pp. 38-40.

72. S.N. Rajaguru, *Inscriptions of Orissa* Vol. III, pt. ii, Bhubaneswar, 1975, p. 309.

73. ibid., Vol. V, pt.i, pp. 118-9.

74. The report of this excavation is published in *Ancient India* No. 5, op.cit. under the heading Sisupalgarh 1948 : An Early Historical Fort In Eastern India. The results of the next year's excavation work are referred in by the excavator in his article, "Sisupalgarh: Some Reflections On Its Layout", published in *Archaeology And Art,* edited by C.P. Sinha and others, (a K. Deva felicitation volume), Delhi, 1990, Vol. II, pp. 589-96.

75. B.B. Lal, "Sisupalgarh: 1948: An Early Historical Fort i In Eastern India", *Ancient India* No. 5 op.cit., p. 67

76. ibid., pp. 67–72.

77. B.B. Lal Sisupalgarh : some Reflection On Its Layout", *Archaeology And Art*, op.cit., pp. 589–96.

78. ibid., p. 590.

79. ibid.

80. A. Das, "Some New Facts About Sisupalgarh", *Orissa*

Historical Research Journal, Vol. XV, Bhubaneswar, 1967, p. 54.

81. ibid., p. 60.

82. ibid., pp. 58-9

83. ibid., p. 59.

84. ibid., p. 60

85. The herostone does not bear any testimony of the continuation of the ancient settlement. It is a late one and by that time the present village habitation would have occupied this site.

86. B.B. Lal, "Sisupalgarh 1948: An Early Historical Fort In Eastern India, *Ancient India* No. 5, op.cit. p. 72.

87. While conducting explorations in Dist. Cuttack, I have come across to such villages which have a suffix of *garha* and associated with some zamindar locally called *rājā*. The house of zamindar is being called *rājabāṭī* (palace) or *garh* (fort), though, it never resembles to a palace or fort.

88. B.B. Lal, "Sisupalgarh: Some Reflections Of Its Layout", *Archaeology And Art,* op.cit., p. 592.

89. ibid.

90. B.B. Lal, "Sisupalgarh 1948 : An Early Historical Fort in Eastern India", *Ancient India* No. 5, op.cit. p. 64.

91. ibid., p. 68.

92. A. Ghosh (ed.), *An Encyclopaedia of Indian Archaeology* Vol. I, op.cit., p. 298.

93. B.B. Lal, "Sisupalgarh 1948 : An Early Historical Fort In Eastern India", *Ancient India* No. 5, op.cit., pp. 76-7.

94. ibid., p. 70.

95. ibid.

96. ibid., p. 70 and plate XXXI, A.

97. Lal guessed it as the royal palace, and compared the whole layout of the fort and of this palace with the principles of fortification laid in *Arthaśāstra* of Kautilya. See, "Sisupalgarh: Some Reflection On Its Layout", *Archaeology And Art,* op.cit., p. 592.

98. B.B. Lal, Sisupalgarh 1948 : An Early Historical Fort In Eastern India, *Ancient India* No. 5, op.cit., p. 78.

99. Ibid.

100. ibid.

101. ibid.

102. ibid.

103. K.C. Panigrahi, *Archaeological Remains At Bhubaneswar,* Cuttack, (2nd ed.) 1981, p. 88.

104. B.B. Lal, "Sisupalgarh 1948 : An Early Historical Fort In Eastern India", *Ancient India,* op. cit., p. 68.

105. B.C. Mazumdar, *Orissa In The Making*, Patna, (reprint) 1984, p. 37. See also, K.P. Jayaswal and R.D. Banarji. "Hathigumpha Inscription of Khāravela", *Epigraphica Indica* Vol. XX, p. 83, note 19 : *Kaliṁga-nagarī* :– The capital of the Kaliṁga country, very likely now represented by Kaliṁgapatanam in the Ganjam district. The ancient capital of Kaliṅga, according to the tradition prevalent in the Chicacole taluk of the Ganjam district, was swallowed up by the sea. Fragments of bas-reliefs of the Maurya period with the well-known protuberance on the head-dress have been however, discovered by R.D.B. in the temples of Mukhaliṅgeśvara and Kurmeśvara in the locality. The coast from Chhatrapuram to Masulipatam is subject to great cyclones during the north-astern monsoon and the majority of ancient sites on this coast are buried in drift sand.

106. K.C. Panigrahi, *History of Orissa*, op.cit., pp. 27-32.

107. D.C. Sircar, *Select Inscriptions* Vol. I, op.cit., p. 213.

108. S.N. Rajaguru, *Inscriptions Of Orissa*, Vol. II, Bhubaneswar, 1960, p. 337.

109. *Indian Archaeology 1978–79 – A Review*, p. 66.

110. S.N. Rajaguru, *Inscriptions of Orissa* Vol. II, op.cit. p. 58.

111. S.R. Balasubrhmanyam, "*Bhuvaneswar*", Orissa Historical Research Journal, Vol. VI, pp. 292-5.

112. B.M. Barua, Hathigumpha Inscription of Khāravela", *Indian Historical Quarterly*, Vol. XIV, op.cit., pp. 483-4.

113. S.R. Balasubrahmanyam, "Bhuvaneswar", *Orissa Historical Research Journal*, op.cit., p. 295.

114. B.M. Barua, "Hathigumpha Inscription of Khāravela", I.H.Q., op.cit. pp.476-7, see fn., 170, 171 and 172 and p. 483.

115. K.P. Jayaswal and R.D. Banarji, "Hathigumpha Inscription of Khāravela", *Epigraphica Indica* Vol. XX, op.cit., p. 87.

116. K.C. Panigrahi, *History of Orissa*, op.cit., pp. 27-32.

117. B.M. Barua, "Hathigumpha Inscription Of Khāravela", *I.H.Q.* Vol. XIV, op.cit., pp. 476–7 and fn. 171.

118. K.P. Jayaswal and R.D. Banarji, "Hathigumpha Inscription of Khāravela", *Epigraphica Indica* Vol. XX, p. 79 :

    "*............ Abhisita-mato cha padhame vase, vāta-vihita-gopura-pākāra-nivesanaṁ paṭi-saṁkhāryati, Kaliṁga-nagarī-khibīra-isi-tala-taḍāga-pāḍiyo cha baṁdhāpayati; sav-ūyāna-paṭisaṁthāpanaṁ cha kārayati panatisāhi sata – sahasehi Pakatiyo cha ram jayati.*"

119. ibid., p. 86: "As soon as he is anointed, in the first (regnal) year (he) causes repairs of the gates, the walls and the buildings (of the city), (which had been) damaged by storm;

in he city of Kaliṅga (he) causes the embankments of the lake (called after) *Khibīra Rishi*, (and) of (other) tanks and cisterns, (also) the restoration of all the gardens (he) causes to be." See also note 20 on page 83: "*Knibīra-isi-tāla* : We have examined these two words very car:fully on the rock, estampages and casts. The reading is perfectly clear but no explanation seems to be possible unless khibīra is taken to be a proper name of a Rishi who excavated the tank or lake or after whom it was named. In that case, the phrase will have to be taken as referring to a particular tank while taḍāga and pāḍiyo will refer to artificial excavations."

120. B.M. Barua, "Hathigumpha Inscription of Khāravela", *I.H.Q.* Vol. XIV, op.cit., p. 474.

121. P.V. Begde, *Ancient And Medieval Town Planning In India*, New Delhi, 1978, pp. 52-3.

122. Line–3 of Hahigumpha Inscription can be interpreted better in the light of structural remains of the said tank. Thus, we can say that Khāravela repaired the portals and the fortification wall of the capital city (proper) of Kaliṅga as well as the cantonment alongwith the tank located therein. See for the settlement provisions for a Śivira.

121. *Śabdakalpadrumah* Vol. V, op.cit. p. 106: *Śivira: (Śerate rājabalānyatra); niviеśah ityamarah; dve āgaṅtuka-sainya-vāse kaṭake ianye, nṛipasya mūlasthāni ityāike iti Bharatah. Atha-Śivira-Lakśhanam:*

Śiviram parikhāyuktamuchchaih prakārāvesṭitam

Yukta dvādaśadvaraścha Siṁhadvārapuraskṛitam

Yuktam chitravichitraischa kṛitribhaischa kapāṭakaih

Nishiddha vṛiksharahitam prasiddhesch puraskrita

Sulakshanam chaṅdravedham prāṅganascha tathaiva cha

Tatra vihita nishiddhakashṭhe pi yathā.

123. B.B. Lal, "Sisupalgarh 1948 : An Early Historical Fort In Eastern India", *Ancient India* No. 5, op.cit., p. 77.

124. A.Das, "Some New Facts About Sisupalgarh", *Orissa Historical Research Journal* Vol. XV, op.cit., p. 60 and pl.III, From the above plate it appears to be a sealing, however, the author has used the term "seal". Another fragmentary sealing bears the letter *ra* which is interpreted as *rājā* by Sri Das.

125. B.B. Dutta, *Town Planning In Ancient India*, Delhi, 1977, op.cit. 39–40.

126. P.V. Begde, *Ancient And Medieval Town Planning In India*, op.cit., p. 52.

127. *Gazetteer Of India And Pakistan*, Vol. II, Director of Military Survey GSGS, Army, India, Delhi, 1950, p. 878.

129. M.M. Mahapatra, *Traditional Structure And Changes In An Orissan Temple*, op.cit., p. 45.

130. R.L. Mitra, *Antiquities Of Orissa,* Cuttack, (reprint), 1963, p. 100.

131. D.C. Sircar and K.G. Krishanan, "Bhubaneswar Inscription of Gaṅga Narsimha", *Epigraphica Indica* Vol. XXXII, Delhi, 1958, pp. 229-38. Also see, S.N. Rajaguru Inscriptions of Orissa Vol V, pt. i, op.cit. pp. 305–14. The inscription is bilingual (in Oriya and Tamil). The relevant line of Oriya version reads :

    श्री कीर्तिवास क्षेत्र सिधेस्वर मठर बड़ नरसींघ देवंकर आशाकामार्थ पूर्व्वे ....

132. S.N. Rajaguru, *Inscriptions of Orissa* Vol. III, pt. ii. op.cit., pp. 336–7.

    मठवेदादीनां द्विजपुरविहारा: प्रतिदिशं बिराजन्ते .....

133. ibid. To the pious Brāhmaṇas he gave a Brahmapura which

was superintended by the Śaiva teacher Vishṇu; as reads verse 31:

आरादब्रह्मपुरं बृहस्पतिपुरस्पद्धि स्मरारेः सदाचार्य विष्णुमभिस्फुर(द) द्विजवर ग्रामाय धर्म्मात्मने ।

दत्तं तेनमुदसाद तमख प्रारबध धूमध्वज स्फूर्ज्जमययन यत्र स कलिब्यालः समुत्सार्यते ।

134. These can be translated as : the locality of Badu–preists, the locality of Panda-priests, the locality of Misra Brahmans, the locality of Confectioners, the locality of Sweeper-class, the locality of Ferrymen the locality of Potters, the locality of Bauri-labours, the yard of Iron-smiths and the turning square near the Carpenters' locality, respectively.

135. During his visit to Bhubaneswar, R.L. Mitra noticed that most of the houses were built with mud or rubble-walls and covered with leaves or straw. See, R.L. Mitra, Antiquities of Orissa, op.ci., p. 100. The gabled roofs made of straw with or without clearstorey are still at Bhubaneswar.

136. *Brahmapurāṇam,* Ch. 41, op.cit. According to the verses 51-53, for the welfare of all people, *Śiva* brought the water drops from all the holy places, rivers, lakes, ponds, tanks, stepped wells, wells and seas of Earth and put here (in this lake) and Lord *Śiva* with all gods and sages (*munis*) created here the holyplace called *Biṅdusara* (literally : Lake of drops).

हिताय सर्वलोकस्य भुक्तिमुक्तिप्रदः शिवः । पृथिव्यां यानि तीर्थानि सरितश्च सरासिं च ॥ 51 ॥

पुष्करिण्यसतडागानि वाप्यः कूपाश्च सागराः । तेभ्यः पूर्व समाहृत्य जलविन्दूनपृथकपृथक ॥ 52 ॥

सर्वलोकहितार्थाय रूद्रः सर्वसुरैः सह । तीर्थ बिन्दुसरो नाम तस्मिन्क्षेत्रे द्विजोत्तमाः ॥ 53 ॥

137. S.N. Rajaguru, *Inscriptions of Orissa* Vol. V, pt. i, op.cit., p. 213, verse: 11:

यस्मिन् बिन्दुसरः सरस्वद्सदृगहक पेयपाथः पत -
पान्थ श्रान्तिहर सुधाजनित निभ्यदद्बपु शम्भवीं ।
यद्विदोरपनानुयन्ति पदवीं तीर्थानि तानिस्फुट
भूतानिग्रहनिर्मित पुरजिता लोकैकशोकापह ॥ 11 ॥

See also *Orissa Historical Research Journal* Vol. XIII, No. 4, pp. 1–14.

138. *Mahābhārata, Vanparva (Tīrthayātrā),* Hiralal Haldar (ed.), (translated by P.C. Roy), C.I.E., Calcutta, pp. 249–50.

139. *Brahmapurāṇam*, ch. 41, op.cit., verse : 11–12

लिंगकोटिसमायुक्तं वाराणसीसमं शुभम् ।
एकाग्रकेति विख्यातं तीर्थाष्टकसमन्वितम् ॥ 11 ॥

एकाग्रवृक्षस्तत्रासीत्पुरा कल्पे द्विजोत्तमाः ।
नाम्ना तस्यैव तत्क्षेत्रमेकाग्रकमितिश्रुतम् ॥ 12 ॥

140. S.N. Rajaguru, *Inscriptions of Orissa* Vol. IV, op.cit., p. 247, and see, ibid, vol. III, pt.i, p. 210

स्वति श्री चोड़गंगदेवस्य प्रवर्द्धमानविजयराज्ये संवत्सरे 67 2 आदत्वाद्विग्रामाय अगु ....

एकाम्ब्रे विनयदेवि अखण्डुर्धकस्य ......

and Vol V, pt. i, pp. 211–20, verses : 9 and 21

तत्र च क्षेत्रमकाम्रमाम्रारामशताश्रितम् ।
एकदेवकुल देवकुलैराकुलमद्भुतम् ॥ 9 ॥

एकाम्राद्वयवेदिते सुमहसि श्री कृतिवासप्रिये क्षेत्रेपुण्यवले श्रुतामृतफले सर्वत्र पुष्पज्वले

प्रसाद पुरूषोत्तमस्य सबल सेषापद वैष्णवं गन्तु मंगलपूर्णकुम्भशिरस श्रद्धाश्रिताचीकरत् ॥ 21 ॥

141. S.N. Rajaguru, *Inscriptions of Orissa*, Vol. III, pt.i, pp. 63 and 169: *Śrīkīrtivāsabhaṭṭārakāya* Vol. III, pt. ii, p. 294 : *Śrīkīrtivāsa pritayae,* p. 298 : *Śrī Kīrtibāsāya,* p. 309 : *Śrī Kīribāsa*, Vol. V, pt.i, pp. 78 and 96 : *Śrī kīrtivāsadevasya,*

p. 97 : (....) *vāsadevasya,* pp. 118–9 : *Bhagvao Jagadīśvarasya kṛitivāsaso devasya,* p. 117 : *Bhagavate kṛittivāsase,* p. 125 : *Śrī kīrtivāsāya Siddhalambodarasya,* p. 128 : *Kīrtivāsāya,* p. 215 : *Śrīkṛittivāsa priyae,* p. 307 : *Śrī kīrtivāsa-kshetra.*

142. P. K. Ray, "The Oriya Inscription of Liṅgarāja Temple", *Orissa Historical Research Journal*, Vol. XV, pt. 1 & 2, Bhubaneswar, 1967, pp. 113–4 :

वीर श्री गजप्ती प्रतापप्रुसोत्तमदेव महाराजांकर वीजे राइजे सम्वत् 14 अंक श्राह भुवनेश्वर

देउले जलसर महान्ती जे बन्दी पड़िथिला ताहाकु एही भुवनेश्वर लींग पारीकला ॥

143. S.N. Rajaguru, *Inscriptions of Orissa*, Vol. III, pt. i, op.cit., p. 67. The text is damaged and gives the reading *Tribhuvanādhi,* which can be restored as *Tribhuvanādhiśvara.*

144. S. Tripathy, "Ekāmr-Bhubaneśvara", *Orissa Historical Research Journal*, Vol. XXXIII, op.cit., p. 149.

145. K.C. Panigrahi, *History Of Orissa*, op.cit., pp. 54–6.

146. P.V. Begde, *Ancient And Medieval Town Planning In India,* op.cit., p. 40.

147. K.C. Panigrahi, *History of Orissa*, op.cit., pp. 191-2.

148. The word "*Kaṭaka*" denotes here a settlement type, though in Sanskrit laxicans it is interpreted variousely; and in recurring usage it is found for *senā* (army), *rājadhānī* (capital), *nagarī* (city) and circle or circular objects. See *Sabdakalpadrumah*, op.cit., Vol. 2, p. 8: Kataka : parvatamadhyabhāga tatparyāyah niambah – ityamarah, mekhalā iti Bharatah (yathā *Raghu* : 16/31 – margairghini sa kaṭakāntar–eshu vaiṅdhyaeshu) senā bahudhā vibhinna valayah, chakraṁ ityamarah; hastidantamaṅḍalam, samudralavaṇaṁ, rājadhānī iti medinī; nagarī iti *Sabdaratnāvalī*, senā iti Hemachandrah (yathā *Hitopadeśe* 1/332 sa cha digvi

jayaka melāgatya chandrabhā–gānadītīre samavesita kaṭako vartate); sānuh parvatabhūbhāgah iti visvah. Girkūteshu durgeshu nānā janapadesu cha, Janākī – rṇeshu deśeshu, kaṭakeshu parevuchah. iti *Mahābhāratam̐* 8/28/12.

The above citation also speaks that "*kaṭaka*" is a particular landscape which is surrounded by a series of hills. Such landscapes proved good encampment sites for the armies; and many cantonments and capital cities emerged in such sites and thus this word coined in such many usage. In Orissa the word "*kaṭaka*" is still popular in both the usage i.e. "*senā*" and "*rājadhānī*" and many places bear the name which has *kaṭaka* as suffix. The modern city of Cuttack was founded by the Gaṅgas bearing the name of Varaṇasīkaṭaka. The Gaṅgas also founded many *kaṭakas* in the hilly surroundings; the ruins of which still survive at many places; e.g. Chudangagarha, Bualigarh, Teligarh and Amrabati kaṭaka etc. In the same line *Kṛittivāsakaṭaka* also emerged as one of the important military and political centre. Though modern city of Cuttack was more important than any other *Kaṭaka*, however, Bhubaneswar which was famous for *Śiva* was also popularly known as *kaṭaka*, as evident from a folk-song in *Avadhī* in Uttar Pradesh:

चली बांकुड़ी रीवॉ जबलपुर, सागर बूँदी को जाती ।
मुल्क चंदेला और ग्वालियर, पूना सतारा ने चकाती ।
अमराती (अमरावती) गुजरात मालवा बुढ़ानपुर तक नपाती ।
झांसी दतिया ढूंढ़ के झन्ना-पन्ना पछताती ।
नासिक हैदराबाद नागपुर, सूरत बम्बई गुलजारा ।
पिया ढूढ़ फिरी मुल्को सारा - सारा ।
प्राग बांदा और लखनऊ चली सुन्दर कम्प्यू म्यानी ।
कोयल कनउज और फर्रूखाबाद कालपी सब जानी ।
मिलो मुरादाबाद भगलपुर सदा आगरा संसारा ।
पिया ढूढ़ फिरी मुल्को सारा - सारा ।
तखत जमुनपुर टांडा बांगला अयोध्या पहुंची नारी ।
बुटबल बेतिया और जनकपुर आजमगढ़ की तैयारी ।

रामपुर बड़हर गोरखपुर बहेराइच बस्ती प्यारी ।
पिया ढूढ़ फिरी मुल्को सारा - सारा ।
तानसेन नेपाल की बेटी भूप चीन जाके ढूंढ़ा
राम नगर कासी नैनागढ़ गाजीपुर को गई अबला ।
अपड़ा - झपड़ा औ दानी पुर औ पटने तक पता नहीं ।
मिलो बॉकुरी सैदपुर कलकत्ते के पूरब दिशा ।
ढ़ाका और बंगाला कटकपुरी में शिव प्यारा ।
पिया ढूढ़ फिरौ मुल्को सारा ।
मेहर भई उस रात नाम पर बहुत दिना में मिले सजन ।
हनूमान दुरगा के चेला चारो दिशा को कतल किया ।
मिर्जापुर दरम्यान शहर जहॉ राधे जी ने ललकारा ।
बना रहे ठाकुर द्वारा पिया बना रहे ठाकुरद्वारा ।
पिया ढूढ़ फिरी मुल्कों सारा ।

149. N.K. Sahu (ed.), *History of Orissa By W.W. Hunter, Sterling, John Beams and Sahu*, op.cit., p. 270.

150. B.B. Dutta, *Town Planning In Ancient India,* op.cit. p. 201 (*Mānasāra,* Ch. 9.1.61.)

151. ibid.

152. K.C. Panigrahi, *History of Orissa*, op.cit., p. 314.

153. S.N. Rajaguru, *Inscriptions of Orissa,* Vol. I, pt. ii, pp. 191-8.

154. K.C. Panigrahi, *History of Orissa,* op. cit., p. 314.

155. ibid.p. 366.

156. ibid.

157. ibid.

158. ibid., p. 368.

159. ibid., p. 367.

160. S.C. De, "Three Ancient Naga Images From Bhubaneswar And Naga Cult In Orissa", *Orissa Historical Research Journal*, Vol. III, No. 1, Bhubaneswar, 1954, p. 1.

161. ibid., p. 4.

162. ibid.

163. K.C. Panigrahi, *History of Orissa*, op.cit. pp. 366–7.

164. ibid.

165. Vasudeva Sarana Agrawal, *Bharatiya kala* (Hindi), Varanasi, (reprint) 1987, pp. 122–3.

166. ibid., fig. 188 (p.124) and plate 78.

167. ibid., pp. 122–30.

168. ibid., p. 126.

गॉव - गॉव को ठाकुर गॉव - गॉव को बीर ।

169. K.C. Panigrahi, *History of Orissa*, op.cit., pp. 356-64.

170. E. Hultzsch, *Corpus Inscriptionum Indicarum* Vol. I, op.cit., p. XVIII.

171. *Brahmapuranam,* Ch. 41, op.cit., verse : 76:

तस्मिन्क्षेत्रेवरे लिंगं भास्करेश्वरसंज्ञितम् । पश्यन्ति ये तु तं देवं स्नात्वा कुण्डे महेश्वरम् ॥ 76 ॥

3

Early Period and Society

1. POLITICAL PHENOMENON

In the early historical period Bubaneswar formed the part of Kalinga *janapada.* Though, we donot have any mention of Kaliṅga in the Buddhist text *Anguttara Nikāya* among the sixteen *mahajanapadas* which existed during the time of Buddha;[1] however, its existence during the sixth century B. C. is well established by its recurring references in *Jātakas* and *Mahābhārat.*[2] Further, another Buddhist' text *Chullaniddeśa*[3] also informs us about the existence of Kaliṅga *Janapada.* In the succeeding centuries, Megasthnese and Pliny also speak about Kaliṅga.[4] When we come to the age of Asoka we get some definite information about Kaliṅga. In his thirteenth rock-edict Asoka mentions the conquest of Kaliṅga. In this version it is said that when king *Devānāṁpiya Piyadasin* had been anointed eight years, Kaliṅga was conquered.[5] Since the known date of Asoka's coronation is 269 BC[6], the date of Kaliṅga war comes to circa 261 BC.[7]

We know that during the reign of Asoka's grandfather Chandragupta Maurya almost all parts of India except Kaliṅga and southernmost tip of the peninsula were included in Magadhan empire.[8] This empire had also included such outside territories as Afganistan and Baluchistan. It had within its orbit Kashmir and Nepal.[9] It seems that whether Chandragupta Maurya or his son Bindusara were unable to subjugate Kaliṅga due to some reasons and maintained peaceful co-existence by some alliance. However, this peace between a vast empire and a small Janapada was broken during the time of Asoka. Probably Magadha required good sea-ports for its naval supremacy and undeterred trade. The existence of Kaliṅga as an independent country would have been detrimental in execution of Asoka's trade policy.[10] Further to win the only independent country lying near his capital would have also tempted a warlike ambitious king Asoka to adopt the policy of aggression on Kaliṅga.

The details of the Kaliṅga war are not available; Asoka has made a brief reference to its results only.[11] He states that the Kaliṅgan war resulted in one hundred thousand people killed, one hundred and fifty thousand carried away as captives and may more died of starvation and disease that followed in the wake of terrible the war. The figures given by Asoka about the number of the slain and the captives indicate that the Kaliṅgan army was a vast one and that the country had a large population.

What sort of government was functioning in Kaliṅga country, is not known from any source. Asoka's edicts do not mention the name of any kings also therefore it seems that there was no sovereign monarch who would have led the Kaliṅgan army, rather the edicts speak the term *kaliṅgah* which would stand for the people of Kaliṅga, thus by implication it can be opined that Kaliṅga was republic during the invasion of Asoka.

There would have been a number of great wars and terrible battles causing large scale genocide but Kaliṅgan war is known as a turning point in the history. This was such a war where the conqueror felt defeated. Generally, a great victory leads a warlike monarch to greater attempts for the greater victories but Kaliṅgan war produced a very different effect on Asoka. He was struck with remorse and became horrified at the immense bloodshed caused in the war.[12] He ended his warlike career and seathed his sword forever. Asoka expressed his feelings in the thirteenth rock-edict as follows:[13]

"In conquering indeed an unconquered country, the slaying, death and deporting that occur there are considered extremely painful and serious by the *Devanaṁpiya*. Even more serious than this is that those who live there whether the *brahmans* or *śramaṇas* or the other sects or the householders etc. to them occurs injury or death or deportation of the beloved ones. And also to the people who are fortunate to have escaped and whose affection is undiminished (by the brutalising effect or war) occur the injury as their friends acquaintances companions and relatives also suffer from the same misfortune. This is shared by all them men and is considered very deplorable by *Devanaṁpiya.*" He further adds,[14] "even the one hundredth part or one thousandth part of all those people who were deported at that time in Kaliṅga would now be considered very deplorable by *Devanaṁpiya.* These are the genuine feelings of a noble monarch who hencefarward replaced *bherīghosa* (the sound of trumpet announcing war) by *dhammaghosh* (announcement of moral path).[15] He not only wanted to win the hearts of the people through the methods of peace and *dhamma* he also exhorted his sons and successors not to indulge in any warfare for the purpose the conquest.[16]

Even though Asoka gave up warfare as a method of conquest after the Kaliṅgan war; but, how he dealt with this newly conquered country is clearly reflected in his special edicts. His special edicts (also known as Kaliṅga edicts) at Dhauli near Bhubaneswar, are addressed to the *mahāmātras* and *kumārāmātya* or prince viceroy of *Tosalī*,[17] while his same edicts at Jaugada in district Ganjam are addressed only to the *mahāmātras* of *Samāpa.*[18] From those two inscriptions of Asoka it is obvious that though Asoka was worst affected by the bloodshed of Kaliṅga war but he retained his victory over Kaliṅga. Instead of granting freedom to subjugated people he wanted to win their faith by good administration. Kaliṅga became a major province of Asoka's empire. It had two administrative centres: one at *Samapa* which was the headquarter of the southern Kaliṅga and other the *Tosali* the vicereagal seat of Northern Kaliṅga. The finds of rock-edicts at *Samapa* indicated that Asoka did not deal Kaliṅga casually, rather he entered interior Kaliṅga and published his motto for better control in the whole province. We have already suggested in the preceding discussions that the fortifications at Sisupalgarh were laid by Asoka to provide firm security to the officers posted at *Tosalī* because it was a strategic need to get stronghold in a newly conquered territory. It seems that political unrest against the conqueror did not cool-down immediately and therefore Asoka imposed tight security arrangements and deployed sufficient armies for suppression of any unwanted agitation and on the other hand he adopted a pacifying and conciliatory attitudes to win the hearts of the subjugated people. In his special edicts he instructs his officers,[19] "In affairs of administration, there might be persons who would suffer imprisonment coercion. In that also might occur accidental death in prison and many also heavily suffer for that. In that case you must strive to deal with all them impartially, the attribute which are malignity, irascibility, harshness, hastiness, lack of practice, indolence and weariness. You all must strive so that these attributes may not be in you. At the root of all impartial dealings lay the absence of anger and avoidance of hurry. The judicial officers of the capital (*nagara i.e. Tosalī*) must strive at all times for this and they should not inflict sudden imprisonment or sudden coercion on people. For this purpose I would be sending on quinquennial tours the *mahamatras* who would not be harsh and irascible and would be soft and gentle in dealings."

Asoka died in 232 BC[20] and some scholars have suggested that Asoka's empire was partitioned by his two successors, *Dasaratha* and *Samprati.* The eastern part of the empire fell in to the share of *Dasaratha* and it may be that Kaliṅga was ruled over by this prince.[21] About the political condition of Kaliṅga the source material is very meagre and therefore exact duration of Mauryan occupation of Kaliṅga cannot be suggested. It seems however, that in 185 BC when *Pushyamitra Suṅga* usurped the Magadhan throne the dissolution of the empire started.[22] The change of the dynasties signified that this dissolution and the independent kingdom like the *Sātavāhanas* and the *Chedis* came into existence. The Mauryan rule in Kaliṅga appears to have ended in about 185 BC.[23]

The veil of obscurity that covers the history of Kaliṅga since the days of Asoka is only lifted by the records found engraved on the walls of some rock-cut caves at Udayagiri hill near Bhubaneswar. These records reveal that the rulers of *Cheti* (Chedi) dynasty, who called themselves *Mahāmeghvāhna* came into power in Kaliṅga. Of the rulers of this dynasty only the names of Khāravela and Kudepasiri (or Vakradeva), together with that of a prince Vadukha, are known from the inscriptions in the caves at Udayagiri. The names of Kudespasiri and Vadukha are found simply a donors of the cells of *Manchapurī* cave (cave-9)[24] while many details are available about Khāravela from his famous inscription engraved on the brow of the rock over Hāthīgumphā i.e. cave - 14 of the same hill. The inscription engraved in seventeen lines, is largely defaced and indistinct, resulting, in incomplete and various controversial readings of the text and its interpretation.[25] Yet, the following facts seem to be well established:

The third king of this dynasty, Kharavela was powerful ruler. As a prince he acquired great proficiency in games and received good education. He was installed as *yuvarāja* (crown prince) when he was sixteen and succeeded to the throne when he has twenty-five. Almost immediately thereafter he launched Kaliṅga on an ambitious career of conquest, leading expeditions far and wide. In the second year of his rule he led his troops to the west without caring for Sātavāhana king Sātakarṇi and reached the river Krishṇā, where he threatened the city of Asika (Musika). In his fourth year he captured the capital of a prince named Vidyādhara and subdued the *Rāshṭrikas* and *Bhojakas* in the north Deccan. Four years later he stormed *Goraṭhagiri* (Barabar hills, district Gaya) and harassed the ruler of *Rājgṛiha* (Rajgir, district Nalanda). A *Yavana* (Indo-Greek) is said to have fled to Mathura out of fear. In his eleventh year he destroyed the city of *Pithuṅḍa* (Masulipatnam region) and next year threatened the rulers of *Uttarāpatha* (north India) and defeated king *Bhasatimita* of Magadha. He brought back with him booty from *Aṅga* and Magadha, including a *Jaina* cult object (*Kaliṅga-Jina*), which had been taken away long ago by Naṅda, the ruler of Magadha. Next, he snatched treasures from the *Pāṅdya* king in the extreme south. Khāravela also attended the job of public welfare and civil works at his capital with full zeal and dedication. In the first year of his rule he rebuilt the gates and walls of *Kaliṅganagara,* his capital, which had been devastated by a cyclone. In the fifth year he enlarged a canal, said to have been excavated by a Nanda king 103 years ago. In the *Kumārī parvata* (Udaygiri-Khandagiri) he excavated, in the thirteenth year, caves for *Jain*-ascetics and erected at enormous cost on the *prāgbhāra* in the neighbourhood of the monastic retreats.

Identification of various kings and kingdoms found in the above statement remained the subjects of controversies as well as the date of Khāravela. Various scholars have assigned Khāravela a period ranging from second century BC to the first century BC. Dr. D.C.Sircar[26] retained only on the paleographic evidences and opined first century BC to first century AD while on the other hand Dr. Barua[27] and Dr. Panigrahi[28] had investigated the internal evidence and Dr. Panigrahi categorically fixed the beginning of his reign in circa 159 BC.[29] Dr. Panigrahi has based his arguments mainly on two points: (1) the *ti-vasa-sata* of the inscription stands for one hundred and three year and not for three hundred years. This interpretation is also accepted by Dr. K.P.Jayaswal and Dr. R.D. Banerji.[30] (2) Therefore Nandarāja of the inscription can not be Mahāpadmananda rather this may fit with Asoka who also hailed from Magadha and is better known for his welfare activities through his rock-edicts and other sources.

Dr. Barua's identification of Asoka and Dr. Panigrahi's arguments for dating Khāravela's reign are more convincing and we may accept them.

Taking the date of Khāravela as datum line we can guess about the preceding rulers of the dynasty. We know that Khāravela was the third king of *Chedi* dynasty; he was a crown prince and he inherited the kingdom of Kaliṅga. Thus if we grant minimum thirty years for two preceding rulers of the dynasty the period of emergence of *Chedi* power in Kaliṅga coincides with dissolution of Mauryan empire.

Hāthīgumphā inscription speaks only about thirteen years of Kharavela's reign suggesting the sudden death of this great hero of Kaliṅga. About Khāravela's successors we lack in definite information. The *Mañchapurī* cave inscription furnish us two names belonging to royal family one Kudepasiri (Vakradeva)[31] was entitled as *mahārāja* suggesting that he was one of the rulers of *Mahāmeghavāhana family*.

His name is recorded as the donor of Mañchapurī cave. At Udaygiri hill Mañchapurī and Svargapurī Cave excavated as the lower storey and upper storey respectively in major rock. The upper i.e. Svaragapurrī bears an inscription recording the dedication on the cave by the chief queen on Khāravela.[32] We may add here that in the building work we proceed from bottom to top while in rock-excavation the system changes and one has to begin from top. This is the only safest way to hewn caves successfully. Therefore, we may opine that the upper storey dedicated by chief queen preceded that of Mahārāja Kudepasiri and thus he was among the successors of Khāravela but not as the predecessors. Similarly, prince Vaḍukha[33] whose name occurs in the Maṅchpurī cave, was also in the succeeding generation. These evidences suggest that the rule of the Chedi dynasty did not come to an abrupt end in Kaliṅga. In this connection it is interesting to note that Guntupally inscription reveals the name of one more king of Mahāmeghavāhan family. It records a gift of maṅḍapa by the recorder of royal message of Mahārāja Kaliṅgādhipathi *Mahāmekhavāhan* who was also the lord of *Māhishka.*[34] According to the amended readings by D.C. Sircar the inscription records the gift of a mandapa by one Chula-Goma, the scribe of the Mahāmeghavāhan Mahārāja Sadā, the lord of Kaliṅga and Māhishka.[35] The city of Pithuṅḍa of Khāravela's inscription is identified with modern Gunupally. Thus, we can say that the boundaries of Kaliṅga empire extended in the Gunupally region by Khāravela and therefore there is no doubt that *mahārāja* Sadā happened in the succeeding generation of Khāravela.

Political conditions of *Kaliṅga / Tosalī* are very much obscure after the rule of king Khāravela. Scholars like D.N.Das[36] toiled to make some meaning of the references about Kaliṅga or Tosali occurring in the Naturalis Historia of Pliny but these references do not help in tracing of ruling dynasties or the proper identification of the territory.

The excavation at Sisupalgarh yielded altogether thirty one coins . It is noticed that in early periods of Sisupalgarh inscribed puncharmed coins were prevalent. From AD 200 and AD 250 to AD 350 lead coins as well as copper coins of Kushāṇa and Puri-Kushāṇa type were more frequent.[37] A legible coin bears the well-known lion type of some of the Andhra coin.[38] Corroborating this piece of evidence Dr. Panigrahi has suggested the supremacy of the Sātavāhnas over Kaliṅga after the fall of the Chedis.[39] In this connection he adds, "Since the mount *Maheṅdra* situated in Kaliṅga, has found place in the list of mountains which Gautamīputra Sātakarṇi is stated to have been the lord, it may not be unreasonable to infer that there was a period of Andhra-Sātavāhana supremacy in Orissa and this supremacy began with the fall of Khāravela's dynasty and ended in about third century AD.[40] On the other hand a copper coin of Huvishka, is datable to the last quarter of second century AD while a gold coin of Kushāṇa type datable to second-third quarter of third century AD (copied from the Kushāṇa king Vāsudeva) informs us about a king Dhamadamadhara ('ma') ('hara') *ja - ra (ja) dhasa dhama damdhar (sa).* Altekar thinks that king Dhamadamadhara might have been a Jaina and belonged to Muruṅḍa family.[41] Dr. Panigrahi also postulated a period of Muruṅḍa supremacy in Kaliṅga.[42] However, frequent occurrence of Puri-Kushāṇa type coins from various regions of Orissa,[43] led Dr. R.D. Banerji to opine a period of Kushāṇa rule over Orissa. At Sisupalggarh, the Puri-Kushāṇa coins continued till the end of settlement and decide uppermost limit of the habitation about the middle of fourth century AD.[44] The coin-moulds of clay found from late level of period III attributable to circa AD 300 indicate that coins were produced in this settlement till this date.[45] It is interesting to note that these coin-moulds were designed to produce in a single operation imitations of coins of a kind which were normally of a series of punches.[46] Further, these finds also confirms that punchmarked coins continued in circulation till as late as circa AD 300.[47] Atleast but not the last it would be suffice to say about the political environment of Bhubaneswar region in the early period that this witnessed the presence of great monarch of the age, viz: Asoka and Khāravela and remained a major or minor seat of government till AD 350. As we have already noticed that currency was minted at the ancient settlement of Sisupalgarh during AD 300, therefore we may say with a high degree of certainty that political importance was still attached with this settlement. The degenerated habitation would have continued for some more time in and around the fortification.

2. PEOPLE AND ECONOMY

Since human settlement in this area flourished mainly due to location of royal court and therefore we may guess that a greater portion of local population would constituted by the people working in government services i.e. both the civil and military. During the reign of Asoka, here was the provincial capital governed by a *Kumāra*. *Mahāmātras* were the subordinate officers to the *Kumāra*. Another important office located in the city was of *nagaraviyohālaka* (town-judge). A class of high officers appointed by Asoka consisted of *lājukas* (*rājuka*), *yutas* (*yuktas*) and *dūtas*.[48] During the period of Chedi dynasty the place raised to the status of imperial capital of Kaliṅga.

Now the number in the imperial service would have increased as a usual sequel to the increased status of the city. The place gained the proud privilege of being the royal abode. The king and his family formed the chief elite population of the city. The ministers and high officer responsible for state affairs would constituted the major elite class. From the dedicatory records in the cave of Udaygiri and a terracotta sealing from Sisupalgarh we get some information about the important personalities of the age and the office-bearers; and a brief description of these known persons is being given hereunder:

(i) Khāravela[49]

He was the third king of the Chedi dynasty. He was handsome and possessed a ruddy body. As a prince he acquired great proficiency in games and received good education. He excelled in royal correspondence (*lekha*), currency (*rūpa*), finance, civil and religious administration, laws (*vyavahāra -vidhi*) and acquired in all other branches of polite learning. He was installed as *yuvarāja* [crown-prince] when he was sixteen and succeeded to the throne when he was twenty-five. Almost immediately thereafter he launched a number of successful military expeditions against the various states lying far and wide. His military compaigns were against mainly Magadha towards north; towards south he subjugated the city of *Pithuṅḍa* and Pāṅḍyan king; he also defeated a confederation of the *Draviḍas. Rāthikas* and *Bhojakas* whose habitats were probably Deccan [(Maharashtra)] and *Berar* [*Vidarbha*] were subdued by him. Towards the west he led his expeditions upto *Asika* [*Mūsika*] and highly destroyed the city.

Khāravela had a deep understanding of politics. First of all he made the proper arrangements of the safety and security of the capital city by repairing the damaged fortification and by other means and only then he proceeded on war. He let to feel the public of capital his active presence among them and therefore he organised the public festivals. He was well- versed in music {*gandharva-vidyā*] and other art and patronised them. He had profound faith in Jainism and he excavated in the thirteenth year caves for Jain-ascetics. Even though he was a *Jaina,* he had an eclectic attitude regarding religions and gods of other sects.

(ii) Chief Queen [Agramahishi] of Khāravela

The chief queen of king Khāravela was the daughter of king Lalāka who was the grandson of Hastisahas. She caused the excavation of *Svargapuri* [*cave-9*] *for the Jaina* monks of Kaliṅga.

(iii) Mahārāja Kaliṅgādhipati Mahāmeghavāhana Kuḍepasiri [or Vakradeva][51]

Probably he succeeded king Khāravela as king of Kaliṅga. At Udayagiri he was donor of the main wing of Mañchapurī-cave.

(iv) Kumāra Vaḍukha[52]

He was a prince of royal family probably brother or son of *mahārāja* Kuḍepasiri [Vakradeva]. He donated the cells of right wing of the *Mañchapurī*-cave

(v) Sabhuti, the Town-judge[53]

This name is known as the donor of Bāghaguṁphā [cave-12]. The name is preceded by the designation of *nagra akhaṇḍa* (sanskrit: *nagrādhyaksha*). Scholars like R.D. Banerji and Barua have opined that he was a judicial officer like the *nagaraviyohālaka* [sanskrit: *nagavyavahārika*] of Asokan edcts. In Kautilya's *Arthaśāstra* both the portfolios viz: *vyavahārika* and *adhyaksha* are dealt *Vyavahārika* was purely judicial officer while *adhyakshs* were given various responsibilities. There fore, the *nagarådhyaksha* at the capital city of king Khāravela probably also looked after the city administration.

(vi) Nākiya the wife of Mahāmada (Minister)[54]

This name of donor appears in the *Jambeśvara* cave. She was the wife of a top minister (*mahāmada - mahāmātya - mahāmātra*?).

(vii) Amātya Sapasanaka[55]

This name is known from a terracotta sealing found at Sisupalgarh. By designation it is clear that Sapasanka was a minister.

Besides, we get the readings as Chullakamas, *Kammas Halakhinaya* at *Sarpagumphā* (cave - 13), *Chulakamas* at *Haridāsgumphā*[56] (cave - 16) and *Pādamūlikas Kusumas* at *Tatwvagumphā* (Khandagiri, cave - 1).[57] They were donors of the respective caves. The word *chulakama* and *kama-halakhinay* are interpreted variously by Dr. Banerji[58] and Dr. Barua.[59] It seem that *chulakama* which may be translated in sanskrit as *kshudrakarman* or *chūḍākarman* were personal attendants to the king and worked as ad-de-camp to the king and his family. It is very likely that each member of royal family had his personal attendants. The faithfulness of such attendants should be of very high degree and probably they were tested for this virtue. Since, they performed menial jobs for their masters therefore, they were *khsudrakarman* but they enjoyed power and status in the society and were capable of making rich offerings. It seems that atleast two *kshudrakarman* (*chulkama*) accompanied the king and other royal persons to *Kumārīparvata* (udaygiri) and they also made their offerings by excavating the caves for *jaina*-ascetics. Word *kama* probably stands for worker or employee (Sanskrit: *karmi*) who was different in status from *chulakama*. In case of *Sarpaguṁphā* it appears that one *chulakama* (royal attendant for personal services) and *halakhinaya* who was also a royal servant jointly caused the excavation of the said cave. For *pādamūlika kusuma*[60] it seems that he was an attentive servant of his master (king). If we take his name in consideration then it can be opined that he also performed the job of flower-carrier for his master.

From the cave-inscriptions we get the name of only those royal servants who believed in the same sect. However, there is no doubt that the capital city was crowded with a great number of people who were running the government machinery; looking after the city and state's administration and security through out the whole period.

Another dominant class of the city population was formed by military forces. The settlement of Sisupalgarh was basically a fortification for defence purpose. We have already stated that this fort was well guarded and there would have been advance military camp being always ready to encounter any attack at capital. Though, Asoka turned to *dhammavijaya* after Kaliṅga war but he would certainly keep sufficient armies to meet any exigency in the newly conquered province and for security of Government's high officer stationed at *Tosalī*. During the rule of Chedi dynasty the importance of military organization was highly increased. King Khāravla had an excellent military career. His armies were organised in four-fold division, viz: cavalary (*haya*), elephant (*gaja*), infantry (*nara*) and chariot-warriors (*ratha*).[61] We may usually infer that there were various ranks and files among the warriors of the each division commanding the troops in different levels. The high military officers, the commanders of each division and the chief commander would have been regarded as the elite of the city. Besides the Warrior classes, there would have been a large number of militia maintaining the stables of the horses and elephants, looking after the food-supply and other like jobs; chariot-drivers and elephant-drivers. The military encampments at Sisupalgarh would have continued more or less till the life of ancient settlement.

The structural remains of Sisupalgarh witness a fully urbanised culture while presence of inscriptions at Dhauli and Udaygiriḥ hills are the evidence of reading and writing. The monastic establishment would have been undoubtedly the great centre of learning and religious discourses. However, we have noticed that king Kharavela mastered in various subjects like royal correspondence (*lekha*), currency (*rūpa*), arithṃatics (*gaṇanā*), law and legislation. Not only that he was also well-versed in music and other fine-arts. Such advance education in the above subjects was not possible without highly qualified teachers of different subjects. Therefore we may say that qualified *āchāryas* were engaged for education of princes as well as youths coming from the houses of elite class. It is also not unlikely that there may have been *gurukula* organizations with specialized teachers in different disciplines including martial-arts which were patronised by the king and the affluent class of the city. Udaygiri hill was a centre of learning, and students practised there writing the *Brāhmī* alphabet. The painted inscription in *Tatwagumphā* no. 1 show the repeated writings of Indian alphabets - "some young monk had used the back-wall of the cell as a copy-book and improved his knowledge of alphabet by writing on it. The characters belong to the first century AD.[62]

For the supply of various articles and consumers' goods to the city-population there would have been an organised system of traders, vendors and craftsmen. Among the craftsmen the potter, the weaver, the carpenter, the ironsmith, the stonecutter, the goldsmith, the beadmaker, the jeweller always formed an important part of the civilized society. The archaeological remains as well as the antiquities from Sisupalgarh yield the evidences that the people of above mentioned classes inhabited the Sisupalgarh and worked in their respective trades. Large-scale constructional activity at Sisupalgarh as well as the rock-cut monuments of Udaygiri are the living testimony that there lived the master sculptors, stone-carvers, masons and stone-cutters. Besides, a number of people worked as labourers and attending menial jobs at the houses of affluent classes. Literary traditions inform that Kaliṅga had many rich trading centres. Excavations proved that Sisupalgarh had trade-links with Rome. Therefore, we may think of the rich traders inhabiting in this city. Some of these *śhreshthins* (traders/business persons) may have been close to the royal-court and enjoyed the elevated position.

Another class of people who were closely associated with trade and commerce were the cart-drivers and navigators. While discussing the geomorphology of Bhubaneswar area we have told that the place was full of deep water-sources yielding the full scope for fishing as livelihood. Orissa is also fish-consuming state therefore, we may say that fishermen also constituted a good proportion of common man. This conjecture is further supported by a fact that still, there is a colony of fishermen locally called Keutsāhī (the hamlet of fishermen/navigators) within a kilometre from Sisupalgarh. These people also perform the job of navigator. From *Hāthīguṁphā inscription* as well as bas-reliefs of Udaygiri cave we get the evidences of dancers and musicians. Such professionals were important part of the society and entertained the people.

The above mentioned classes of people belonged to the *grihastha* order. While the *Kumārī* hills, *Ekāmra kānana* and Dhauli hills were inhabited by monks and ascetics.

To feed such a large city population an established agrarian system was essential. The vast stretches of fertile alluvial tract towards south and east would have supplied the sufficient quantity of foodstuffs. Besides, the land-revenue would have served the major source of state-income. Probably *rājukas* (keeping the rope for measurement of land) were the officers for measuring the cultivated land and to decide the king's share. It can be observed that while monastic settlements were located at hills and likewise and *Ekāmra* in the laterite, both unfit for agriculture, but the civil settlement was located in the fertile alluvial tract. This location provided a convenient collection of agricultural produce. Looking for the location of the fortification amidst the fertile land it is plausible to opine that there might have been some land under the direct cultivation of royalty.

Being a capital city this place would have proved a good market for traders, therefore, tarrifs, custom-duties and octroi etc. from traders and craftsmen would have been the another important source of state's income. Besides, great booties collected from the conquered countries also increased the royal treasure.

The use of currency is well established by documentary as well as antiquarian evidences. It seems that only required quantity of food-stuff was collected in kind as a part of land revenue while the remaining porticn was collected in the form of cash-currency. Similarly, the tariffs, custom-duties and the octroi were collected in both cash and kind.

Government officers and military forces were paid in cash; however it is likely that some of them were also granted lands as permanent assets so that they could earn their livelihood in distress. A great consumer class yielded the scope for traders, vendors, and craftsmen to earn more and more and flourish. While poor people and labourers earned their livelihood serving all above people. Thus an interchain of money circulation would have evolved from collection of revenue and consumers' requirement which created jobs for large population. Further, during the powerful ruler like Kharavela money circulation also increased due to state's various welfare projects and providing more chances of jobs for various professionals and craftsmen. But such favourable conditions would have not lasted long. Since the annexation by Asoka Kaliṅga almost remained under monarchy and under the weak rulers the kingdom fell in chaotic conditions and while being unable to protect the life and money of the public, the unsafe people moved to safer places. Such phenomenon would have caused greater harm to the economy of this settlement during its end.

3. WAY OF LIFE

The preceding discussions evidenced that the early settlements developed in Bhubaneswar area of highly urbanized nature. The *Hāthīguṁphā* inscription also discriminates *paurāṇ* and *jānapadān*. The *paura* were probably the people inhabiting in the *pura* (city) while the *jānapadān* were the peasant population living in the surrounding villages to feed the urban settlement. Being the place of royal court there would have developed a standard of social behaviour, manners and etiquette and we may expect that people inhabiting here were accustomed of *nāgara* (refined of city like) behaviour. Asoka and Khāravela brought the local people in wide contacts of northern as well as southern India. Besides, inland trade links and maritime trade made this place a centre of advance culture. People were versed with the reading and writing *Brāhmī* script and were speaking *Prākrit*. The knowledge of Sanskrit - language cannot be ruled out. The use of currency also confirms that common people knew the science of counting and related arithmetic. The *Hāthīguṁphā* inscriptions mentions that Khāravela was anointed as king after completing twenty-four years of his age i.e. in the twentyfifth year. This factor suggests that the *āśhrama-vyavasthā* was in vogue. The religious fanaticism did not develop during this period as Asoka[65] and Khāravela[66] both claim themselves as *sarvapāsaṅḍapūjaka* (worshipper of all stones i.e. idols of all kinds of gods) and *sarvadevāyatana-saṅskāraka* (conservator of all type gods temples).

Jainism was a predominant religion, the monastic establishment of *Jain*-sect was patronised by royal and public charity. Image-worship is not found in this period. From the narrative reliefs it appears that the worship of symbols was in vogue at that time likewise among the followers of Buddhism. Cave-5 (*Jaya-Vijaya guṁphā*) of Udayagiri and cave-3 (*Anaṅtaguṁphā*) of Khandagiri, for example, depict a tree worshipped by devotees with floral offerings. In cave-9 (*Maṅchapurīguṁphā*) of Udayagiri it is shown that an unidentifiable symbol being worshipped probably by a royal family. Some symbols forming part of eight auspicious objects (*ashtamaṅgalas*) often worshipped by *Jains* appear in carvings; e.g. on the backwall *Anaṅtaguṁphā* a *naṅdipada* is shown placed upon a pedestal flanked on either side by a set of three symbol, a triangle headed one (*bhadrāssna or sthāpana*), *śrīvatsa* and *svāstika*. Buddhism and Saivism were other important religions. Besides, *nāga* and *yaksha*- worships were also in vogue as attested by the antiquarian remains of the period.

In the relief art of Udaygiri the contemporary social life, dress and ornaments are faithfully delineated. The chief drapery, as in the reliefs, consisted of a *dhotī* or *sārī,* often diaphanous in the case of women, tied by a waist-band, with folds of the ends hanging in front. Both men and women usually were without any garment to cover the part of the body above the waist, but scarves were used in ceremonial occasions. Men used to bear a turban, the rich often with crest jewels. Veils covering the head but not the face were sometimes worn by woman. The coiffure of woman was varied and often bedecked with ornaments. In contrast to the scantiness of dress, ornaments were profuse. Both women and men of rank were decorate with ornaments e.g. ear ornaments, necklaces and bangles; women having girdles, anklets and head ornaments in addition. The ear ornaments were of two type ; rings and studs. Furniture, like the bedstead, stool, table and seat ; utensils like the bowl, plate pitcher, umbrella fan, toilet-tray and caskets are depicted.

NOTES AND REFERENCES

1. R.C.Mazumdar (ed.), *The Age of Imperial Unity,* (Bhartiya Vidya Bhavan) Bombay, 1951, p. 1, "The same political condition is reflected in the stereotyped list of sixteen great countries. (*solasmahājanapada*) which according to Buddhist text; flourished shortly before the time of Gautama Buddha. They are as follows according to *Aṅguttara Nikāya*:- Aṅga, Magadha, Kāsi, Kosala, Vriji (Vajji) Malla, Chedi, Vatsa (Vamsa), Kuru, Pāñchala, Matsya (Machchha), Surasena, Asvaka or Asmaka (Assaka), Avanti, Gāndhāra and Kāmboja, each being named after the people who settled there or colonised it. (Footnote - 1: *Aṅguttara Nikāya* I, 213; IV 252, 256, 260; *Mahāvastu,* I, 34 II, 3; *Vinaya* Text II, 146 fn; *Niddesa* II, 37.)."

2. H.C.Raichaudhary, *Political history of Ancient India,* op. cit., pp. 87-88, "the kingdom of Kaliṅga, too was in existence in the time of Nimi and his contemporaries of *Brāhmaṇa* period. The evidence of *Jātaka* (Kuṁbhkāra *Jātaka*) is confirmed by that of *Uttar-adhyana sutra.* The *Mahāgoviṅda Suttanta,* makes Sattabhu, king of Kaliṅga, and of Dhataraṭṭa or Dhṛitarāshṭra, king of Kāsī, mentioned in the *Śatapatha Brāhmaṇa.* There

can be thus no doubt that Kaliṅga existed as an independent kingdom in the time of which the *Brāhmaṇas* speak. It is mentioned both by Pāṇini and Baudhāyana. The latter regards it as an impure country, but evidently not unfrequented by Aryans. According to epic tradition it comprised the whole coast from the river Vaitarani (*Mahābhārata*-III, 114, 4) in Orissa to the borders of Andhra country. The southern boundary of the *janapada* was not well defined. It reached Yellamanchili and Chipurupalli in the Vizagapatam district and at times even Pishṭapur or Pithapuram, north-east of the Godavari, but not the river itself which flowed through the Andhra country. Pargiter says that Kaliṅga as settled kingdom appears to have consisted properly of the plain between of the Eastern Ghats (Mahendra range) and the sea. But its kings seem to have exercised suzerainty over the jungle tribe which inhabited the hills, far inland, far the Amarkanatak range, in which the Narmada rises, is said to be in the western part of Kaliṅga. That large tracts of the country were covered with forests appears from reference *Kaliṅgāraṇya* in *Pālī* texts. The windows of the capital city in the days of Kalidas looked out on the sea, and the deep roar of the waves drowned the sound of trumpets. In the days of Yuan Chawng Kaliṅga Wu-tu (Orda) and kung-yu-to (kāṅgoda) in the north and An-to-lo (Andhra or Vengi) in the south and seems to embrace the parts of Ganjam and Vizagapathnam districts."

3. R.C.Mazumdar (ed.) The Age of Imperial Unity, op. cit., pp. 1-2.

4. Cf. K.C. Panigrahi, *History of Orissa,* op, cit., p. 4 ; see also, N.K.Sahu, Historical Geography of Orissa", *Souvenir, Orissa Sahitya Akademi*, Silver Jublee - 1982, Bhubaneswar, 1982, pp. 1-2.

5. E.Hultzsch, *Corpus Inscriptionum Indicarum,* Vol. I, op. cit., pp. 66-68: When king Devanampriya Priyadarsin had been

anointed eight years, (the country of) the Kaliṅga was conquered by (him).

अठवष अभिसितम प्रिअद्रशिस रजो कलिंग विजित।

(extract from 13th rock-edict, Sahabazagarhi)

6. R.C. Mazumdar (ed.), *The Imperial Age of Unity,* op. cit., pp. 88-9. And V.A. Smith. *Asoka, The Buddhist Emperor of India,* New Delhi, (reprint) 1970, pp. 19-20 and 73.

7. V.A. Smith, *Asoka, The Buddhist Emperor of India,* op. cit., pp. 24 and 73.

8. R.C. Mazumdar, *The Age of Imperial Unity,* op. cit., pp. 61-2; and V.A. Smith, *Asoka, The Buddhist Emperor of India*, op. cit., pp. 75-81.

9. V.A. Smith, ibid, pp. 76-7.

10. K.C. Panigrahi, *History of Orissa,* op. cit, p. 11; There is a story prevalent in Orissa, particularly among the fishermen of the eastern coast, that Asoka fell in love with a beautiful fisherwoman named Kāruvākī and wanted to marry her. In order to gain his object he wanted to acquire the homeland of this beautiful woman by force and this led to the Kaliṅga war.

    The name Kāruvākī referred above appeared in the Asokan Minor Pillar Edict of Kosambi where Kāruvākī (Chāruvākī) is said to have been the second queen of Devānaṁpriya and the mother of Tīvara (*dutiyaye deviye ti Tīvala-mātu Kāluvākiye*), see Corpus Inscriptionum Indicarum Vol. I, op. cit., p. 158. However, Panigrhi himself admits that this tradition is of very late origin and therefore cannot be relied upon.

11. E. Hultzsch, *Corpus Inscriptionum Indicarum,* Vol. I, op. cit., pp. 66-8: One hundred and fifty thousand in number were deported thence, one hundred thousand in number were those who were slain there, and many times as many who died.

दिअढमचे प्रणशतसहस्रे ये ततो अपवुढे शत सहस्र मचे तच हते बहुतवतके व मुटे।

12. Ibid. pp. 66-8: After that, now that (the country of) the Kaliṅgas has been taken, Devanāṁpriya (is devoted) to a zealous study of morality, to the love of morality and to the love of morality and to the instructions (of people) in morality.

ततो पच अधुन लधेषु कलिगेषु तिव्रे ध्रमशिलन ध्रमकमत ध्रगनुशास्ति च देवनप्रियस सो अस्ति अनुसोचन देवनप्रिअसि विजिनिति कलिगानि।

13. Ibid., pp. 66-9: For this is considered very painful by Devanampriya that while one is conquering an unconquered (country), slaughter, death and deportation of people (are taking place) there.

But the following is consideration even more deplorable than this by Devanāṁpriya.

(To), the Brāhmaṇs or Sramaṇas, or other sects or householders, who are living there .... to these then happen injury or slaughter or deportation of (their beloved ones).

Or, if there are then incurring misfortune, and relatives of those whose affection (for the latter) is undiminished, this (misfortune) as well becomes an injury to these (persons) themselves.

This shared by all men and is considered deplorable by Devanāṁpriya.

अविजितं हि विजिनमनो यो तच वध व मरणं व अपवहो व जनस तं बढ़ं

वेद नियमतं गुरूमतं च देवनं प्रिअस इदं पि चु ततो गुरूमततरं दंवनं प्रिअस ये तत्र

वसति ब्रमण व श्रमण व अंजे व प्रषंड ग्रहथ........, तेष तच्च भोति अपग्रथो

व वधो व अभिरतन व निक्रमणं येष व पि सुविहितनं सिही अविप्रहिनो इदं पि चु

ततो गुरूमततरं देवनं प्रिअस ये तत्र व तेष मिच संसतुतसहयजतिक वसन प्रपुणति त

च तं पि तेष वो अपधषो भोति प्रतिभगं च एतं सव्रमनुशनं गुरूमतं च देवनं प्रियस

14. ibid.

Therefore even the hundredth part of the thousandth part of all those people who were deported at time in Kaliṅga, (would) now be considered very deplorable by Devanāṁpriya.

सो यम चो जनो तद् कलिगे हतो च मुटो च अपवुढ च ततो शत भगे व सहस्र

भगं व अज गुरूमतं व अज गुरूमतं वो देवने प्रियस

15. ibid.

And this conquest is considered the principal one by Devanāṁpriya, viz. the conquest of morality.

अयि च मुखमुत विजये देवानं प्रिअस यो ध्रम विजयो

16. And for the following purpose has this rescript on morality been written, (viz.) in order that the sons (and) great-grandsons (who) may be (born) to me, should not think that a fresh conquest ought to be made, (that), if a conquest does please them, they should take pleasure in mercy and light punishments, and (that) they should regard the conquest by morality as the only (true) conquest.

एतये च अठये अयि ध्रमदिपि निपिस्त किति पुच पपोच मे असु नवं विजयं म

विजेतविअ मजिषु स्यकस्पि यो विजये क्षतिं च लहुदंडत च रोचेतु तं च यो विज मअतु

17. ibid, pp. 92-100.

18. ibid., pp. 100-18.

19. ibid., pp. 96-7 (from first Separate Rock-Edict of Dhauli):

It happens in the administration (of justice) that a single person suffers either imprisonment or harsh treatment.

In this case (an order) cancelling the imprisonment is (obtained) by him accidently, while (many) other people continue to suffer.

In this case you must strive to deal (with all of them) impartially.

But one fails to act (thus) on account on the following dispositions: envy, anger, cruelty, hurry, want of practice, laziness, (and) fatigue.

(You) must strive for this, that these dispositions may not arise to you.

And the root of all this is absence of anger and the avoidance of hurry.

He who is fatigued in the administration (of justice), will not rise; but one ought to move, to walk and to advance.

He who will pay attention to this, must tell you: See that (You) discharge the debt (which you owe to king) ; such and such is the instruction of Devanāṁpriya.

For the following purpose has this rescript been written here (viz.) in order that judicial officers of the city of may strive at all times (for this), (that) neither undeserved fettering nor undeserved harsh treatment are happenings to (men).

And for the following purpose I shall send out every five years (a Mahamatra) who will be neither harsh nor fierce, (but) or gentle actions, (viz. in order to ascertain) whether (the judicial officers), paying attention to this object .... are acting thus, as my instruction (implies).

नितयं एक पुलिसे पि अथि ये बधनं वा पलिकिलेसं वा पापुनाति तत होति अकस्मा

तेन बधनंतिक अंने च .... हु जने दविये दुखीयति तत इछितविये तु फेहि कितं

मझं पटिपादयेमा ति इयेहि चु जातेहि नो संपरि पजति इसाय अनुलोपेन निहूलियेन

तूलनाय अनावूतिय आलसियेन किलमथेन से इछितविये कितिं एते जाता नु हुवेवु ममा

ति एतस च सवस मूले अनामुलोपे च नितियं ए किलं ते सिया न ते उगछ संचलित

विये तु वटितविये वा देवं मेव च देवं च देवानां पियस अनुसथि

एतायं अठाये इयं लिपि लिखित हिद एव नगल वियोहालका

सस्वत समयं यूजेव ति .... नस अकस्मा पलिबोधे व अकस्मा पलि किलेसे

व नो सिया ति एताये च उठायं हकं .... मते पंचसु पंचसु वसे सु निखामयिसामि

ए अखखसे अचंडे सखिनालंभे होसति एतं अठं जानितु .... तथा कलंति अथ मम अनुसथी ति

20. On the basis of Puranic statements R.K. Mukharji dated the end of Asoka's reign in 236 B.C.; however, corroborating the various evidences contained therein the edicts of Asoka the end of Asoka's reign is computed 232 B.C.; for 236 B.C. see, V.A.Smith, *Asoka,* op. cit., p. 74 and Romila Thapar, *Asoka and the Decline of the Mauryas,* Delhi, 1973, p. 51; and for the 236 BC. see R.C. Mazumdar (ed.) *The Imperial Age of Unity,* op. cit., p. 88-9.

21. K.C. Panigrahi, *History of Orissa,* op. cit., p. 17.

22. ibid. See also K.A. Nilakanta Sastri, *Age of The Nandas And Mauryas,* Banaras 1952, p. 246. In some other works it is 187 B.C., see *The Age of Imperial unity*, op. cit., p. 90 a and pp. 95-7. V.A.Smith mentioned it as 188 B.C., see, op. cit., p. 74.

23. K.A. Nilakanta Sastri, *Age of The Nandas And Mauryas,* op. cit, pp. 246-7: Pushyamitra Śuṅga usurped at the heart of kingdom while the Chetas of Kaliṅga and the Sātavāhanas succeeded in other portions of the empire.

24. R.D. Banarji, "Inscriptions in The Udayagiri Caves", *Epigraphica Indica* Vol XIII, op. cit., pp. 160 1. See also, B.M. Barua, "Minor old Brahmi Inscription in Udayagiri and Khandagiri, *Indian Historical Quarterly* Vol. XVI. op. cit., pp. 158-66.

25. K.P. Jayaswal and R.D. Banarji, "The Hathigumpha Inscription of Khāravela", *Epigraphica Indica* vol. XX, op, cit., pp. 71-88; and B.M. Barua, "Hathigumpha inscription of Kharavela", op. cit., *Indian Historical Quarterly* vol. XVI, op. cit., pp. 459-85.

26. R.C. Mazumdar (ed.) *The Age of Imperial Unity,* op. cit, p. 216; and D.C.Sircar, *Select Inscriptions,* vol. I, op. cit., p. 213. ff.

27. B.M. Barua, "Hathigumpha Inscription of Kharavela", *Indian Historical Quarterly* Vol. XIV, op, cit., p. 261.

28. K.C. Panigrah, *History of Orissa,* op. cit., pp. 27-8.

29. ibid., p. 29.

30. K.P. Jayaswal and R.D. Banarji, The *Hathigumpha* Inscription of Khāravela", *Epigraphica Indica,* Vol. XX, op. cit., p. 87 and fn. 7 on the same page, "the natural explanation is 'the one hundred and three of king Nanda". See p. 84 also.: "The date in this line apparently refers to an era founded by king Naṅda."

    Even, if we accept it an era of some king of Naṅda dynasty, then also the date of excavation of original canal referred in the above inscription falls some where in the reign of Asoka, which Barua and Panigrahi stressed in their writings that the said canal was excavated by Asoka; see, K.C. Panigrahi, History of Orissa, op. cit., p. 28.

31. R.D. Banar ji, "Inscription In the Udayagiri and Khandagiri Cave", *Epigraphica Indica* Vol. XIII, op. cit., p. 161. (The name Vakradeva is suggested by D.C.Sircar, see, *Select Inscriptions,* op. cit., p. 222.)
32. ibid., pp. 159-60.
33. ibid, p. 161.
34. Cf. D. Das, *Early History of Kaliṅga,* op. cit., pp. 79-80.
35. ibid.
36. ibid, pp. 80-3.
37. B.B. Lal, "Sisupalgarh 1948: An Early Historical Fort in Eastern India", *Ancient India* No. 5., op. cit., p. 97.
38. ibid.
39. K.C. Panigrahi, *History of Orissa,* op. cit., pp. 41-2.
40. ibid.
41. B.B. Lal, "Sisupalgarh 1948; An Early Historical Fort in Eastern India", *Ancient India,* No. 5. op. cit., pp. 97 and 100-101.
42. K.C. Panigrahi, *History of Orissa,* op. cit., pp. 40-3.
43. R.D. Banarji, *History of Orissa* Vol. I.
44. B.B. Lal, Sisupalgarh 1948: An Early Historical Fort In Eastern India", *Ancient India* No. 5., op. cit., p. 72.
45. ibid., p. 99.
46. ibid.
47. ibid.
48. E. Hultzsch., *Corpus Inscriptioum Indicarum* Vol. I, op. cit., pp. 4-5. The third rock-edict: yuta cha rājuke cha etc.

49. K.P. Jayaswal and R.D. Banar ji, The Hāthīgumphā Inscription of Khāravela" *Epigraphica Indica* Vol. XX. op. cit., pp. 71-89.

50. R.D. Banarji, "Inscriptions in the Udayagiri And Khandgiri Caves", *Epigraphica Indica,* Vol. XIII, op. cit., pp. 159. 60.

51. ibid, 161.

52. ibid.

53. ibid. 163.

54. ibid., pp. 163-4.

55. A. Das, "Some new Facts About Sisupalgarh" *Orissa Historical Research Journal* Vol. XV. op. cit., p. 60.

56. R.D. Banar ji, "Inscriptions in the Udayagiri and Khandagiri Caves", *Epigraphica Indica* vol. XIII, p. 162.

57. ibid, p. 164.

58. ibid.

59. B.M. Barua, "Minor Old Brahmi Inscriptions in Udayagiri - Khandangiri Caves", *Indian Historical Quarterly* Vol. XIV, 165, fn. 20, 21 and 22.

60. ibid, p. 165, fn. 20: *Pādamūlika* is not a place-name. It is evidently the same official designation as *rājapādamūlika* mentioned in the *Pālī Asadisa-Jātaka* (Fausboll No. 522); and fn. 23: Banar ji taken *pādamūlika* in the sense of a menial, lit. "a server of the feet". According to the *Jātaka* usage, a *pādamūlika* was a personal attendant, a body-guard.

61. K.P. Jayaswal and R.D. Banarji, "The Hāthīgumphā Inscription of Khāravela", *Epigraphica Indica* Vol. XX, op. cit., pp. 79 and 86-87; "*Dutiye cha vase achitayita Sātakaṁnim pachhima-disam haya-gaja-nara-bahulas daṁḍaṁ paṭhāpayati*".

62. R.D. Banarji, "Inscriptions in Udaygiri And Khandagiri Caves", *Epigraphica Indica* Vol. XIII, op. cit, p. 165.

63. The expenditure-cost of various works is stated in currency, that was "*paṇa*" in the Hāthīguṁphā Inscription of Khāravela e.g. *panatisahi sata-sahasehi* (at the cost of thirty-five-hundred-thousand)", see, The Hāthīguṁphā Inscription of Khāravela", *Epigraphica Indica* Vol. XX, op. cit., pp. 79-89.

64. ibid, From the frequent mention of use of currency coins in the State's expenditure in the Hāthīguṁphā Inscription it appears that currency system dominated in exchange than the barter system.

65. In the Seventh Rock-Edict it is said that king Devānampriya Priyadarsin desires (that) all sects may reside everywhere. See, *Corpus Inscriptionu Indicarum,* Vol. I, op. cit., pp. 13-4 and 89.

66. K.P. Jayaswal and R.D. Banar ji, "The Hāthīguṁphā Inscription of Khāravela", *Epigraphica Indica* Vol. XX, op. cit., p. "*guṇa-visesa-kusalo-sava Pāsaṇḍa-Pūjako sava-de (vāya) tana-saṁkhāra-kārako.* See translation, p. 89, ... accomplished in extraordinary virtues, respecter of every sect, the repairer of all temples.

4

End of Sisupalgarh Culture and Transition to Religious Settlement of Ekamra-Bhubaneswar: A Narrative

In the foregoing chapter we have noticed that after the fall of the Chedis the empire of Kaliṅga became a subordinate territory of the Sātavāhans and subsequently of the Muruṅḍas. On the other hand the Allahabad Pillar Inscription of Samudragupta indicates that he defeated a number of rulers of the south (dakshiṇāpatha).1 The Gupta emperor is said to have conquered Maheṅdra of *Kosala,* Vyāghrarāja of *Mahākāntāra*, Maṅṭarāja of *Kaurāla,* Maheṅdragiri of *Pishṭapura,* Svāmidatta of *Koṭṭūra,* Damana of *Eraṅḍapalla,* Vishṇugopa of *Kāñchī,* Nīlarāja of *Avamuktaka,* Hastivarman of *Veṅgī,* Ugrasen of *Pālakka,* Kubera of *Devarāshrṭra* and Dhanañjaya of *Kusthalapura.*[3] The phrase, "*grahaṇa moksha-anugraha*" used in the Allahabad Pillar Inscription about the Kings of *dakshiṇāpatha* (south), implies that the Gupta emperor later liberated these kings whom he first defeated and captured.

Of these territories *Koṭṭura* has been identified with modern Kothoor in the Ganjam district of Orissa and *Pishṭapura* with modern Pithapuram in the Godavari district of Andhra Pradesh. *Erańḍapalla* and *Devarāshṭra* are located in modern Erandapalli (Vizagapatam district) and Yellamanchili (Vizagapatam district) of Andhra Pradesh respectively These identifications suggest the inclusion of the territories concerned in ancient Kaliṅga, which extended in the littoral area upto the river Godavari. It appears that, in the days king Samudragupta (AD fourth century) or rather on the eve of his invasion, Kaliṅga was divided into a number of kingdoms and was not under one single sovereign.[5]

While listing the names of the kings alongwith their kingdoms of south the composer of the above inscription used a word *prabhriti* after the name of Dhanañjaya. The meaning of this word is 'etcetra' which implies that there were more such principalities but they were insignificant therefore the composer had not taken pain to mention their names. Among such least important ruling families the mention may be made of the Matharas who are known to us from their copper plate grants.[6] On paleographic considerations their records appear to have belonged to roughly fifth-sixth century AD.[7] They were at first living as minor chiefs in remote part of Kaliṅga with their capital at Śhrīpura, and probably therefore remained unnoticed by Samudragupta.[8] It is also possible that after hearing the unabated victories of Samudragupta they surrendered in person or by their envoys without any resistance hence remained undisturbed and emerged as a greater power subsequently. Soon after the expedition of Samudragupta the Matharas gradually amalgamated most of the small principalities [which were already weakened in battles with Samudragupta] into a united kingdom in the region around the mountain Mahendra.[9] One Umāvarman was the first king of this family who assumed the title of *Kaliṅgāḍhipati* and preferred *Siṁhapura* to the capital of Kaliṅga.[10] After Umāvarman all other kings of the dynasty used the same title. By the time of Śaktivarman as recorded in the Ningandi copper plate of his son Nandaprabhañjanvarman the Māthara kingdom extended from the river Krisṇa in the south to the river Mahanadi in the north and being the ruler of this vast land Nandaprabhañjanvarman introduced the title of sakalakaliṅgadhipati.[11] By the middle of sixth century AD, however, the Māthars were overpowered by one Pṛithvī Mahārāja of the *ŚriRām-Kāśyapa-gotra.* Who captured *Pishṭhapura* after defeating them in Kaliṅga.[12] In his attempt to annexe the country north of river Mahanadi he reached upto *Viraja*[13] [modern Jaipur] but apparently he was defeated by the Vigrahas[14] and retreated to *Pishṭhapura* where his successors continued to rule until they were overpowered by the Chālukyas of Badami during the first quarter of seventh century AD.[15]

The country on the north of river Mahanadi, extending as far north as *Tāmralipti* (Tamluk of modern Midanapur district in West Bengal) was divided into two sections in the mid-sixth century AD which were known as *Dakshiṇa Tosalī* and *Uttara Tosalī,* ruled over by the Vigrahas and the Mudgals respectively, with both divisions bifurcated by the river Vaitarni.[16] Some of the copper plate grants of these rulers are recorded in the Gupta-era therefore it is likely that they served initially as the feudatories under the Imperial Guptas. This is further corroborated by the Sumandal copper plate of Dharmarāja issued in the Gupta-year

250 (AD 569) starting that Dharmarāja is a subordinate kin under Pṛithvīvigraha the ruler of the *Kaliṅga-rāshtr* which formed a part of the Gupta-kingdom.[17]

When and how this part of Orissa came under the Gupta is still enveloped in the darkness. Since, Samudragupta did not include in his *divgijaya* (victory expedition); it seems that this came into control on the Guptas as soon as the Muruṇḍas were overpowered at *Pataliputra* by the formers. Getting automatic control over the territories earlier under the Muruṇḍas the Gupta-kings would have appointed their regional chiefs (they may be also from the local dissidents to the Muruṇḍas) and this Gupta-era came into use in Kaliṅga/Tasolī region.

The preceding discussion on the political conditions of fourth-fifth centuries lead us to opine that the old city of *Tosalī* and fortified settlement of modern Bhubaneswar area were constantly falling into background due to rise of a number of new political centres in the land of erstwhile Kaliṅga-empire. At first the status of this imperial capital would have decreased to a regional administrative centre under the Sātavāhanas and the Muruṇḍas respectively. These circumstances would have certainly decreased the local population due to migration to government staff to the new political centres which might have acted adversely on traders and craftsmen. During circa AD 350 we find that small but independent and flourishing kingdom already emerged in the banks of rivers Godavari, Nagavalli and Vamsdhara in the district of East-Godavari, Srikakulam and Ganjam. These principalities were not far from the Bhubaneswar region particularly that lying in the district Ganjam. These new kingdoms would have yielded more scopes of employments for literates and working masses as well as warrior-class. The traders and craftsmen also would have found better markets in these centres. Thus, all these conditions and perspectives would have led large-scale migration of population from Bhubaneswar which was basically depended upon court-patronage.

On the other hand we notice that the northern part of Kaliṅga was divided into independent kingdoms known as *Dakshiṅa Tosalī* and *Uttara Tosalī.* The territory of *Dakshiṅa Tosalī* extended from river Mahanadi to river Vaitarani while the region from river Vaitarni to *Tāmralipti* was known as *Uttara Tosalī.* In the Kanas copper plate of Lokavigraha (dated Gupta-era 280 i.e. AD 600) he is told to be master of the eighteen *Tosalīs.*[18] Scholars suggested that from the eighteen *Tosali* we should mean that there were eighteen part of the kingdom.[19] However, it is more probable that apart from the major kingdom of the Māthāras (subsequently under Pṛithvī Mahārāja of *Srīrāma-Kāsyapa-gotra*) and the *Dakshiṇa Tosalī* of the Vigrahas there were also a few small principalities scattered in between the both. These petty rulers would have been taken into subordination and their estates were consolidated into the main kingdom. The purpose of pointing out the above probability is to show that the northern region of Kaliṅga as also under dissolution which would have proved a major setback to the desertion of great city-settlement at Bhubneswar. The city of *Virajā* emerged as the Capital of *Dakshiṇa Tosalī* during the Vigrahas[20]; there is no doubt that this city would have become the centre of political activities during the time of the Gupta; and its development would certainly have deprived the Bhubaneswar region of political importance. Similarly, we see that the Māthars claimed themselves as the master of the whole Kaliṅga (*sakalkalingādhipati*) having the extension of their territorial boundaries upto the river Mahanadi in north, but their political centres remained confined within the range of about sixty miles (say 100 km.) from Srikakulam.[21] Thus, in this period Bhubaneswar region lay politically neglected and was deprived of any political patronage.

Owing to the causes dealt above the elite class, the traders and the craftsmen gradually abandoned the long cherished city of Kaliṅga leaving behind the *śaiva*-ascetics, the Jains and probably the Buddhist monks for their undisturbed penance and *sādhanā*. The settlements of the alluvial plains now became the simple peasants' locality The monastic settlement of the *Jainas* at Udayagiri hills also would have suffered in want of royal patronage, however, they continued to exist. In the laterite tracts and Dhauli hills probably a confrontation took place between the Śaivites and the Buddhist during the fourth-fifth century AD in which Śaivites ultimately succeeded to establish their supremacy ousting the Buddhist from that area.[22] In this confrontation the Śaivites would have certainly backed by the favourable political condition. We see the Māthara were followers of Śalivism and claimed themselves as *māheśvaras*.[23] Likewise, the Vigrahas were also the followers of neo-brahmanical pantheon.[24] Sasāṅka the king of *Gauda* who was also the overlord of *Kaṅgoda* (a part of modern Orissa), was the staunch devotee of *Śiva* and noticed for his anti-Buddhist activities.[25] The rulers of the Sailodbhava dynasty who ruled over *Kaṅgoda* (comprising of modern Puri and Ganjam districts of Orissa) earlier as subordinates of king Sasāṅka and later on independently were all *Śaiva*[26] but they also patronised Jainism.[27] Thus, the royal factor was in support of Saivism at Bhubaneswar. The rulers of the Bhaumakara dynasty who succeeded the Śailaodbhavas in Orissa were Buddhist though some later kings of this dynasty were Śavites.[28] They patronised Buddhism in Orissa and a number of Buddhist centres flourished around their capital Jajpur, viz : Ratnagiri, Udayagiri, (identified as Mādhavapur mahāvihāra), Lalitagiri, Bajrarigi, Solampur and at other various sites all lying in district Cuttack. But, the Bhaumakaras did not disturb the supremacy of Śaivism at Bhubaneswar.

When and how the Saivite settlement the *Ekāmra-Bhuvaneśvara* came into prominence, is a point of further discussion. In this regard some of the existing temples of Bhubaneswar confirm that the temple-building activity at *Ekāmra* was going in full swing during the early seventh century AD or late sixth century AD, hence, the sanctity of the place was well acknowledged by the people in that period. The literary sources e.g. *Mahābhārata* and some *Purāṇas* also support that *Ekāmra* was a celebrated *tīrtha* of the age of their compilation. Among these sources the *Brahampurāṇa* is worthy to note which assigned a separate chapter for *Ekāmra* and has given a vivid picture of this settlement with the description of the holy places, rivers, tanks, flora, fauna, people and the religious cults and auspicious days.[29] The discription mentions a few landmarks which provide clues to assess the antiquity of the place. Thus verse-50 and 73 mention the existence of the *svyaṁbhū* (self-existant) *kṛittivāsas*[30] and verse-60 confirms that there was temple on that *svyaṁbhū-liṅga* of *kṛittivāsas*.[31] The verses-11, 19 and 83 indicate that there were eight main holy centres[32] and temples;[33] and various *liṅgas*[34] either in worship or in defunct condition located at different spots, e.g. in the forests, lanes, crossways and cemeteries. Verses-51 to 59 ½ eulogise the lake *Viṅdusaras* (Biṅdusāgara),[35] while verse-76 and verse-91 mention, the *liṅg* called Bhāshkareśvara[36] and *Kapilatīrtha*[37] respectively. About the Biṅdusāgara we have already noticed that it existed prior to seventh century AD while the temple of Bhāshkareśvara and Kapileśvara are of much later date i.e. of the Gaṅga and the Suryavaṁsī periods respectively. But, a noteworthy point here is that the *Brahmapurāṇ* mentions only the Bhāshkareśvara *liṅga* and the *Kapilatīrtha* and not the temples on those. The Bhāskareśvara *liṅga*, which was originally a remnant of Asokan pillar was conceived as *liṅga* during the fourth-firth AD.[38] The colossal size of the *liṅga* would naturally have attracted the visitors and therefore became popular, thus got mentioned in the *Purāṇa's* compilation. Further, regarding the *Kapilatīrtha* is an established fact that the existing temple was constructed at the site of some earlier temple, the remnants of which are datable to circa seventh century AD are reused in some structures at the place.[39]

Now, the earliest date of the compilation of *Brahmapurāṇa* is fixed to circa AD fifth century.[40] Therefore, we can say that when this *purāṇa* was compiled i.e. during the fifth century AD the *Ekamra* was recognised as an established *Śiva-kshetra* (sacred land of Śiva), and it was a properous settlement.

On the other hand some landmarks of the description of the settlement of the *Ekāmra* in *Brahmapurāṇa* creates a complicated situation by mentioning the existence of a fortified settlement that was well-guarded with a moat, a fortification wall and the armies.[41] In the area of *Ekāmra* we get the erstwhile existence of the two fortification, first is of the Sisupalgarh belonging to the early historical period while another is of the *Kṛittivās-kaṭaka* of the Gaṅga and Suryavaṁsi periods. Now the question is that which may be corroborated with the description of *Brahmapurāṇa.* For the *Kṛittivās-kaṭaka* there are no such remains which can give an idea of its architecture, though, there may have been the fortification walls and high portals and military forces, however, it is certain that it was not surrounded by any water-moat and therefore cannot withstand to the description of *Brahmapurāṇa.* On the other hand all the characteristic features, viz : fortification wall, portals and moat are available with Sisupalgarh and the said description very much fits with it. Thus *Brahmapurāṇa* obviously mentions the citadel that lay in revenue mauza of Sisupalgarh. The expansion of the *Ekāmrakshetra* is given in the *Brahmapurāṇa* which extend to two and a half *yojanas* from the *Kṛittivās liṅga* on all cardinal directions.[42] Obviously this distance covered the fortification at Sisupalgarh within the boundaries of *Ekāmra.* The above corroboration again takes back the antiquity of Śavite occupations of *Ekāmra* i.e. to the period of fortified settlement of Sisupalgarh. Though, the *Brahmapurāṇa* was compiled during fifth century AD and by that time the citadel would have fully deserted or reduced to merely the status of village but it included an old tradition about the prosperous citadel which would have in vogue through ancient fables and folk lores. The *Brahmapurāṇa* has maintained an ancient tradition regarding the citadel at *Ekāmra* is also evident from its some expressions, e.g. *iti śrutam* and *purakalpe.*[43]

To fix a certain date to the origin and celebacy of *Ekamra*- the Śaivite settlement in early historical period is a vexed problem. The two important epigraphic records, viz: Asokan Rock Edicts at Dhauli and Hāthīgumphā inscription of Khāravela did not make any mention of *Ekāmra* or *Svayambhū* or any other epithets of that *linga*; however, their claim of being generous to all religious sects with these expressions such as *savvapāshanḍa-pūjaka* and *savvadevā yatana-samkhāraka* imply the presence of different religious sects, and therefore, it may possible that there would have been a minor shrine (*devāyatana of* Hāthīgumphā inscription) at the vast laterite stretches of the present Bhubaneswar area.

In the *Vanaparva* of *Mahābhārat* a tradition is recorded about the location of *Svyambhū-vana* in the *Utkala* measuring its distance from the *Virajā-tīrtha*;[44] no doubt this *Svayambhū-vana* (forest of self-existant) is referred to for the same *Ekāmravana* or *purāṇas*. This reference of *Mahābharāta* takes the antiquity of *Svayambhūlinga* of *Ekāmra* to the pre-Mauryana date. But the scanty description again suggests that the Śaivism at this time was not well propagated on sectarian grounds and rather remained a folk religion with the *linga* as an object of the cult worship. In the *Vaishṇavakhanda* of *Skandapurāṇa* it is stated that the *rājā* Indradyumna (the builder of original shrine of *Purushottama-Jagannātha*) had worshiped *Śiva* at *Ekāmra* before building of the shrine of *Purushottam* at Puri[45]. Professor K.S. Behera has established the date of original shrine of *Purushottama-Jagannātha* and his worship at Puri in the third century AD.[46] Therefore, the

Svayaṁbhūliṅga was being worshipped popularly before the third century AD.

In the post-Mauryan period Śaivism received wide popularly in India under the patronage or various ruling families of native and foreign origin, viz. the Sātavāhanas, the Kadambas, the Ikshvāku, the Udumbaras, the Yoḍheyas, the Bhāra Śivas and other tribal kingdoms as well as the Śakas and Kusāṇas[47] Sātavāhanas were the earliest benefactors of *Śiva*-worship in south India[48]. It can be, therefore, opined that Śaivism would have received impetus also at Bhubaneswar during their rule in Kaliṅga after the end of Chedi dynasty.[49]

The development of *Śiva*-sectarianism played an important role in the rise of Saivism in early historical period. Among the *Śiva*-sectaris the *Pāśupat*-system became most prominent in the early centuries of Christian era as evident from the *Purāṇas* and epigraphic records. Lakulīśa, the first *Pāśupat* teacher was born in the first century A.D. or earlier at Kayarohan in Gujrat and was considered to be the last incarnation of *Śiva* and that, of his four disciples Kūśik established himself at Mathura and Garga at Somnath in Kathiawad.[50] At Bhubaneswar frequent depictions of Lakulīśa in the early temples confirm the influence of the *Pāśupat*-sect in the Śaivism practised during that time. Further, verse-71 of the *Brahmapurāṇa* refers to the existence of the *Pāśupat-Yoga* at Ekāmra,[51] which suggests that Lakulīśa-*Pāśupat* sect had propagated Śaivism at Bhubaneswar zone and ultimately succeeded to place it in the superior position among all the contemporary religions being practised in this area.

In the light of the preceding discussions the following factors can be ascertained for the evolution of Śaivite settlement of Bhubaneswar :

(i) The *Svayaṁbhūliṅga* was a simple cult object of folk worship during the Mauryan and the Chedi period.

(ii) As the Jainism and Buddhism were dominant religious during these periods which flourished under royal patronage but Saivism did not receive any particular favour from the royal house.

(iii) With the end of Chedi family the great Jaina centre of Udayagiri-Khandagiri received a major setback though Buddhism still had its stronghold.

(iv) *Śiva*-worship would have received favour from the royal house of the Sātavāhanas and the Muruṇḍas in the succeeding periods and the *Svayaṁbhūliṅga* would have arose above from the mere status of cult-object.

(v) The advent of the *āchāryas* of the *Lakulīśa-Pāśupat* sect opened new vista of Śaivism by propagation of the advance doctrines of their philosophy which attracted the attention of the public and ultimately placed Śaivism in the firm grounds.

(vi) The shifting of the political centre from the citadel of Sisupalgarh made that area less attractive while being the place of cult-worship of the common folk duly supported by the new charm-worship of the *Pāśupat-Yogins Ekāmra* was attracting masses of people.

(viii) Ultimately the royal patronage of the Matharas, the Vigrahas and king Śaśāṅk of *Karṇa-Suvarṇa* hastened the development of the *Ekāmra-Bhuvaneśvara* as the temple-city of Lord *Śiva*.

NOTES AND REFERENCES

1. B.C. Chhabra (ed. & revised by D.R. Bhandarkar), *Corpus Inscriptionum Indicarum* Vol. III (Inscriptions of the Early Gupta Kings), New Delhi, 1981, pp. 212-7.
2. ibid.
3. ibid. pp. 13-17.
4. G. Jouveaw-Dubreuil's *Ancient History of Deccan,* p. 60, Cf, S.N. Rajaruru, *Inscriptions of Orissa* Vol. I, pt. ii, op. cit., pp. 58-60.
5. S.N. Rajaguru, *Inscriptions of Orissa* Vol. I, pt. ii, op. cit., pp. 58-93.
6. ibid., 58-73.
7. ibid., p. 59.
8. ibid., pp. 61-2.
9. ibid., p. 62.
10. ibid., pp. 62-63.
11. ibid, pp. 45-7 and 61.
12. ibid., pp. 63-6.
13. ibid.
14. ibid.
15. ibid.

16. A.K. Rath, "A Note On The Vigraha Dynasty", *Orissa Historical Research Journal* Vol. XI, Bhubaneswar, 1962, pp. 54-9.

17. S.N. Rajaguru, *Inscriptions of Orissa* Vol. I, pt, ii, op. cit., pp. 113-6.

18. ibid., p. 123.

19. ibid., p. 151.

20. ibid., 152.

21. ibid., p. 59.

22. K.C. Panigrahi, *History of Orissa,* op. cit., p. 314.

23. S.N. Rajaguru, *Inscription of Orissa* Vol. I, pt., ii, op. cit., pp. 32 and 36.

24. ibid., pp, 113-27. The donees in their copper place grants are Brahman of different gotras.

25. B.N. Srivastava, "Śaśāṅka, king of Gauda", *Bulletin of The U.P. Historical Society* Vol. X, No. 3 (1965), pp. 43-5; Cf. T.E. Donaldson, *The Hindu Temple Art of Orissa* Vol. I, Lyden, 1985, p. 26.

26. S.N. Rajaguru, *Inscriptions of Orissa* Vol. I, po. ii, op. cit., pp. 157-247. It is evident from the titles and invoking prayers in their copper plate grants.

27. ibid., pp. 223-8.

28. K.C. Panigrahi, *History of Orissa,* op, cit., p. 311; and see, T.E. Donaldson, *Hindu Temple Art of Orissa* Vol. I, op. cit., pp. 69-1.

29. *Brahmapurāṇam* (Ch. 41, Ekamra-kshetra-māhātmya-kathanam), op. cit., pp. 253-60.

30. ibid, see, verse-50 and 73 :

    नानाजलाशयैः पुण्यैः शोभितं तत्समन्ततः।

    आस्ते तत्र स्वयं देवः कृत्तिवासा वृषध्वजः।। 50 ।।

ये च पश्यन्ति तं देवं कृत्तिवाससमुत्तमम्।
विमानेनार्कवर्णेन शिवालोकं व्रजन्ति ते ।। 73 ।।

31. ibid., "Sambhoṛgṛham" implies the temples of Siva, see, verse-60 :

पितृणामक्षयां तृप्तिं ते कुर्वन्ति न संशयः ।
ततः शंभोगृहं गत्वा वाग्यतः संयतेन्द्रियः ।। 60 ।।

32. ibid., verse : 11 :

लिंगकोटिसमायुक्तं बाराणसीसमं शुभम् ।
एकाम्रकेति विख्यातं तीर्थाष्टकसमन्वितग् ।। 11 ।।

33. ibid., verse : 19 :

देवतायतनैर्दिव्यैः प्राकारोद्यानमण्डितैः ।
पूजाविचित्ररचितः सर्वत्र समलंकृतम् ।। 19 ।।

34. ibid., verse- 82 & 83.

तस्मिन्क्षेत्रवरे पुण्ये लिंगं यद्दृश्यते द्विजाः ।
पूज्यापूज्यं च सर्वत्र वने रथ्या न्तरे पि वा ।। 82 ।।
चतुष्पथे श्मशाने वा यत्र कुत्र च तिष्ठति ।
दृष्ट्वा तल्लिंगमव्यग्र श्रद्धया सुसमाहितः ।। 83 ।।

35. ibid., verse - 53 :

सर्वलोकहितार्थाय रूद्रः सर्वसुरैः सह ।
तीर्थ बिन्दुसरो नाम तस्मिन्क्षेत्रे द्विजोत्तमाः ।। 53 ।।

36. ibid., verse - 76 :

तस्मिन्क्षेत्रवरे लिंगं भास्करेश्वरसंज्ञितम् ।
पश्यन्ति ये तु तं देवं स्नात्वा कुण्डे महेश्वरम् ।। 76 ।।

37. ibid. verse - 91 :

स्नात्वा च कापिले तीर्थे विधिवत्पापनाशने ।
प्राप्तोत्यभिमतान्कामजिछवलोकं स गच्छति ।। 91 ।।

38. K.C. Panigrahi, *History of Orissa,* op, cit., p. 369.

39. D. Mitra, *Bhubaneswar,* op. cit., p. 71-2.

40. F.E. Pargiter *Ancient Indian Historical Traditions,* Delhi, (reprint), 1972, p. 49.

41. *Brahmapurāṇam,* Ch. 41. op. cit., verses - 14, 15 and 16 :

गृहगोपुरसंबाधं त्रिकचाद्वारभूषितम् ।
नानावणिक्समाकीर्णं नानारत्नोपशोभितम् ॥ 14 ॥
पुराट्टालकसंयुक्तं रथिभिः समलंकृतम् ।
राजहंसनिभैः शुभ्रैः प्रासादैरूपशोभितम् ॥ 15 ॥
मार्गगद्वरसंयुक्तं सितप्राकारशोभितम् ।
रक्षितं शस्त्रसंघैश्च परिखाभिरलंकृतम् ॥

42. ibid, verse - 75 :

देवस्य पश्चिमे पूर्वे दक्षिणे चोत्तरे तथा ।
योजनद्वितयं सार्ध क्षेत्रं तद्भुक्तिमुक्तिदम् ॥ 75 ॥

43. ibid, verse - 12 :

एकाम्रवृक्षस्तत्राऽऽसीतुरा कल्पे द्विजोत्तमाः ।
नाम्ना तस्यैव तत्क्षेत्रमेकाम्रकमिति श्रुतम् ॥ 12 ॥

44. *Mahābhārat,* op. cit., pp. 249-50.

45. *Skaṅdapurāṇa,* Vaishanavakhanda,

46. K.S. Behera, "The Antiquity of Purushottama-Jogannatha"..., Daitari Panda and S.C. Panigrahi (ed.) *The Cult And Culture of Lord Jagannatha,* Cuttak, 1984.

47. P. Banerjee, *Early Indian Religious,* Delhi, 1973, pp. 47-54.

48. ibid., p. 44.

49. Supremacy of the Sātavāhānas, Kushāṇas and Muruṅḍas over Orissa is already discussed in Chapter III, they were followers of Śaivism.

50. P. Banerjee, *Early, Indian Religions,* op, cit., pp. 55-9:

The other Siva sectaries that deserve a special mention here are the *Pāśupatas* and the *Māheśvaras.* The *Mahābhārata* refers, as we know, to the *Pāśupata* doctrine as a contemporary religious system with the *Sāṁkhya, Yoga, Vedāraṇyaka* (or Veda) and *Pāñcharātra.* As stated there, *Kapila* formulated the *Sāṁkhya, Hiraṇyagarbha* the *Yoga, Apamtaratma* was the teacher of the *Veda Śiva-Srīkanṭha* declared the *Pasupata and Vishṇu* the *Panchatra.*

About the *Pāśupata* system which is the subject of our present enquiry, we derive more specific information from the *Purāṇas.* In the *Vāyupurāṇa,* Chapter 23, and *Liṅgapurāṇa,* Chapter 24, *Śiva* is represented as having told *Brahmadeva*:

During the 28th repetition of the *Yugas* when Vasudeva would be born as the son of Vasudeva at the time Krishna Dvaipayana, I whose essence is *Yoga,* will by the magic power of Yoga assume the form of a *Brahmachārī* and entering an unowned corpse thrown out into a cemetry, will (according to *Vayu*) and Kayavatara (according to *Linga*) will become famous as a sacred place. And there will be born my sons (disciples) the ascetics Kusika, Garga, Mitra and Kaurushya; these *Pāśupatas,* their bodies besmeared with ashes, having attained the *Māheśvara Yoga,* will depart to the world of Rudra, hence it is difficult to return.

From the above it is quite clear that the Pasupata doctrine, thought is said to have been formulated by *Śiva* himself in the *Mahabharata,* originated with Lakuilisa, a human teacher, who was later on regarded as an incarnation of Siva.

While writting on this subject in the journal of Bombay Branch of the Royal Asiatic Society (Vol. XIII 1950), Bhandarkar made some important observations regarding Lakulisa's date which may be summed up as follows.

Lakulisa and his disciples are mentioned in the *Vāyupurāṇa* which, in its account of the great dynasties of India, mentions at the end the Guptas reigning over Saketa, Magadha, and along the Ganges as far as Prayāga. Applying to the early and imperial Gupta dynasty, this can only refer to the territories of Chandragupta I (c. 320-335 A.D.) before the expansion of the Gupta kingdom by Samudragupta. This shows that the *Vāyu* was completed shortly after the reign of Chandragupta I. Hence, the incarnation of Siva as Lakulisa, mentioned as an object of faith in the *Vāyupurāṇa* in the early fourth Century A.D. must already have been known, probably not later the first century A.D.

About 25 years he published the above theory regarding Lakulīśa, Bhandarkar edited the Mathura Pillar Inscription of Chandragupta II, Gupta Era 61, in the Epigraphical Indica XXI, and in the light of contents fixed Lakulisa's date between 105-130 A.D. The epigraph in question credits Uditacharya, a Mahesvara teacher, with the establishment in the teachers' shrine, of two *Liṅgas* "Upamiteśvara" and "Kapileśvara" named after his deceased preceptor and preceptor's preceptor, Upamita and Kapila respectively. Uditāchārya as described in the epigraph is tenth in descent from Kusika (and hence, eleventh from Lakulisa). His own teacher was Upamita. The latter again was a pupil of Kapila, and Kapila, a pupil of Pārāśara. "If we now allot" observes Bhandarkar, "twenty five years to each generations, we have to assign Lakuli to A.D. 105-130. This agrees pretty closely with the view I expressed 25 years ago (J.B.B.R.A.S., Vol. XXII, p. 151 ff) that Lakuli has to be placed as early as the first century A.D. My conclusion was then based merely on the mention in the *Vāyupurāṇa* of Lakuli as the first incarnation of Siva. Evidence of this type will always remain of somewhat conjectural nature. Epigraphic evidence, on the other hand, is more accurate. We may, therefore, take it now as well nigh proved that Lakuli flourished in the first quarter on the second century A.D., about half a century later than the time so long ascribed to him.

The date 105-130 A.D. assigned to Lakuli by Bhandarkar on the basis of his interpretation of Mathura Pillar Inscription Gupta Era 61, does not seem to be satisfactory one for various reasons. The *Mahābhārata* (*Śāntiparva*), we have seen, regards the *Pāśupata, Sāṁkhya* and *Pāñchrātra,* etc., as contemporary religions. The Puranic tradition asserts that the founder of *Pāśupata* system was Lakulīśa, who was an incarnation of Śiva and a contemporary of Krisṇa-Vāsudeva and Krisṇa-Dvaipāyana. Even if it is presumed that the Lakulīśa and the *Pāśupata* system can not be given such an early date, yet they can not be as late as 105-130 A.D. That the *Pāśupata* system is older than the Christian era is evident from the point of close similarity between the *Śiva-Bhāgavatas* and the *Pāśutpas* in respect of their doctrinal practices. The very characteristic reference in Pantañjali's *Mahābhāshya* to the adoption of violent means by the *Śiva-Bhāgavatas* reminds distinctly, as J.N. Banarjea points out, of the fourth Vidhi or means to which the *Pasupatas* restored to reach their desired goal, i.e. *Dukhāṅta* (termination of ills or final liberation), The fourth *Vidhi* of the *Pasupatas* as Mādhavāchārya tells us, includes certain extreme practices, like loud laughter, song, dance, muttering of *aum* and pious ejaculations. This led R.G. Bhandarkar to describe the *Pāśupata* School or schools with two of its offshoots, the Kāpāla and Kālāmukha, as atimārgika schools, that are away from the common path or go astray, and are spoken of *Śambhudeva* as revealed by Rudra." These discussions show beyond doubt that the *Śiva-Bhāgavatas* and the Pāśupataś were allied sects, some of the practices common to both were known to Patañjalli and that they flourished earlier than this i.e. 200. B.C.

Further, the *Pāśupatas,* who were also known as Māheśvaras, were an important sectarian organization in the middle of the first century A.D. Vima Kadphises, a celebrated Kushāṇa ruler of this time was an adherent of this sect as is evident from one of his appellations "Mahesvara" on the reverse sides of his coins. The generally accepted date of Kanishka is 78 A.D. Hence Vima, who was his predecessor, must have belonged to the period between 50-77 A.D. Thus, the rise of the Lakulīśa and the *Pāśupata* sects must also be placed earlier than the first century A.D. If we accept Bhandarkar's date of Lakuli, it is difficult to explain the presence of a *Māheśvara* or *Pāśupata* sect in Vima's time.

51. *Brahmapurāṇam,* Ch. 41, op. cit., verse - 71 :

जायते योगिनां गेहे चतुर्वेदी द्विजोत्तमाः ।
योगं पाशुपतं प्राप्य ततो मोक्षमवाप्नुयात् ।। 71 ।।

5

Evolution of Society from Sixth Century AD to Sixteenth Century AD

1. POLITICAL BACKGROUND

The mountain Mahendra and Rushkulya valleys as well as coastal plains in the north, which all once formed a part of ancient Kaliṅga empire, were gradually warming up in political arena of Orissa in the sixth century AD. We have already noticed the rise of the Māṭhara family in the area around the mountain Mahendra. Around the close of fourth century AD or in the early part of fifth century AD one Śaktivarman of this family is said to have been the lord of whole Kaliṅga. (*sakaliṅgādhipati*). His domain extended from the river Krishna in the south to river Mahanadi in the north. The Māṭharas ruled this kingdom till the mid-sixth century AD and at this time they were defeated by one Pṛithvī Mahārāja of *Srīrāmkāsyapa-gotra.* Obviously, Bhubaneswar remained under the control of the Māṭharas in the said period and subsequently annexed to the kingdom of Prithvi Maharaja.

It appears that Kaliṅga and Tosalī which once formed the united state of Kaliṅga, now separated into two respective parts. The boundaries of Kaliṅga have already been mentioned above;

however, the known area of Tosali region extended from river Mahanadi to Tāmralipti (modern Tamaluka in dist. midanapur of W.B.). From the contemporary copper plate grants we come to know that this region was divided in two sections; namely: *Dakshiṇa Tosalī* and *Uttara Tosalī*, ruledover by the Vigrahas and the Mudgals respectively with both divisons bifurcated by the river Vaitarni.[1]

In the Vigraha family we know the names of atleast two rulers: Pṛithvīvigraha and Lokavigraha. One Dharmarāja was a subordinate king under Prithvīvigraha. Since in his Sumandal copper plate grant (dated in the Gupta era 250 i.e. AD 569).[2] Dharmarāja has mentioned Prithvīvigraha as the ruler of *Kaliṅga-rāshṭra* which formed a part of Gupta-kingdom, therefore, it seems that Pṛithvīvigraha was in possession of Kaliṅga as well as Tosali. His supremacy over Kaliṅga would have been a result of subjugating the Prithivī Mahārāja of *Sriramkāsyapa-gotra* who had become *sakalakaliṅgādhhipati* (master of the whole Kaliṅga) after defeating the Māṭhars. In one of his copper plate grants,[3] Pṛithivī Mahārāja of *Srīrāmkāsyapa gotra* is said to have been reached to *Virajā* and some was camping there in connexion with an ongoing war. This advance of his was obviously against the Vigrahas in an attempt to annex the area north of the river Mahanadi. While corroborating the statements of these copper plate it appears that though Mahārāja of Srīrāmkāsyapa gotra beseiged the city of Virajā (the capital of Vigrahas). but he was defeated and had to retreat Pishathapura in very humiliating terms i.e. leaving a considerable part of *Sakalakaliṅga* in possession of the Vigrahas.

On the other hand the Vigrahas have had also rivalry with the Mudgalas of *Uttara Tosalī.* In the Kanas copper plate, Lokavigraha, the successor of Pṛithvīvigraha is said to be overload of eighteen *Tosalīs* which implies that this kingdom was greatly expanded.4 This expression would have also caused on the cost of the Mudgalas apart from the other minor states. In the Soro copper plate of Mahārāja Sambhuyasas of the Mudgala family (AD 580), it is revealed that the king was involved in a war with Lokavigraha which probably continued upto the death of Lokavigraha sometimes between AD 600-602.[5] From the Patikella copper plate of Śivarāja, a subordinate ruler of Sambhuyasas, it is evident that the Mudgalas had ousted the Vigrahas from *Daksjiṇa Tosalī* by this date.[6]

During the last quarter of this century it seems that the rulers of the Sailodbhava dynasty had also made their home in the *Kāṅgoda* country (district Ganjam and southern parts of district Puri in Orissa). Though, we do not find the copper plate grants of any of the rulers before Mādhavarāja II who was earlier a subordinate ruler under king Śaśāṅka but later on asserted his independence. From the geneology mentioned in his copper plate grants we can guess that the earlier ruler of this family : namely : Araṇabhitā, Mādhavarāja I and Ayasobhīta would existed in between AD 575 to AD 605.[7] But the ruling family Śailodbhava did not seem to rise in much importance during this period. The Vigrahas and the Mudgalas remained the main political figures and ultimately the latter surpassed the former.

In the wake of seventh century AD, we find that the mudgalas were defeated by king Śaśāṅka of *Gauda,* who remained the overload of this area alteast upto AD 619 as is evident form Ganjam copper plate grant of Mādhavarāja II of Śailodbhava dynasty (dated in Gupta year 300 i.e. AD 619) in which Śailodbhavas are said to have been the feudatories under Śaśāṅka.[8] The newly conquered territory was apparently divided into two parts by Śaśāṅka for administrative purposes, the southern half that was *Kāṅgoda* remained under the viceroyality of Mādhavarāja II and the northern division that was *Utkala* with *Daṇḍabhukti* was governed by Somadatta of the Datta family as indicated by the latter's Midanapur copper plate grant.[9] Śaśāṅka was a staunch devotee of Śiva and his contribution in the rise of Bhubaneswar has already been mentioned in the preceding chapter.

The Śailodbhavas

With the submission of Śaśāṅka by Harshavardhan and Bhāsakaravarman, sometimes in between AD 619 and 624, his sovereignty apparently confined to *Gauda,* the feudatory Śailodbhava and Datta families appear to have asserted their independence as evident from the later copper plate grants of both the families viz: Puri copper plate grants of Mādhavarāja II[10] and Soro Copper plate grant of Somadatta.[11] Orissa at this time as described by Huien Tsang was thus divided into three distinct parts; *Odra, Kāṅgoda* and *Kāliṅga. Odra* making up the northern part of the country, consisted of the modern district of Midanapur, Balasore, Cuttak and Puri north the marshes along the northern fringe of Chilika whereas *Kāṅgoda* began south of these marshes. Kaliṅga probably extended from the southern border of *Kāṅgoda* (Ganjam and Vizagapatam) to the banks of the river Godavari.[12] It is noteworthy feature that the land which was known as Tosali in the preceding century now referred to by terms like *Utkaladeśa* and *Odra.* Whereas Kaliṅga receded to further south, the Kāṅgoda middled the Orda and Kaliṅga. Bhubaneswar area though forming the southern extremes of *Odra* probably annexed by Mādhavarāja II of Śailodbhava dynasty as guessed by the statements in his Puri copper plate grant where it is recorded that he performed the *Aśvamedha* sacrifice after reducing the sovereign influence of his enemies over his own statal circle of kings.[13] From Khurda copper plate grant we know that he succeeded in possessing Kaliṅga[14] also though this possession remained for a short period as evidenced by his later inscriptions suggesting that probably Gaṅgas rivalled the expansion of the Śailodbhavas towards south.[15] From the accounts of Huein Tsang we know that the fort of the Śaildbhavas was naturally strong, that they maintained a powerful army which kept the neighbouring countries in awe and they had no powerful enemy.[16] These statements imply that the Dattas were weakened and the Śailodbhavas expanded towards north.

In the early seventh century India was dominated by three great powers, Harshavardhana of Kannauj in the north the Chālukya king Pulkeśin II in the south and west; and Śaśāṅka in the east. With the death of Śaśāṅka Orissa became the centre of conflict between Harsha and Pulkeśin II. After conquering Veṅgī the Chālukyas invaded Kaliṅga and Kāṅgoda but they returned south after leaving the area in chaotic condition.[17] In AD 643 Harshavardhana led an expedition as far as Kāṅgoda, though, he does not seem to have annexed this territory but reinstated the defeated king Mādhavarāja II in his own kingdom.[18]

Mādhavarāja II ruled for a period of atleast fifty years (AD 620-670).[19] He was succeeded by his son Ayashobhīta alias Maḍhyamarāja I who reigned for atleast twenty six years.[20] Upon the death of Madhyamarāja a war of succession was fought between Mādhava (the second son of Madhyamarāja) and Dharmarāja (the eldest son of Madhyamarāja). In this war Dharmarāja defeated Mādhava and banished him from the country.[21] It was at about this time that Tīvararāja of *Dakshiṇa Kosala* extended his reign over *Utkal* and was threatening the northern gate of Kāṅgoda.[22] He was foiled in his attempt to occupy Kāṅgoda, however, by espousing the cause of Mādhava against his brother. Dharmarāja succeeded in defeating the combined strength of Mādhava and Tīvararāja in a second battle with Tīvararāja subsequently retreating even from Utkala. It was probably after this crushing victory over Tīvararāja of Dakshiṇa Kosala, that Dharmarāja claimed the proud titles of *mahārāja* and *paramabhaṭṭārakha*.[23] Dharmarāja was the last great ruler of this dynasty. After him very little is known about the rulers of this dynasty except for their names. The dynasty ended in or before circa AD 736.

The late sixth and early seventh century constitute an important epoch in the cultural history of Orissa. It is during this period that the Orissa temple first emerges, the earliest example of which are at Bhubaneswar. Despite the ravages of war and conquests, the Orissan temple evolves into a peculiar type apart from other artictural styles with little apparent external influence. Although the Śailodbhavas of Kāṅgoda annexed the area around Bhubaneswar there is no evidence whatsoever that any of the Śailodbhava rulers were responsible for the construction of any temple at Bhubaneswar, though, it is evident that many temples were constructed at this time. All that can be said with certainly, as Panigrahi points out, is that the earliest surviving temple belongs to the period of their rule.[24]

SAILODBHAVAS

1. Pulindasen

|

2. Sailodbhava

|

3. Arnabhita (c. A.D. 575)

|

4. Madhavaraja I alias Sri Sainyabhita I (c. AD. 590)

|

5. Ayasobhita I (c. A.D. 605)

|

6. Madhavaraja II alias Sri Sainyabhita II (c. A.D. 620-670)

|

9. Dharmaraja alias Manbhita (c.A.D. 705-725) — 8. Madhava

|

10. Madhyamaraja II — 11. Allavaraja

12. Yuvaraja Tailapanibha (ended in or before A.D.736)

(Adopted from S.N. Rajaguru, *Inscription of Orissa, Vol. I, pt. i,* Bhubaneswar, 1958)

The Bhaumakaras

It seem plausible that the breakup of the Śailodbhava kingdom was primarily due to emergence of the Bhaumakaras in the northern Tosalī. Their capital was *Guhādevapāṭaka* identified with modern Guhirtikra about 8 km. from Virajā or Jaipur on the river Vaitarni. The earlier rulers of the dynasty seem to have ruled in north (Tosali) contemporaneously with the Śailodbhavas in the south (Kāṅgoda), but from copper plate grants of the later members it becomes evident they eventually extended their sway to the region once occupied by the Śailodbhavas.[25] The hegemony of the Bhaumakaras over the area around Bhubaneswar is proved by the inscriptions in the rock-cut cave at Dhauli and at Udaygiri which refer to the reign of the Bhaumakara king Śāntikara (AD 829).[26] According to the astronomical calculations worked out by S.N. Rajaguru the foundation of the Bhauma dynasty lasted for nearly two hundred years. Though the later part of their is characterized by political turmoil and disunity.

Śivākaradeva I alias Unmattasimha (AD 756-786) the son and successor of Kshemakara, greatly expanded the boundaries of the kingdom. He conquered Kāṅgoda and northern parts of Kaliṅga.[28] He is credited for defeating the king of *Rāḍha* (south-west Bengal) in the battle and forcing him to give his daughter in marriage. Probably in his time autographed manuscript of the Buddhist work *Gaṅdavyūha* was sent to contemporary Chinese emperor Te-tsang.[29] Śivākara I was succeeded by his son Śubhākara I. Śubhākara I was succeeded by his eldest son Śivākara II who is described in Chaurasi copper plate grant as the lord of both the *Uttara* and *Dakshiṇa Tosalī* and addressed with the epithets of *paramabhaṭṭāraka, mahārājādhirāja* and *parameśvara.*[30] Śubhākara I and Śivākara II ruled between AD 790 and 829. It is during this period that the Rashtrakuts of Deccan and Palas of Bengal claim in their Inscriptions to have invaded Orissa. Govinda III (AD 793-814) of Rāshṭrakūṭa dynasty is said to have conquered *Kosala, Kaliṅga, Veṅgī, Dahala,* and *Odraka.*[31] during the reign of Śivākara II Devapāla (AD 810-850) of Pāla dynasty invaded Orissa and appears to have succeeded in bringing it under his political hegemony. The loss of prestige and power as well as distress and disorder caused by these invasions appear to be reflected in the Dhekanal plate of Tribhuvanamhādevī I (AD 846), where the sad plight of Orissa prior of her accession is described.[32]

Śivākara II was succeeded by his younger brother Śāṅtikara I alias Lalitahara (AD 829). He married Tribhuvanamahādevī, the daughter of Rājamalla I of the western Ganga dynasty of Mysore. According to statements in Dhekanal copper plate grant of Tribhuvanamahādevī, it was her father Rājamalla who put an end to the Rāshṭrakūṭa and the Pāla dominion in Orissa.[33] The accession of Tribhuvanamahādevī I to the throne in about AD 846, after short reigns by Śubhākara II and Śubhākara III, thus marked the revival of the Bhauma-kingdom after a period of external dominion. She was succeeded by her grandson Śāṅtikara II. He had two sons Śubhākara IV and Śivākara III who ruled in succession.

The death of Śivākara III was followed by a struggle for succession to the throne. Though, Śivākara III had two sons: Śāṅtikara III and Śubhākara V, the throne was usurped by Tribhuvanamahādevī II (AD 894) the queen of Śubhākara IV with the aid of her father, the Somavaṁsī - king Janme jaya Svabhavatuṅga, who probably killed Śivākara III.[34] Tribhuvanamahādevī was eventually overthrown by Śivākara's (III) son Śāṅtikara III, the husband of Dharmamahādevī. He was succeeded by his younger brother Śubhākara V who married Gaurīmahādevī and Bakulamahādevī. Gaurimahādevī succeeded her husband and she was followed by her daughter Daṅḍīmahādevī. Dandimahādevī's known dates are the Bhauma years 180 and 187 (AD 916 and 923). Her supremacy over *Yamagartamaṅdala* and *Dakshiṇa Tosalī* is testified by her copper plate grants.[35] She was succeeded by her step-mother Bakulamahādevī though the circumstances of this succession are not known. It is possible that Vakula secured aid from her paternal family in dethroning Dandi. Vakulamahādevī was succeeded by Dharmahādevī, the wife of Śāṅtikara III and was probably the last ruler of the dynasty.

Bhaumakara rule marks a period of cultural changes in Orissa which Bhubaneswar also shared. The dominant form of religion of the early Bhaumakaras appeared to have been *tāṅtrika*-Buddhism, though, they were tolerant of all sects. Although, the Buddhist monasteries were in existence during the Śailodbhava hegemony in Orissa, it is not until the mid-eighth century that Buddhist images and motifs, as well as terrifying aspects of Śāktism appear on temples at Bhubaneswar. Śaivism, Śāktism and Tāṅtrism seem to have inseparably mixed together at this time with the *Kapālikas* being one of the most extreme sects emerging from this amalgamation. Their influence is noticed on many temples, beginning with the Vaitāla deula, and terrifying images come to be one of the more popular motifs decorating the Orissan temples.

BHAUMA KARAS

1. Kshemankaradeva (AD. 736)

|

2. Śivakaradeva alias Unmattasimha (AD. 756 or 786)

|

3. Subhakaradeva I (AD. 790)

|

4. Śivakararadeva II (AD. 809)

|

6. Subhakaradeva II (AD. 836)

5. Santikara alias Lalithara =

9. Tribhuvan Mahadevi AD. 846

|

7. Subhakaradeva III alias Kusumhara (AD. 839)

8. Santikara I alias Lonabhara I

|

10. Subhakaradeva IV alias Kusumahara II (A.D.881)

12. Tribhuvana-Mahadevi II (A.D. 894)

11. Śivakaradeva III alias Lalitahara II (A.D. 885)

|

13. Santikaradeva III alias Lavanabhara

14. Subhakaradeva V

15. GauriMahadevi

17. VakulaMahadevi (A.D. 940)

16. DandiMahadevi (A.D. 916, 923)

18. Dharma Mahadevi (A.D. 949)

(Adopted from K.C. Panigrahi, *Chronology of The Bhauma-karas and Sovavamsis of Orissa,* Madras, 1961.)

The Somavasis

During the last quarter of ninth century when Bhaumakara dynasty of *Odra* was leading towards dissolution, the Pāṅduvaṁśīs of *Dakshiṇa Kosala* were probably forced to leave their homeland by Kalachuris' conquest of that area.[36] The Pāṅduvaṁśīs apparently took refuge in the Sambalpur tract, where, under Jamne jaya I Mahābhavagupta, (c. A.D. 882-922) they established as independent kingdom, with the capital at *Suvarṇapura* (Sonpur), changed the name of their dynasty to Somavaṁśī. Jamne jaya also known as Svabhavatuṅga had matrimonial relations with the Bhaumakaras. He had intervened in the dispute for succession in the Bhauma kingdom and succeeded to install his daughter Pṛithvīmahādevī alias Tribhuvanamahādevī in the throne after killing the ruling king of *Odra.* Thus, it seems that he marched upto the capital of the Bhaumas. Traditionally, he is said to have performed *aśvamedha*-sacrifice at *Kaṭaka* (Chaudwar) new modern Cuttak as well as constructing Uttareśvara temple. All these events witness that he had much influence though indirectly in the Bhauma-kingdom and which paved the path of their conquest of *Utkal* in future.

The next king of the dynasty was Yayāti I who founded a new capital, named *Yayātinagra,* on the bank of river Mahanadi near Baudh.[37] In the ninth year of his reign he granted the villages of *Chandagrama* in *Marāda-vishaya* suggesting that he conquered a portion of flat plains of Orissa in AD 931.[38] The amalgamation of *Kolasa* with *Kāṅgoda* and *Utkala* must have taken place sometime after the death of Vakula whose last record is dated to *samvat* 204 (AD 940). Yauti I is credited with building another capital known as *Abhinavayayātinagara* which identified with modern Jaipur in *Utkala* where he performed ten *aśvamedha*-sacrifices, to which ten thousand *brāhmans* were invited from *Kannauja,* probably to celebrate his conquest of *Utkala.*[39] From the evidence of their copper plate grants it appears that the Somavaṅśī-kings during this early period ruled their vast kingdom *Yayātinagara* in upper Mahanadi valley while *Abhinavayayātinagara* in *Utkala* was their second headquarter. After Yayāti I the Somavaṁśī kingdom was ruled over by Bhīmaratha Mahābhavagupta II (c. AD 955-980), Dharmaratha Mahāśivagupta II (c. AD 980-1005) and Nahusha Mahābhavagupta III (c. AD 1005-1021). Bhīmaratha is said to have defeated a king name Ajapāla from whom he received thirty-two war elephants.[40] Dharmaratha had taken aggressive action against *Gauda* and Andhra.[41] On the other hand during the later part of the rule of Nahusha his kingdom was invaded by the Paramāras, the Cholas and the Kalachuris. The invasions may have resulted in death of dethronement of both : Nahusha and Iṅdraratha (he was stationed at Abhinavayayātinagara), apparently caused chaotic conditions in the Somnavaṁśī kingdom.[42] To save the kingdom from anarchial conditions Chaṅdīhara Yayāti Mahāśivahgupta III, referred to as Yayāti II, was chosen as king. Yayāti II was an efficient ruler, he freed both the *rāshṭras* (*Kosala* and *Utakala*) from enemy chiefs. In driving out the foreign enemies Yayati II (c. AD 1025-1040) restored internal peace and order in the vast territories roughly comprising the whole of modern Orissa.[43] It is stated that he also fought against the kings of *Karanāṭa, Lāṭa, Gurjar, Kāñchī, Gauda* and *Rāḍha.*[44]

Yayāti was succeeded by his son Uddyotakesari (c.AD 1040-1065) who was the last great monarch of this dynasty. The disintegration of the Somavaṁśī-kingdom began from the time of Jamne jaya II (c. AD 1065-1085), the son and successor of Uddyotakesari. Janmejaya II was succeeded by his son Purañjay (c.AD. 1085-1100) and when Purañjay's younger brother Karṇadeva (c.AD. 1100-1110) ascended the throne the Somavaṁśī kingdom must have been confined only to the coastal tract comprising the present districts of Puri, Cuttak and Balasore appears to have been a bone of contention in between the Pālas and the Gaṅgas. And it was Anaṅtavarman Choḍagaṅgadeva of the Ganga dynasty who ultimately brought *Utkala* in his possession.

The rule of the Somavaṁśī dynasty is the most noteworthy period in the history of Orissa. The Somavaṁśī kings are called the builder of modern Orissa. The whole land of Orissa was brought under the cultural and linguistic bond during the rule of this dynasty. They had a cosmopolitan outlook in religion though they patronised Śaivism at Bhubneswar and in the whole kingdom. Many important temples including the most important one i.e. the Liṅgarāja temple were constructed at Bhubaneswar during the reigns of this dynasty. Bhubaneswar, particularly enjoyed the royal patronage under the kings of this dynasty and flourished. It is also probable that some of the kings of this dynasty would have camped here during the construction of temples and thus Bhubaneswar had enjoyed the status of their religious capital. Several new trends in the field of temple architecture were also introduced this period.

SOMAVAMSI-KESARIS

1. Janme jaya I Mahabhavagupta I alias Svabhavatunga
c. AD. 882-922

2. Yayati I MahaŚivagupta I
(c. AD. 922-955)

Vichitravira

3. Bhimaratha Mahabhavagupta II
(c. AD. 955-980)

Abhimanyu

4. Dharmaratha Mahasivagupta II
(c.AD. 980-1005)

5. Nahusha MahaŚivagupta III
(c. AD. 1005-1021)

6. Yayati II Chandihara MahaŚivagup III
(c. AD. 1025-40)

7. Uddyotkesari Mahabhavagupta IV
(c. AD. 1040-1065)

8. Janme jaya II MahaŚivagupta IV
(c. AD. 1065-1085)

9. Puran jaya Mahabhavagupra V
(c. AD. 1085-1100)

10. Karnadeva Mahasivagupta V
(c. AD. 1100-1110)

(Adopted from K.C. Panigrahi, *Chronology of The Bhumakaras and The Somavaṁśīs of Orissa,* Madras, 1961)

The Gaṅgas

The Gaṅga-rule in Kaliṅga commenced in of about AD 626-7, after which they introduced their family era.[45] Regarding the original place of the Gangas before they came to Kaliṅga, the epigraphical evidences are silent. But in connection with the western Gangas of Mysore B. Lewis Rice writes thus.[46]

"According to the inscriptions, the progenitor of Kaliṅga-Gaṅgas was Bhāgadatta, between whom and his brother Srīdatta their father Vishṇugopa divided his kingdom giving Kaliṅga to the former, and the ancestral kingdom, with the elephant, to the latter."

On the other hand S.N.Rajaguru opined in the light of various other evidences that the Gangas had not come Kaliṅga from the South. He assigns Ghumsur taluk in Ganjam district and its adjoining district of Phulbani as the homeland of the Kaliṅga-Gaṅgas. He writes thus.[47]

"It is, therefore, not conclusive that the Gangas had come to Kaliṅga from the south. Now, according to our assumption, namely that Kurāla of the time of Samudragupta might subsequently be transformed into Kolahalapurapattana, our attention is attracted to the regions of Ghumsur taluk in Ganjam district and its adjoining district of Phulbani, (situated in between the countries of Mahākāṅtāra and the Maheṅdra-mountain) to identify the same with the kingdom of Kurāla of the fourth century AD. That was probably the original home-land of the Gaṅgas, who afterwards called it as Trikaliṅga, From the inscriptions of the Somavaṁśī-kings of Dakshiṇa Kosala, who ruled the country from Suvsrṇapur (the modernpur Sonpur in Balagir district), we learn that they assumed the title of Trikaliṅgādhipati for a considerable long time. Presumably they could capture the Phulbani and Ghumsur area as these were contiguous to their territory."

The dynasty variously termed as Eastern Gaṅgas or Kaliṅga Gaṅgas by the historians, is known from a number of copper grants. The rulers of the Gaṅga dynasty are again categorised as the Early Gaṅgas and the Greater Gaṅgas of Imperial Gaṅgas by the historians basing upon their territorial extension. The rulers of the Early Gaṅga dynasty established their rule in the parts of Kaliṅga when the Śailodbhavas were ruling in the Kāṅgoda. We know that Māṭhars-rule in Kaliṅga came to an end sometime in the sixth century AD, when another family called Srīrāma-kāsyapas captured the southern portions of Kaliṅga and established its capital at Pishṭapur near the river Godavari. They seem to have advanced towards north but repelled by the Vigrahas of Tosalī. The Vigrahas of Tosalī were declined when king Śaśāṅka of Karṇa-Suvarṇa (Bengal) was the overload of the Śailodbhavas in Kāṅgoda. Then there was a fight between Harsha of Kannauj and Śaśāṅka on one side, and Harsha and the Chālukya-king Pulakeśin II of Veṅgī on the other. The result was that Śaśāṅka was killed and the Śailodbhav-ruler Mādhavarāja II became independent in Kāṅgoda who also once declare himself the *Sakalakaliṅgādhipati.* But, due to invasion of Pulakeśin II he had to lose the status of the *Sakalakaliṅgādhipati.* After this Kubja-Vishṇuvardhana I, the brother of Pulkeśin, became the king of Veṅgī and the first Gaṅga-king assumed the title of *Trikaliṅgādhipati.*[48] Accordingly the country of *Sakalakaliṅga* or greater Kaliṅga was divided among the eastern Chālukya and the Gaṅgas in or about the first quarter of AD seventh century.[49]

It seems that the family of Śrīrāma-kāsyapas declined in the southern Kaliṅga due to emergence of the Gaṅgas. The early Gaṅga-kings Indravarman I (Ganga era 39 - AD 665) and his successor Sāmaṁtavarman (Gaṅga era 64-AD 690) possessed the southern portions of Kaliṅga.[50] They used the title *Trikaliṅgādhipati* but not of the *Sakalkaliṅgādhhipati*.[51] *Mahārajā* Hastivarman whose name occurs after Sāmaṁtavarman in the Gaṅga geneology, and who issued two charters in Gaṅga era 79 and 80 (AD 705 and 706 respectively) used the title of Sakalakaliṅgādhipati for the first time in the family which was incorporated in the Gaṅga-*prasasti* and used by each king of the family in conventional form.[52] Most probably he was a contemporary king of Śailodbhava-king Dharmarāja (AD 705-720).[53] We know that the Śailodbhava-king Mādhavarāja II (AD 620-70) held the title of '*Sakalakaliṅga....*' for a very short period and none of his successors bore this title. As such, we may say that Hastivarman of Gaṅga family could manage to subdue the Śailodbhavas of the Kāṅgoda-*maṇḍala* during the first quarter of eight century AD, After this he shifted his capital to Kaliṅganagra, on the banks of river Vaṁśadhārā near Mukhaliṅgam of Srikakulam district.[54]

Thus, the contemporary political conditions show that the family of Śrīrāmkāsyaps would have been evicted from southern Kaliṅga by the rise of the Gaṅgas in that area while the Śailodbhavas were crushed to decadence and the northern half on the Kaliṅga came into possession of the Gaṅgas though Kāṅgoda was not retained in Kaliṅga any longer, but annexed by the Bhaumakaras of Tosalī/Utkala later on. Most probably during this conflicted period a colleteral branch of the Gangas occupied the country of *Śvetaka,* which extended to the north of Mahenḍra-mountain. It formed a buffer state between of Kaliṅga and the Kāṅgoda.[55]

The capital of Kaliṅga before the Gangas was at *Siṁhapura*. The Gaṅga did not prefer to have their headquarters there. So, for the purpose of administration they kept an outpost at Siṁhapur by appointing a member of royal family as their subordinate.[56] The Gaṅgas had their capital at *Daṅtapura* at the beginning of their rule.[57] *Mahārāja* Iṅdravarman issued his grant in the Ganga era 39 (AD 665) from *Daṅtapur.* Although, after Hastivarman, Kaliṅganagra was made the capital headquater of the Gaṅgas, even though Dantapura continued as an important administrative centre and in many copper plate grants the residence of the king is stated to be at Daṅtapura.[58]

Beginning from Ad 626-7 the Gangas ruled over the country of Kaliṅga uninterruptedly for more than four hundred years from the capital city of Kaliṅganagara. During the last part of the rule of the Early-Gaṅgas, troubles arose on account of contest for power between the then existing colleteral branches at their family. Towards the end of the tenth century AD, another line of the Gaṅgas rose into power and became the over-lord of Kaliṅga by unifying five separated parts of Kaliṅga, hitherto divided among five branches of the family and it was accomplished by Vajrahasta IV alias Aniyaṅkabhīma (AD 981-1016).[59] It is quite significant to note that in the neighbouring country of Veṅgī there followed a period of anarchy for 27 years i.e. from AD 972 to AD 999. With our present knowledge regarding the Gangas the relation between the Veṅgī and Kaliṅga is not known during that conflicting period, but the political condition in both the countries appear to be almost the same.[60]

As different colleteral aspirants in Veṅgī fought with each other weakened the resisting power of that country so also in Kaliṅga the different branches of the Gaṅgas might have fought with each other only to create a field for outside powers to march into their land. We find that Ammaraja I and Chālukyabhīma (AD 925) of Veṅgī had used the glorious title of *Trikaliṅgādhipati.*[61] But afterwards that family made an alliance with the Gaṅgas probably through matrimonial relationship. That was probably a period of weakness, ensued in territory of Kaliṅga, when a hostile Somavaṁśī-king, named Yayāti, could easily capture the province of Trikaliṅga and used the glorious title of Trikaliṅgadhipati.[62]

However, the political conditions were going adverse to the Somavaṁśīs towards the later part of eleventh century AD and the Gaṅgas were progressing to set up an empire annexing the neighbouring territories; and very soon they became the Imperial or Greater Gaṅgas when they conquered the Utkala country from the Somavaṁśīs. The Somavaṁśīs were slowly loosing their northern dominions to the Pālas and their southern dominions to the Gaṅgas. The southern coastal area was most likely lost during the invasion of the Gaṅga-king Rājarāja I Deveṅdravarman (AD 1070-80) who is credited, in an inscription found in the village of Dirghasi (Ganjam dist.), with destroying the troops of the king of Utkala[63]. Trikaliṅga, the hilly tracts of Ganjam, Kalahandi and Koraput may have been lost even earlier as the Gaṅga-king Vajrahasta III Anaṅtavarman (AD 1038-70) of Kaliṅganagara, when ascending the throne assumed the title of *Trikaliṅgādhipati* as stated in his Nadagam inscription of AD 1058.[64]

By the second half of the eleventh century AD the Gangas of Kaliṅganagara captured some portions of southern coastal region (district Ganjam) of the Somavaṁśī kingdom. However, final jolt to the Somavaṁśī rule in Utkala was given by Gaṅga king Anaṅtavarman Choḍagaṅgadeva. The subjugation of Utkala alongwith the capitulation of Suṁha (Maṅdāra) and Veṅgā apparently took place sometimes between AD 1110 and 1112 as this great victory of Choḍagaṅga is recorded in the Korni copper plate grant of AD 1112 as well as in a grant issued from Sinddurapura in AD 1118.[65] That Utkala was included within Choḍagaṅga's empire by the second decade of the twelfth century AD is further supported by a stone inscription in the Liṅgarāja temple datable to AD 1112,[66] an inscription in the Khileśvara temple in the village of Khilore, district Puri dated to AD 1113[67] and inscriptions in the Garteśvara temple in the village of Algum (Puri dist.)[68] and the Mārkaṅdeśvara temple at Puri.[69] It is also this time that Choḍagaṅga transferred his capital from Kaliṅganagara (Mukhalingam) to a site near the present city of Cuttak. During his long reign of seventy years he built a series of forts at strategic points and dense forests; remains of which are still extant, which accounted for his military success.[70] Choḍgaṅga was by far the strongest and most powerful king of his age in north-eastern India. From inscriptions we know that Choḍagaṅga ruled over a vast empire extending from river Gaṅga in the north to river Godavari in the south, this area referred to as Trikaliṅga in later part of his rule with a large group of feudatory chiefs posted on the borders.[71]

The rule of the Gaṅgas lasted for more than three centuries and the dynasty had given most notable and powerful monarchs

IMPERIAL GANGAS

Anantavarma Chodaganga (A.D. 1077-1147)

Kamarnavadeva (A.D.1147-56)	Raghavadeva (A.D.1156-70)	Rajaradeva II (A.D.1170-90)	Anangabimadeva II (A.D.1190-98)

Anangabimadeva II (A.D.1190-98)
|
Rajarajadeva III (A.D.1198-1211)
|
Anangabhimadeva III (A.D.1211-1238)
|
Narasimhadeva I (A.D.1238-1264)
|
Bhanudeva I (A.D.1264-1278)
|
Narasimhadeva II (A.D.1278-1308)
|
Bhanudeva III (A.D.1308-1327)
|
Narasimhadeva III (A.D.1327-1352)
|
Bhanudeva III (A.D.1352-1377)
|
Narasimhadeva IV (A.D.1377-1413)
|
Bhanudeva IV alias Nisanka (A.D.1413-35)

(Adopted from T.E. Donaldson, *Hindu Temple Art Of Orissa* Vol. I, Lyden, 1985)

who not only maintained peace in the country but also increased the territorial extent as well as the economic prosperity. They were great patron of religion as well as of art and architecture. After Choḍagaṅga fourteen successive kings of the dynasty ruled the empire till AD 1434-35. Choḍagaṅga was succeeded by four of his sons, Kāmārṇava (AD 1147-56), Rāghava (AD 1156-70), Rājarāja II (AD 1170-90) and Aṅagabhīma III (AD 1190-98). After Anaṅgabhīma II the following kings of his lineage continued to rule : Rājarāja III (AD 1198-1211), Aṅangabhīma I (AD 1211-38), Narasiṁhadeva I (AD 1238-64), Bhānudeva I (AD 1264-78), Narsiṁhadeva II (AD 1278-1308), Bhānudeva II (AD 1308-27), Narsiṁhadeva III (AD 1327-52), Bhānudeva III (AD 1352-77), Narsiṁhadeva IV (AD 1377-1413), Bhānudeva IV alias Niśaṅka (AD 1413-35).

During the closing years of twelfth century AD the political condition of northern and eastern India was drastically changed due to fall of the great Rājapūta kingdoms to the Muslims. The Muslims had advanced as far as Chunar and Maner near Patna at the time when Anaṅgabhīma II died and the conquest of western Bengal which followed next year opened the road to Orissa. The first Muslim invasion of Orissa occurred in AD 1205 during the reigns of Rājarāja III and he repelled it.[72] A second invassive attempt of Muslims in AD 1211 was also likewise repelled by the Gaṅga army.[73] Rājarāja III also invaded Tumman and defeated the Kalachuris[74] but it appears that the annexation of Sambalpur tract was concluded during the time of Anaṅgabhīmadeva III. We know a stone inscription originally attached to the Mageśvara temple, Bhubaneswar that his daughter Chaṅdikadevi was given in marriage to Pramāḍideva a member of Kalchuri ruling family who was also appointed the commander-in-chief of the Gaṅga armies, therefore, it seems that some amicable settlement was made establishing matrimonial relations between both the families effecting the possession of the Sambalpur tract in favour of the Gangas.[75] During the reign of Narasiṁha I an aggressive policy was followed against the Muslims. He frequently invaded the Muslim territories and plundering them reached upto Lakhanavati or the capital of Gauḍa.[76] The period of relative internal security and peace in the country extended to the reigns of Bhānudevaa I (AD 1266-78) and Narasiṁha II (AD 1278-1308) though the latter was continuously at war with Muslims during the later part of his reigns. However, Orissa received a major setback during the reign of Bhānudeva III when Firoz Saha Tughlaq, the *sultan* of Delhi. combined with Siquandar Shah of Bengal penetrated upto fort of Sāraṅgagarha (Chaudwar) plundering the towns falling on their way. The army of *sultan* also plundered and massacred in the capital city of *Abhinava Varānasi* (Cuttack).[77] Firoz Shah returned after receiving twenty-two elephants from the Gangas as tribute.[78] In the contemporary copper plate of Bhānudeva III, no mention of this invassion is found. Bhānudeva III, in fact granted rewards to some of his Generals for their meritorious services.[79] His glory sung by the Court-poet which suggests that though the armies of Delhi *sultan* reached upto the heart of the Ganga-kingdom but they were ably repelled and Bhanudeva III continued to rule. But, the kings of Orissa could not sit in peace. Narasiṁha IV, the son and successor of Bhānudeva III was also continuously at war to maintain sovereignty throughout Orissa against the Reddis. This time the kingdom was invaded by the

Bahmani *sultan* Tazuddin Firoz Shah and Maliq-us-Sharq of Jaumpur and the king was compelled to pay tribute.[80] The last Gaṅga king (Niśaṅka) Bhānudeva IV likewise had to struggle with the Hindu and Muslim foes. Though, he succeeded in suppressing the Reddis and the *sultan* of Bengal but he had to purchase his liberty from *sultan* Hosangashah of Malwa.[81]

Upon the death of Bhānudeva IV the Gaṅga dynasty ended as Kapilesvara, the minister and the general of the kingdom usurped the throne in AD 1434-35 and became the king, thus, he founded the new dynasty known as the Sūryvaṁśī.

During the Gaṅga rule sanctity of Bhubaneswar was frequently honoured by royal dedications. Though, after construction of the Jagannātha temple at Puri by Choḍagaṅga, Puri became the main centre of royal patronage, however, it is evident from the existing monuments and the royal grants that Bhubaneswar was never ignored by the royal house. The location of imperial capital at Cuttack and the extensive fortifications of Chudangaraha near Barang and at Bualigarha near Chhandka and Kṛittivās-Kaṭaka at Bhubaneswar itself made politically sensitive the surrounding regions of Bhubaneswar and brought drastic demographic changes which variously acted upon the social, economic and religious conditions of Bhubaneswar. A sacred place like Bhubaneswar attached with greater sanctity would have invited the faith of masses. Thus, becoming more and more popular it may have received pilgrims in increasing numbers. On the other hand we noticed that most of the iconoclastic invasions were from northern side which were repelled before the invaders reached Cuttack, thus, Bhubaneswar was lying in a safer zone. These political circumstances were conclusive in the uniterrupted growth of Bhubaneswar during the Ganga rule.

The Sūryavaṁsīs

Kapileśvara ascended the throne in the AD 1434-35, though, his first year may have been spent in suppressing the rebellion princes of the Gaṅga family, he rapidly gained possession of the Telugu country including the eastern edge of the Deccan plateau, became the undisputed master of the whole Telangana with his victory over the Bahmani forces at Devakonda in AD 1458.[82] Two inscriptions at Mannur in South Arcot district prove that he was in possession of atleast the entire seaboard of the Vijayanagara empire.[83] In an inscription in the Jagannātha temple at Gopinathpur, Kapileśvara is eulogised as saviour," when the world was sinking down under mud and when a civilization was completely vanishing through the attrocities of *Mlechchas,* the king Kapileṅdradeva appeared on the scene and saved the world. He saved his country *Udra* and acted like *Kalkī* and the boar incarnation of Vishṇu.[84] In an inscription at Bhubaneswar, Kapileśvara is called, "the conqueror of Gulbarga, the destroyer of Malwa and Delhi and one who crushed Gauda (Bengal).[85] Though, these claims are obviously exaggerated, Kapileśvara did revive the ancient glory of Orissa and succeeded in conquering the entire eastern coast of the Bay of Bengal from river Hugaly in the north upto Trichinapally in south and prevented the Muslims from over running the eastern India.

Kapileśvara was succeeded by his son Purushottamadeva (AD 1467-97). He had to struggle with the Bahmanis of Bidar and Sāluā Narasiṁha of Vijayanagara empire for maintaining sovereignly in the southern regions of the kingdom, though, he had to loose the area towards south of river Kṛishṇa in favour of Vijayaganara, but he regained the doaba (basin) of river Kṛishṇa and Godavri, expelling the Muslims.[86] He is also credited for conquering Kanchi as per traditions.

Purushottama was succeeded by his son Pratāparudradeva (AD 1497-1540). During his reign, Krishṇadevarāya of Vijayanagara frequently invaded Orissa and captured area lying towards south of river Godavari. After the defeat of Pratāparudeadeva by Krishṇadevarāya, the political power of Orissa rapidly disintegrated, the king apparently being more concerned with religious affairs than political ambitions.

After the death of Pratāparudradeva and the murder of his two sons, the minister Govinda Vidyādhara ascended the throne. It was a crucial period in the history of Orissa as here independence was threatened not only from the North-East and North by the Muslim-rulers of Bengal and Delhi, but also from the South by the kings of Golkunda. Vidyādhara's reign (c. AD. 1540-49) was marked not only by incessant attacks by the Muslims but also by the internal rebellions. The absence of an effective and recognized central authority continued until the emergence of Mukuṅdadeva Haichaṅdana, a Telanga by birth, who ascended the throne in AD 1559-60. Though, northern Orissa had fallen to *sultan* Ghiasuddin Jalal Shah of Bengal, Mukuṅdadeva succeeded in driving the muslims out and reoccupied the northern districts upto river Bhagirathi. In the eighth year of his reign he even advanced as far as the river Ganga and concluded an alliance with emperor Akbar. Some of his feudatory chiefs, particularly Rāmachaṅdra Bhaṅja rebelled against him and the muslims of Bengal taking the advantage of the situation launched a two-pronged deep attack into the heart of Orissa. It was during the moment of confusion and anarchy with Mukundadeva being absent from the capital that a renegade soldier named Kalāpahāḍa, a converted Hindu occupied the royal fort of *Bārābātī* in Cuttack and conducted a iconoclastic activities throughout the country raiding even the Jagannātha temple at Puri. In the civil war that ensued Mukundadeva was killed by Rāmchandra Bhañja. The death of Mukundadeva in AD 1568, brought the solidarity of the country to an end.

The above narration shows that the emergence of the Gajapati Kapilendra extended the security and the internal peace for two hundred years. *Kṛittivās-Kaṭaks,* the royal camp at Bhubaneswar was active in political stirs and played decisive role in the fall of last weak rulers of the Ganga dynasty as evident from the fact supplied by the *Mādalāpañjī* that Kapilendra announced himself the king of Orissa from Kṛittivās -Kaṭaka where his coronation took place. In this usurpation the *gurus* [religious teachers] and the priests supported Kapilendradeva alias Kapileśvara. No doubt Kapilendra satisfied the priestly class by fabulous endowments. Thus Bhubaneswar received royal patronage from the house of Gajapati-Sūryavaṁsīs, the temple of Jagannāth at Puri remained the object of supreme faith. Though, the period after Kapilendra was one of great turmoil which affected the contemporary society.

The temple of Lord Lingaraja was very much instrumental in installation of Kapileśvara in the throne of the Orissa. Kapileśvara had made many declarations against his rivals and rebellions for punishments and rewards to those who would surrendered. He also said that his these declarations were with the concurrence of *rājaguru* at Bhubaneswar. Thus, active involvement of priestly class in the politics of usurpation of the unworthy members of the ruling family of the Gaṅgas would have annoyed the supporters of the Gangas and they might have adversely reacted to the priestly classes of Bhubaneswar. It appears that for few years the environment of Bhubaneswar would have been tense with the political stirs and military activities simultaneously. There would also have a period of pilgrimage bleakness. However, as stated earlier that it took no time to command the situation for Kapileṅdra and peace and stability was restored very soon. Thus, Bhubaneswar would have also enjoyed the fruit of king making.

2. RELIGIONS IN PRACTICE

The sixth and the seventh centuries constitute an important epoch in the cultural history of Orissa. In the contemporary copper plate grants, we notice that *brāhmaṇa*-donees were frequently obliged by royal charity,87 which imply that Brahmanical -Hinduism was at a great pace. On the other hand the accounts of Chinese traveller Huien-tsang inform that the majority of inhabitants of Odra were Buddhists while in Kāṅgoda they were Hindus and Jainas who predominated; and in Kaliṅga there were adherents of numerous different sects but few Buddhists.[88] The religious make-up of the land of Orissa at the time of Huien-tsang's visit thus consisted of various religious sects.

At Bhubaneswar area Śaivism was a dominant religion in practice; however, other religious practices were also in existence. Jainism had a pioneering establishment at Udayagiri hills in the earlier period. There are ample evidences that Jainism co-existed at these hills with the Śaivism at laterite tracts of Ekāmra. Mādhavarāja II of the Śailodbhava dynasty, in his Banpur copper plate grant , endowed a piece of land in Thorana-Vishaya in favour of a Jain-āchārya Darahadāchārya and his disciple Prabuddhachaṅdra[89] and Huien-tsang also states that the Jaina people were inhabiting the Kāṅgod country, therefore , we may say that the premier Jaina centre of Udayagiri hills would have been kept in survival by the monks and the followers of the particular sect. The place was worthy for dedication and received pilgrims from remote places such as Virajā [Jaipur]. This is attested by an inscription at Ganesagumpha datable in the reign of Bhaumakara king Santikara in the AD ninth century.[90] During the rule of Somavaṁśī king Uddyotakesari the Khandagiri hill which was called *Kumaraparvat,* received special attention. In the fifth year of his reign a few of the old cells were converted into sanctuaries by carving the reliefs of the *tīrthaṅkaras* and the *sāsanadevīs* on the walls.[91] This period unlike the early period is marked with the image worship of *Jaina-tīrthaṅkaras* and *sāsanadevīs,* is evident not only from the inscriptions recording the setting up of the images of twenty four *tīrthaṅkaras* but also by the discovery of a large number of nude chlorite-images of different *tīrthaṅkaras.*[92] The enormous architectural fragments lying in some areas of hill suggest the period also witnessed the construction of structural temples. The crude reliefs of the *tīrthaṅkaras* on the walls of cave-9 [Trisulagumpha] considered of the period of the Gangas and their successors the Sūryavaṁśīs, is suggestive of the prolonged *Digambara* association of Khandagiri caves.[93]

Buddhism was a popular belief of the age as evidenced by the statement of Huien-tsang which is again attested by the discovery of a number of Buddhists monasteries mostly concentrated in the hill-spurs of the Asia range in district Cuttack, viz : Ratnagiri, Udayagiri, Lalitagiri, Olangiri, Vajragiri and Solampur as well as in other parts of Orissa. At Bhubaneswar, the finds of the images of Buddha, Lokeśvara and Tārā suggest that *Mahāyāna* Buddhism was in practice here in the early medieval period[94]; but unlike the Jainism it did not co-exist with the Śaivism much longer. Dr. Panigrahi has opined on the various grounds that Buddhism came in confrontation with Śaivism and after a violent struggle between both the religious groups Buddhism had to leave the place.[95] It is also noticeable that during the early part of seventh century AD, this part was controlled by the king Śaśāṅka of Gauda who was a devout worshopper of Śiva and caused the erection of a temple in the shrine of Tribhrvaneśvara. He is known for his anti-Buddhist activities and therefore he would have played a major role for the eviction of Buddhists from Bhubaneswar and popularising Śaivism.[96] The Saivites proves so strong at Bhubaneswar that Buddhism became almost traceless here and even in the succeeding centuries when Buddhism was a court religion during the Bhaumakaras it could not make its place in the land of Ekāmra .However, its influence in the sculptural art of the temples can be seen in this period. In the synthesis of different religions which was a trait of little later period, Buddha was adopted in Hindu mythology as one of the incarnations of Vishṇu and his images appeared in the walls of temples in the same context.

Śaivism as a popular faith had predominated the whole Orissa by the early seventh century AD. It is noticed earlier that Bhabaneswar was a pioneering centre of Śaivism. It appears that *Lakulīśa-Pāśupata* cult propagated Śaivism in a great pace. The early temples of Bhubaneswar teem with prolific representations of the deified teacher *Lakulīśa*.

The study of narrative reliefs of some early temples reveals that there were atleast two broad groups of Śaivites; the one who strictly adhered to Śaivism with *Śiva* being the supreme power and insisted the frequent depiction of *Śiva* and *Śiva*-myths exclusively, the other deities were represented as subordinate and only in some context of *Śiva*'s lore . Thus at Paraśurāmeśvara temple we get all the representations of Śiva such as *Ravaṅāugrahamūrti, Naṭarāja*, dancing *Ardha-narīśvara*, *Bhikshāṭanamūrti, Śiva-paarvatī, Harihara* and *Lakulīśa.*[97] Besides, busts of *Śiva* are frequently placed in niches or *chaitya-gavākshas.* The *pārśva-devatā* niches contain the images of *Gaṇésa* and *Kārtikeya* on the sides and the central niche on the backside would have been enshrined with image of *Pārvatī* as usual which is now missing. The other deities of Brahmanical pantheon are *Yama, Varuṇa, Gaṅgā* and *Yamunā Saptamātṛikā* and *Mahisāsuramardinī* and *Gajalakshmī* (on lintel). *Mahisāsuramardinī* and *Saptmātṛikās* are *Śākta* representations and this cult was deeply associated with Śaivism as we see further. The depiction of *Yama* and *Varuṇa* became common as *dispālas* while *Gaṅgā-Yamunā* and *Gajalakshmī* used to be depicted in the doorjambs and at door-lintels respectively as a set pattern of architectural embellishments. The deities namely, *Agni, Brahmā* and *Sūrya* and *Vishṇu* have found place in a depiction of the marriage of *Śiva* and *Pārvatī*.[98] Thus in the whole mythological depiction *Śiva* figures in the centre and seems that the carvers were directed to take the stories only from the *Śaiva* literature for depiction of religious theme.

On the other hand the narrative reliefs on the walls of Svarṇajāleśvara temple speak that while selecting the themes for carvings a liberal attitude was adopted. Here, apart from the *Śaiva*-lore the stories from epics like *Rāmāyaṇa* and *Mahābhārata* have also found the place. Thus, we find the depiction of *Śiva-Pārvatī* flanked by flying figures, the worship of *liṅga* and the marriage of *Śiva-Pārvatī* as well as the scenes like conference between Rāma and the monkey chief Sugrīva, Rāma killing the golden deer, Bāli's death at the hands of Rāma (all from *Rāmāyaṇa*) and the fight between *Śiva* in the guise of *Kirāta* with Arjuna (from *Mahābhārata*). Not only that, Vishṇu with his distinctive attribute had been originally enshrined in the central niche of the southern facade and it was removed to the Orissa State Museum by late Dr. K.C. Panigrahi.

Though both of these temples belong to same period and dedicated to Śiva but on the basis of the facts revealed above it appears that each temple was separate institution and was very rigid in his faith while the other was having cosmopolitan outlook for Brahmanical pantheon though believed in Śiva as Supreme power.

In the succeeding period i.e. eighth-ninth century AD the temples dedicated to Śiva as well as independent Śākta shrines were constructed. The Śaivism, this time fully amalgamated Saktism is evident from the depiction of Śiva, Gaṇeśa and

Kārtikeya in the Śākta shrines while Śaiva temples were carved with the figures of Mahisāsuramardinī. During this period four Śākta shrines sprang up on the four sides of Bindusarovara, viz. Vaitāl, Mohinī, Uttareśvara and one on the east bank the name of which is not known.[99] The shrines of Vaitāla and Mohinī are dedicated to goddess *Chāmuṅdā*. In Vaitāla temple *Chāmuṅdā* being the central deity in sanctum, is accompanied with six other *mātṛikās* with *Gaṇeśa* and *Vīrabhadra.* Though, it is a *Śākta* shrine, however, the walls are decorated with the bold representation of *Naṭarāja, Ardhānarīśvara, Harihara,* Śiva-Pārvatī, Lakulīśa and Sūrya. On the side of Vaitāla deula stands the Śiśireśvra a shrine of Śiva.

The name of Vaitāla has been derived from the word Vetāla (spirit) with the help of which the Kapalaikas and the Tāṅtrikas wanted to attain their *siddhis.*[100] It is stated in the *Svarṇādri Mahodaya* that the venerable goddess Chāmuṇḍā garlanded with skulls existes on a spot on the west, not far from the Bindusarovara, and that is of terrific from and is known as Kāpālinī. Evidently the shrine of Vaitala is referred to in this passage.[101] Kāpālinī was the deity of worship of Kāpālikas. Though, the Kāpālikas were the worshipper of *Śiva,* the deity of their worship was very often Chāmuṅdā. The Kāpālikas used to sacrifice human beings to the goddess Chāmuṅdā or Kali[102]. The temple of Kāpālikas, now known as Vaitāla was a shrine for sacrificing human beings and other animals. The basement of a stone-yūpa which was utilized for the sacrifice, is still to be found infront of the door of Vaitāla deul.[103] Unlike the cells of other temples where light and shade intermingle to create a serene atmosphere in the sanctum of Vaitāla deul, it is intensely dark.

The seriousness of the aweful esoteric rites that were once performed here, must have been hightened by the darkness of the cella and by the presence of some images in their most terrific form. Among the terrific figures carved on the walls occurs a Bhairava depicted in most hideous and terrific form that human imagination can ever conceive. This had been another deity of worship, for the Kāpālias take Bhairava to be the creator, protector and destroyer of the world and believe that all other gods are subservient to him.[104] The form of Śāktism that we find from the study of Vaitāla deula is not an independent one but it certainly allied itself with Śaivism, Buddhims and Tantrism.[105]

Besides the four *Śākta* shrines on the four sides of Bindussarovara at Bhubaneswar, there are also other smaller shrines at the places in which the deity Chāmuṅdā is still being worshipped. One such shrine known as Bhuasuni, is in the neighbourhood of Sisupalgarh and the other is near the house of Satyapriya Mohanti. All these images bear the characteristics of the same age and all are depicted in terrific forms.

In the side of Vaitāla deula, stands the Śiva temple known as Śiśireśvara which was also constructed in the same age. The temple is notable for depiction of Lakulīśa bearing the stamp of Buddhims. Another reclining figure of a lady recalls the depiction of Māyādevī though its proper explanation is difficult. In the amalgamation of Śaivism and Śāktism it appears that *Mahiśāsuramardinī* and *Pārvatī* were taken as one and the same as evident from this temples where in the Pārśvadevatā niches though *Kārtikeya* and *Gaṇeśa* are in their respective places but *Pārvatī* is replaced by Mahisāsuramardinī. Śiśireśvara is also remarkable for the depiction of scenes from *Rāmāyaṇa* in the edges of clerestory roof of Jagmohana. It is noteworthy feature of this period that much stress has been given on unification of Śaivism and Śāktism but other deities like Vishṇu and Brahmā representative of major cultic worship in other places were depicted in humiliating position as attested from a depiction in Mārkaṁdeśvara temple where Brahmā and Vishṇu are shown standing folded hands paying homage to *Śiva*. In the representation of Harihara at various temples the motto was to prove that Vishṇu is a part of *Śiva* and not the supreme. In the representations of *Saptamātṛikās, Vaishṇavī* and *Vārāhī,* which are the counterparts of *Vishṇu*, are shown as allied deities to *Chāmuṇḍā* and or *Mahisāsuramardinī.*

From a study of the existing temples of Bhubaneswar it appears that Śaivism inseparably mixed up and formed a strange amalgam in the early medieval period. But, this phenomenon is not limited to Bhubaneswar only rather extended to all parts of India as scholars believed.

The Somavaṁśī period is remarkable for a cosmopolitan religious outlook. The Somavaṁśī rulers performed *Aśvamedha* sacrifice at Chaudwar, worshipped *Saptamātṛikās* at Jaipur and Puri. They revived the *Vaishṇava* shrine of Jagannātha at Puri and constructed the grand and majestic temple of Liṅgarāja and other various temples at Bhubaneswar. Simultaneously they patronized Jainism at Udaygiri-Khandagiri hills. Such liberal attitude of royal house had its cumulative bearing in the religious practices at Bhubaneswar. The Bhubaneswar area had now become a place of frequent visit of royal people and common pilgrims, the *liṅga*-worship was highly popular and crowding folks worshipped the lingas at numerous places as evident from the frequent depiction of *linga*-worship scene in most of the temples of this age at Bhubaneswar. We also notice that Jainism had got much impetus during this time at Udaygiri. The believers of non-violence would have supported the eviction of cruel practices like sacrifices of human beings and innocent animals, though *Śākta* images in their terrific forms continued to be sculpted in the walls of temples, as evidenced by a few appearing in Brahmeśvara temple, but the only *Śākta* image that served as a presiding deity of the Gaurī temple near Kedāreśvara temple, assignable to this period, is found in pacific form. It is notable that a temple of *Chausaṭha Yoginī,* datable to AD tenth century, related with violent Śakti cult or Tāntrism, is located at Hirapur which is only 18 km. far from Bhubaneswar. The existence of such temple at a near distance from Bhubaneswar, suggest that such practices were not favoured at the sacred site of Bhubaneswar and the *sādhaka* (practitioners) of these sects had to migrate from Bhubaneswar being disregarded by people, and remained in isolation.

On the other hand, we notice that representation of deities of Brahmanical pantheon increased in the temples e.g. *ashṭadikpāla* have received a certain place in the walls of temples. Kṛishṇalīlā scenes were depicted in the Lingaraja temple. The image of Lakulīśa, though depicted frequently in Mukte›vara temple, is not found on the other temples built in the Somavaṁśī period, except only one i.e. the Liṅgarāja temple. Śaivism in this period is restored to purer form being more adherent to Brahmanical pantheon and separated from Tāntrism.

During the Gangas and the Sūryavaṁsīs various temples dedicated to *Śiva* were constructed as Bhubaneswar was regarded as the sacred site of Kṛittivāsas i.e. *Śiva*. However, Vaishṇavism, which started with casual representations of *Kṛishṇalīlā* scenes and incarnations of *Vishṇu* have interfered much with the practices of Śaivism in the guise of Jagannātha cult.

Vaishṇavism did not occur much earlier in the coastal plains of Orrisa as a popular religion. The earliest temple dedicated to *Vishṇu* in circa eighth-ninth century is located at Village Gandharadi of district Phulbani lay in hilly regions. The two huge rock-cut images of *Vishṇu* dated in circa AD ninth century in district Dhenkanal also lay in the hilly tracts. In the coastal area, the *Pāñchu-Pāṅdava* temple in the village Ganeśwarpur of district Cuttack datable to circa AD tenth century can be assigned the earliest temple where *Vishṇu* is enshrined as presiding deity. The worship of *Prushottama* at Puri is being dated back to circa AD third-fourth century. Some stray references of his worship in the circa AD eighth-ninth century at Puri have also traced. The temple was revived by Somavaṁśī-rulers, but it came in prominence only after the construction of the present temple of Jagannatha by Anaṅtavarmana Choḍagaṅgadeva in circa AD twelfth century. The deity of Jagannātha is treated by the scholars as the development of a separate cult which combines the diverse religious faiths. Thus, it is said that Lord Jagannātha is not only the composite form of various Hindu religious sects, viz : Vaishṇavism, Śaivism and Śāktism but also of Jainism and Buddhism.[106] Whatever may have been the contributions of diverse religious sects in the origin and development of the composite form of Jagannātha, however, it is obvious that Vaishnavism prevailed in the practice of Jagannātha cult from the very beginning to the present day. It was Vaishṇavism which assimilated diplomatically other sects establishing its supremacy in the guise of Jagannātha.

At Bhubaneswar, the stray depictions of Vishnu and stories relating his other incarnations, viz. : *Rāma* and *Krishṇa* in the temple already referred to. In the Gaṅga period we notice that Vaishṇavite representations were more frequent in *Śiva* temples. The Magheśvara temple built by Svapneśvaradeva, commander-in-chief of the Ganga army towards the close of twelfth century AD, contains on its southern facade a small panel of sculpture depicting *Godhanaharana.* In the compound of Paramguru family at Bhubaneswar (near Vaitāla deula), on the walls of a small rekhā temple (datable to late Ganga period) the ten incarnations of Vishnu appear, in which the last two the Buddha and the Kalkī incarnations have also been found. The Anaṅta Vāsudeva, a Vaishṇava temple was built in AD 1278 on the eastern bank of Vindusarovara by the royal house of the Gaṅgas. In a commemorative inscription earlier attached to this temple, it is stated that the temple was built for *Srīkrishṇa* and *Vala-Vāsa* (*Balarāma*) on the bank of Vindusarovara by Chandrikā devī, the daughter of Anaṅgabhīma III, in the Śaka year 1200 corresponding to 1278. It is further disclosed that Chandrikā devī was married to Pramāḍideva, the ornament of the Haihaya family. Pramāḍi after having successfully fought with the enemies of Narasiṁha I ultimately went to heaven. After his death the widow-lady built the temple of Anaṅta Vāsudeva for *Achyuta* (*Vishṇu*) for whom she had developed a devotion from the childhood.[107] The figures of *Rāma, Lakshmaṇa, Sītā* and *Hanumāna* appears on the balustrades of the window on the northern side of this temple, obviously proving that *Rāma* was considered to be an incarnation of *Vishṇu* and as such, he alongwith his associates, found place on a temple meant for *Śrīkṛishaṇa* and his brother. The free-standing images are also enshrined in the southern and northern niches of the temple respectively. On the southern side of the temple there were also five images of the balustrades of the window consisting of *Śrīkṛishaṇa* and *Gopīs*, from here three of the balustrades were detached sometime and now can be seen fixed in a miniature shrine standing on the eastern bank of Bindusarovara. A few other cult images like *Lakshmī* and *Sarasvatī* appear on the *Jagmohana*. Though it is a solitary temple of *Vaishṇava* faith but it became on of the most important places of worship and equally shared the religious ceremonial celebrations of Bhubaneswar as attested by the traditional literature as well as the current religious practices.

During this period Vaishṇavism also penetrated inside the shrine of Lord *Tribhuvanśvara* or *Liṅgarāja*. The *naṭamaṅdira* and the *bhogamaṇḍapa* of the *Lingarāja* temple are additions of the Gaṅga period and it is noteworthy that in the southern doorjamb of naṭamaṅdira the images of *Vaishṇava dvarapalas Jaya* and *Vijaya* were relieved in the place of *Chaṇḍa* and *Prachaṇḍa.* These images holding in the upper two hands *śaṅkha* (conch) and *chakra* (wheel) and in the lower two a *gadā* (mace) struck to the ground. As pointed out by Dr. Panigrahi the *vṛisha-staṁba* which now stands in front of *Bhogamaṅdapa* would have been the work of that period[108].

This stambha (pillar) is crowned by a bull the mount of *Śiva* together with *Garuda* the mount of *Vishṇu* which had evidently been meant to make the synthesis to two cults visible to a visitor at his first approach to the shrine. In order to bring home to the general public that the shrine belongs to both the sects, the *āudhya* (weapon) crowning the sanctum tower which must originally have been a *triśūla* (trident), was replaced by one consisting of half a disc and a trident. It is said that this change was brought about by a daring man who climbed to the top of the spire in the dead of night under the orders of a Ganga king, pulled down the original *āyudha* and replaced it by this new and composite device. This man is said to have been granted rent-free lands and given the title of *Niśanka malla* (the fearless hero or wrestler) which is still born by his descendents living in the Nuapalli village (now under the notified area of Bhubaneswar)[109]. Besides, Vaishṇavism tried to establish its priority to *Śvayambhū-linga* by enshrining the *Vaishṇava* deities inside the temple complex of Lingarāja, viz.: Jagannātha to the west of *garbhagṛiha* and *Lakshmīnārāyaṇa* on the northern side of *bhogamaṇḍapa* so that a devotee while making the circumbulation will first meet these Vaishṇvite deities before he enters the main shrine.[110] Among the subsidiary temples standing within the compound of Lingarāja temple, there are two shrines to the south of the main temple, one enshrines the *Vārāha,* (boar incarnation of *Vishṇu*) and the other contains the images of *Ananta, Vāsudeva* and *Eknāmsā.*[111] The influences of the Jagannātha cult are discernible in the daily worship of *Lingarāja* in the *mantras* with which he is invoked, in the offerings which are given to him, in the surroundings in which he is worshiped, in the festivals which are held in his honour and the orthodox literature in which he had been extolled and modes and merits of his worship propounded.[112] It is now an accepted fact which the priests here say that *Liṅgarāja* is the combination of both *Vishṇu* and *Śiva*. In other words he is Harihara, and not Hara alone. A natural line that exists in that *Śvayaṁbhū-liṅga* is pointed out by them as the line demarcating *Śiva* and *Vishṇu*-portions of the same deity which is invoked as *Harihara*.[113]

No doubt that Jangnnātha cult became the symbol of sovereignty during the rule of the Gaṅgas and the Sūryavaṁśīs and later the rājās of Khurda in Orissa. However, it is also a noteworthy fact in the dedicatory records of the Gangas and the Sūryavaṁśīs the deity of Liṅgarāja was never regarded as Harihara rather always referred to as Kīṛtivāseśvara or Kīṛtivāsas (Kṛittivāsas). Therefore, it does not seem more plausible that any royal hand was active in introducing some Vaishṇavite features and influences in the temple of Liṅgarāja. It is a well known fact that the whole India was under the sway of Vaishṇavism in the medieval period. The fifteenth and the sixteenth centuries of the Christian era are remarkable for the preaching of Vaishṇavism by a number of great saints and the masses were attracted by them. The great saint Chaitanya brought a revolution in eastern India in the sphere of Vaishṇava religion during this time. He also stayed at Puri in devotion of Lord Jagannātha and highly influenced the whole society through his *kīrtana* (singing of the devotional songs).

The propagation of Vaishṇavism highly endangered the popularity of Śaivism. The cultic disputes in between Śaivas and Vaishṇavas as well as the efforts for synthesis of both are well discernible in the contemporary orthodox literature.[114] Therefore, it appears that looking into the faiths of masses the religious teachers and priests of Bhubaneswar would have to compromise with the Vaishnaṇavism and the Jagannātha cult by setting aside the cultic dispute and by synthesis of both the religions to keep in demand their institutions. Not only the Śaivism but Śāktism had also to compromise; the influence of Vaishṇavism was so intense that no fearful representation of Śākta deities was made even in the Devī-temples as evident from Pārvatī temple and Chitrakāriṇī temple, rather *Kṛishṇalīlā* scenes found place in the walls of these temples. However, Śaivism was so deep-rooted at Bhubaneswar that Vaishaṇavism could not hurt much the religious status of Śiva. Bhubaneswar was always regarded as the land of Śiva and Śaivite temples were being constructed without any break even in the days of great popularity of Vaishṇavism. In the fold-songs of *Hindī-Avadhī* Kaṭaka (probably Kṛittivasa Kaṭaka) was always identified as the place of *Śiva.*

Thus from he preceding discussions we come to know that there were various religious practices going on at Bhubaneswar area. In the early medieval i.e. in the sixth and seventh centuries Śaivism was practised in its purer and canonical form It also co-existed with Jainism though Buddhism was losing its hold in this area. In the succeeding centuries *Śakti* and *Tāntrika* cults crept and amalgamated with the Śaivism which acted upon variously in the society . In this period Buddhism was patronised by the royal house in the whole Orissa and though it receded from Bhubaneswar even it left a cumulative effect in the temple art. Further, it was Vaishnavism and Jagnnatha cult which stood in parallel to Śaivism and both were synthesised ultimately.

3. CLASSES OF PEOPLE AND THEIR FEEDING ECONOMY

The human settlement at Bhubaneswar developed mainly due to greater religious sanctity of this place. The site would have been initially occupied by the wandering monks, ascetics practicing pennance and by the *Śaiva* and *Jaina āchāryas* preaching the religion to sectarian followers as well as lay devotee. These occupants would have increased the grace of this holy place resultantly inspiring public and royal endowments for construction of temples enshrining of lingas and the deities as well as the monasteries for the shelter of monks, religious students (*brāhmanas*) and their teachers belonging to different sects. As soon as such construction took place the religious population did grew under organised institutions; but simultaneously a lot of secular people also would have settled in connection with the construction work or earning their livelihood by other ways at this growing settlement of frequent public visit. Thus the classes of people inhabiting the *Ekāmra*-Bhubaneswar (in its notional boundaries) during the period under review can be classified under the following groups:

(i) Wandering population and free ascetics practising penance at the sacred area.

(ii) Religious population associated with particular deity or sectarian monasteries.

(iii) People working in temple construction.

(iv) Various professional-castes (*pāṭaka*) seeking their livelihood by practising their specialized crafts and business, mainly related with religious phenomenon.

(v) Pilgrims

Now, we shall discuss each class of people mentioned above with their feeding economy as well as their contribution in formation of the social groups which crystallized later in a complicated manner and their adherence with the present Liṅgarāja temple for various kind of services.

(i) Wandering Monks And Free Ascetics Practising Penance At The Sacred Area

The extensive laterite tracts and hill-spurs full of trees and plants yielding soothing shed and edible fruits, with plenty availability of water from different sources, namely, great lake, river Gandhavati and perennial springs, finally attached with the greater sanctity of Lord *Svayaṁbhū Liṅga* and the sacredness of *Jaina* tīrtha, the site had become a favourable resort for the ascetics practising penance for divine powers (*siddhi*) or liberation. The retired *gṛihastha* who were running in the stage of *Vānapṛastha* and *sanyāsa* following the order of four *āshramas* (*āshrama-vyavasthā*) would have also resided here in order to perform their *tapa* and *sādhanā*. The existence of such a class of people is still in vogue. In many *tīrthas* whether in casual days or some special festival occasions they can be seen performing various activities of *hathayoga*. However, at *Ekāmra*-Bhubaneswar the existence of such a class is testified by contemporary description appearing in the grants of Śailodbhavas where king Mādhavarāja II is said to have been attained diviṅe peace due to his great qualities while for the attainment of the same ascetics performed various painful practices:[115]

"In order to attain the supreme heavenly peace the hermits (*munis*) were performing meditation in several ways (namely) by following the functions of wild deers having been lived with them by looking attentively the thousand-hot rays of the burning sun which caused mutilation of their faces, and some of by wearing natural cloths (viz. *balkala* and *ajina*). For attainment of divine peace, people left their abodes and practised *'yoga'* in caves. Some of them inhaled smoke others lived taking air, fruit and water, while others lived without taking anything". Though the whole description appeared only to prove the superior virtue of the said king, however, this reflects the contemporary society of ascetics. The cave-dwelling and other practices closely fits with the *Ekāmrakshetra.* We know that during the period of the said king Bhubaneswar was under the spell of Śaivism and Śāktism with the co-existence of Jainism. It was during this period that various yogika and *tāntrika* performances were being practised as attested by existence of *Śākta* temples, namely Vaitāla deula, Mohinī, Uttareśvara and other associated places. *Jaina* also practise very hard penance. In various copper plate grants of Madhyamarāja the Verses describing the above phenomenon have been repeated.[116] In the copper place grants of his successors also while giving the geneology the king Madhyamarāja was introduced by these verses.[117] It seems that Madhyamarāja was a lay devotee of *Śiva* and liked the simple form of worship, therefore, he proved these practices of lesser value and not favoured. However, such ascetics and hermits would have been a great attraction for the pilgrims visiting the place and would have also received their food in the monasteries and temples during the days of general offerings of cooked food.

(ii) Religious Population Associated With Temples Or Monasteries

The monumental relics as well as inscriptions show that there were a large numbers of temples enshrining the deities known by different names and mathas associated with the particular branches of learning. From the available records it appears that maṭhas were intimately related with some deity and the *tapasvins* or students as the case may be received shelters in these places. In the early medieval period the number of temples as well as of monasteries (*maṭhas*) was not more but during the time of the Somavaṁśīs, the Gaṅgas and the Sūrvaṁśīs the number of these institutions increased. In the sixth century AD, we come to know a monastery (*maṭha*) existing at *Ekāmra* which was associated with deity *Maṇināgeśvara Bhaṭṭāraka.* In this monastery the students of *Taittirīya śākhā* (school) of *Yajurveda* were studying the particular knowledge.[118] In the temple of Parasurāmeśvara the *tapasvina* of Pārāśara school of Śaivism stayed.[119] The temple of Uttareśvara was headed by a *Śaiva* devotee (*Śaivārādhaka*) who was well-versed in *Sāmaveda.*[120] The Siddheśvara *maṭha* (monastery) was endowed for the Śaivite students of particular sect (*sampradāya*) coming from Kāñchīdeśa, Choḍadeśa and Pāṇdydeśa.[121] From the Megheśvara temple inscription of Svapneśvaradeva we come to know that the monasteries (*matha*) for the study of Vedas were founded by Svapneśvaradeva, the Commander-in-Chief of the Gaṅga army.[122] He also founded a settlement for *Brahmans* named Brahmpur which was vying with the city of *Bṛihaspati* (*Bṛihaspatipur*) in the field of knowledge. The *Brahmpur* was superintended by a great *Śaiva* teacher Vishṇu āchārya who also consecrated the shrine of Megheśvara.[123] From one inscription we come to know that the head of a temple was called *sthānapati* as mentioned in dedicatory inscription of Liṅgarāja temple that *Tapodhana Sāmavedī Soma* was the *sthānapati* of the shrine of lord Uttareśvara. The same designation was born by the head of Siddheśvara monastery (*maṭhas*). The *sthānapati* (Oriya : *Thānapati*) of this monastery was Taporāja Mahāmuni succeeded by Tapaschakravartin. The *sthānapati* of Liṅgarāja temple would have been a sage of very high calibre and learnings. Thus, the religious people associated with temple and monasteries were the brahmans and the tapasvins.

Each matha/temple was headed by a *sthānapati* (head of the institution) who would have been appointed in order of merit and seniority in the particular institution. Under him were the other teachers well-versed in their subjects. The students termed as chhātra or sometimes as tapasvin, vaṭu and brahmachārin were consisted the part of religious population.

Besides, priests attending the ritualistic performances in the service of presiding deity of the temples were the important people being an integral part of the religious population. The people performing the worship rituals in the temples are usually called *pujārī,* however, in an inscription we also come across such a term.[124] In a temple or monastery a *pujārī* (priest) would have performed alteast the following jobs:

(1) Burning of perpetual lamp (this is most important offering to the deity as we notice that most of the

dedicatories grants were made by the donors for the maintenance of perpetual lamps i.e. *akhaṇḍadīpa*).

(2) Cooking of food for offerings to the deity and to be partaken by the dwellers of the temple of monastery.

(3) Making/offering of cooked food to the deity maintaining all the paraphernalias.

(4) Receiving the offerings made by pilgrims and devotees for the deity.

The *sthānapati* or the endowers of the temples would have appointed their subordinate teachers or disciples for this job. In the big temple like Ligaraja which was receiving more number of devotees, there may have been engaged many more numbers of priests.[125]

Though, there were other people who were closely related with the deities by their different services, but they did not form the part of priestly class and will be discussed later. However, one important class of people associated with temple was formed by the *devadāsīs* performing ritualistic dances in the precincts of temples. Probably they lived in temple premises. The existence of this class at the temples of Bhubaneswar is attested by atleast two inscriptions. From the inscription of Kolāvatī the queen-mother of the Somavaṁśī-king Uddyotakesarī, we know that she had offered beautiful damsels in the service of lord Brahmeśvara.[126] Similarly, the inscriptions of Svapnaśvaradeva, the Commandor-in-Chief of the Ganga army, mentions that the builder endowed beautiful girls in the service of lord Magheśvara.[127] The provision of the naṭamaṁdira in the Lingaraja temple, Anantavāsudeva temple and Yameśvara temple, speaks that these temples were provided with the *devadāsīs* and ritualistic dances were performed at natamandira before the presiding deity in usual routine.

The feeding economy of the religious population residing in the sacred area was essentially based on the charity by royal houses or benevolent people. There is no doubt that the creator of a monastery or founder of a temple would have also endowed with the institution some permanent landed property or treasure, yielding the seasonal crops or interests as the case may be for the maintenance or the ritualistic offerings as well as food and clothings for the dependent brāhmaṇs and other priestly people. The original endowments were being increased by the fresh grants from new donors. At Bhubaneswar most of the dedicatory inscriptions of public and royal charity are found at Liṇgarāja temple. In these dedications the donor had made arrangements for burning of perpetual lamp (in most of the cases) or for making daily offerings of *Nitya-Naivedya,* sometimes curd, ghee (clarified butter) and Vyañjan etc. to the deity. Only in the Paraśurāmeśvara temple inscription whose donor was *Śaiva* teacher, we find that the arrangements were made for the food-supply to the tapasvina attached with the temple.[128] Similarly, in the Liṅgarāja temple a dedication made by Guru Bhāva Sadāśiva in favour of the students of grammer (*vyākaraṇa*).[129] Therefore, it appears that the *brāhmanas,* priests and devadāsīs associated with a particular institution were maintained by the income of permanent endowment affiliated with that institution. The priest would also have received *dakshiṇā* while placing the offerings of pilgrims to the deity. In Paṇḍā-system such services are still in vogue.

While the *āchāryas* and preceptors earned some money which was donated by Guru Bhāva Sadāśiva for the students of grammer, was earned by alms (*bhikshopārijita*).[130] Sometimes, abots (*sthānapati*) of monastery also found asking loans from other people pledging the land in connection with some *anusṭhāna* (like feeding the medicants or followers of a particular sects). Thus, we may cite the case of Siddeśvara maṭha as revealed by a stone inscription[131] that the *sthānapati* had pledged twelve *vātīs* land in the Vāghamārā village an endowment by the Ganga king Narasiṁha against a loan of 150 gold *māḍhas* and thirty *pauṭīs* (an old measure) of rice from Uttareśvara Nāyaka in connection with to feed the *Māheśvaras* (followers of *Śiva* probably of a *sampradaya* from south India, in an *anusṭhāna* of *Ekādaśa Rudra Bhikshā*. But, the borrower died without repay. The new successor of the monastery was the friend of the lender Durgābhaṭṭa Āchārya and he had written off the loan from his account and also paid the loan borrowed from Uttareśvara Nāyaka, thus, freed the late abot (*sthānapati*) from the sin of not repaying of loan. The mortgaged land was re-endowed for continuation of *Ekādaśa Rudra Bhikashā* and for the study of student coming from Kāñchīdeśa, Choḍadeśa, and Pāṇḍydeśa. The event suggests that Durgābhaṭṭa Āchārya appearing in the inscription was also a *Śaiva* and was capable of lending such a huge amount and writing off the same without repayment. This fact suggests that some of the people of priestly class were very rich. We know one more such person from priestly class who donated five gold *māḍhas* (*suvarṇa māḍha*) for the perpetual lamp before the deity *Kṛittivāsa* (modern Liṅgarāja). He was Bālakañchchhaṭika, the devotee of *Śiva* and was also the royal preceptor (*rājaguru*). His father *Tapodhana*

Sāmavedī Soma was the *sthānapati* (abot) of the Uttareśvara temple.[132]

It is also a noteworthy point that only the material required for offerings to the deity was supplied to priests. While making the grants in currency, it was deposited with some trust with witness of some important person and the trustee or the particular professional had to supply the desired material. For example the rājan Pramādi, the younger brother of Choḍagaṅg deposited five gold *māḍhaa* with the *Pradhāna* of Nāgagarabhā village in the knowledge of other villagemen, now it was the duty of *Pradhāna* to provide the required quantity of oil for perpetual lamp at Kedāreśvara temple (probably from the interest of the deposited amounted).[133] The similar example is of the dedication of Meḍamadevī (in the reign of Rāghavadeva) when the five gold *māḍhas* were paid to the villagemen of Gandhavan village for the same purpose at *Kṛittivāseśvara* (Liṅgarāja).[134] Similarly, the land grants were also given in the name of deity but lying far away from the place of deity, which was tilled and managed by the local people.[135] Thus, appears that the priestly class was provided only their daily livelihood but not arrangements were made for his recurring source of income other than the priestly job.

(iii) People Working In Temple-Construction

The cluster of temples, tanks and embankments of Bindusagara at Ekāmra - Bhubaneswar bespeak of continuous construction-activity. Such constructions would have certainly involved the huge number of workers at quarry-site as well as at construction site. We may well conjecture about

(1) The labourors working in quarry site for quarrying stones,

(2) The stone-dressers,

(3) Transporters of stones to the site of construction,

(4) The sculptors and stone-dressers,

(5) The masons working in building work,

(6) The labourors making the ramp or working platform for the masons,

(7) The black-smiths (*kammāra*) preparing iron-dowels and girders,

(8) The carpenter (*baḍheī*) engaged in wood-work.

Further, we can also think of the supervisors at various stages and sites, the scribes maintaining the account of expenditure and payment of wages etc.

The early temples at Bhubaneswar are though small in dimension and therefore less number of workers would have worked in these constructions, however, temples like Liṅgarāja, Rājārānī, Brahmeśvara, Megheśvara Yameśvara, Sāri deula, Chitrakariṇī, Anatavāsudeva, Pārvatī and some other temples were gigantic projects and there would have worked hundreds and thousands of people in unskilled, semi-skilled and highly skilled jobs. These people would have camped or settled as per the availability of jobs at Ekāmra-Bhubaneswar earning their wages. Some of the labour stock would have been from the neighbouring village. At present Bauris, a class of people inhabiting in their own hamlet called Baurīsāhī, usually work as labourer and quarry laterite blocks. They are also in the lists of traditional servants of Liṅgarāja temple. We may guess that these people would have been working in temple construction at Bhubaneswar in the past. The stone work was done by the people known as *Mahārāṇās*. They are locally called *Pathuriā* (worker in stone). In the canonical writings they are also known as *sthapati.* They were the semi-skilled and highly skilled workers in the temple-construction. They acted in high capacities like Architect as well as in a lower job of petty stone dresser. While performing the job of engraving an script in stone or copper they were mentioned as *sūtradhāra.* At present only a few traditional families stay at Bhubaneswar and their present localities are found in the interior villages in the vicinities of hills where they are quarrying stones and fashioning stone-equipments for domestic use, but, once they formed the integral population of *Ekāmra*-Bhubaneswar. Obviously they earned their livelihoods through wages received for their work in temple-construction. Besides, they would also have sold the stone-equipments to the pilgrims.

Over and above the workers in the temple-construction, there were also some temple-servants who rendered various physical services during daily ritual cycle as well as in special ceremonies. They were given food and wages from temple management as evident from present practices in the Liṅgarāja temple. Such servants were also maintained by some kind donor at many times. One such example is furnished by an inscription engraved in the Pārvatī temple (standing inside the Liṅgarāja temple complex). In this inscription the donor had made arrangements of food supply for a certain physical worker.

Though this Oriya inscription is interpreted by B.V. Nath with detail notes on the name appearing in the text[136] but his interpretation appears confused; in fact Umādevī to whom Sri Nath understood the goddess was the queen Umādevī and donor of the record as identified by S.N. Rajaguru.[137] Thus, the inscription records that Yasodhara Mahārāṇa (an officer of the temple) was assigned the responsibility by Umādevī (by depositing some money as was customary in those days) to provide the *pisṭaka* (cakes) in two *āṭikā* (i.e. an earthen pot which can contain the quantity of cooked rice of one *kaṭaki seera* or *105 tōlā* say near about 1¼ kg. or little less) to one Bhīma who was a (palanquin-) carrier (*bāhika*).

(iv) Various Professional Castes (Pāṭaka) Seeking Their Livelihood By Practising Their Specialized Crafts and Business:

The temple-city of Bhubaneswar proved a good market for traders and craft-professionals in ancient times. The large number of pilgrims gathering during auspicious days and festivals yielded a good scope for traders and craftsmen to sell their goods. The gatherings during auspicious days had given rise to various regular fairs in the annual cycle and the sanctity of temple yielded scope for a good daily market. Previously traders and craftsmen would have coming from different neighbouring places but in due course they would have settled at Bhubaneswar. No doubt these people would have earned their livelihood by selling their articles to casual pilgrims-customers and other settlers of the area. However, the permanent endowments made by the donors with traders and craft-professionals for the supply of different articles, viz.: oil, pots, lime, flowers betels etc. for the purpose of temple-rituals, were the permanent source of those people.

In the present society of Bhubaneswar we find various cast-professionals who claim their traditional association with the Liṅgarāja and other temple institutions. Though, at present their economic setup is changed, however, still a lot of such people are engaged in their traditional business. Dr. M.M. Mahapatra has listed the various castes termed as '*pāṭaka*' and described their role in the temple rituals.[138] We also encounter various references regarding some of these caste-people in dedicatory records. Therefore, we would like to peep in to the past of these caste-people of Bhubaneswar listed by Dr. Mahapatra in their present context:

(i) Abadhāna (the Astrologer)

He is also called '*Jautisha*' or *Graha-Vipra*' (lit. *brāhmaṇa* of planets). The *Abadhānas* are the non-*brāhmaṇa* caste, can be found in many villages. Their parallels, viz. : Bhaḍḍaris are found in Uttar Pradesh. Astrology or *Jyotisha* was a well-established subject in ancient curriculam. But, in a certain stage it came to be practised as a profession. Initially, at *Ekāmra*-Bhubaneswar, *Abadhānas* would have practised their knowledge in various occasions such as telling the auspicious day for foundation and consecration of temples. They would forecasted the duration of particular auspicious occasion on particular *tithi* (date) for abulation and offerings. Besides, they would have served the residents of the place on different occasion as it is still in vogue. At present the *Abadhāna* is associated with Liṅgarāja for his professional duties as well as other jobs namely beating of bells etc.

(ii) Prakaraṇa (the Scribe)

At present he is assigned various jobs relating accounts and other clerical works in the temple. However, in the past we know that the Scribes who were '*Karaṇa*' by caste, had a lot of business to do. They would have maintained the expenditure-accounts during the constructions of temples, paid the wages to the workers on behalf of builders, recorded the dedications by royal and public donors, and documented the deeds of monetary transactions taking their own fees, e.g. we can quote a transaction between the *Sthānpati* of Siddiheśvara maṭha and Durgābhaṭṭa āchārya, where two *suvarṇa māḍhas* were taken by the scribe for recording the deed.[139]

(iii) Gauḍa (the Milkman)

As a milkman they supply the milk and curd for the services of the deity. Besides, they also carry the palanquins of the deity in festive days. In one dedicatory record we get the references of *dadhi* (curd), *ghrita* (clarified butter) and *vyañjana* (cooked food delicacies). Therefore, the Milkman's relation with the settlement of *Ekāmra*-Bhubaneswar is well-established. However, in Orissa this caste is also stamped as load carriers (*bhāro boibā loko*). They carry the load on *'bāhuṅgī'* (*kāṅvara,* in which loads are tied or placed in a roap-cage and further tied with either ends of a flexible bamboo-stick forming the shape of a balance. People carrying such loads are also depicted in temple-art of Koṇārka as well as Bhubaneswar. Whether these depictions are of the Gaudas, is not sure in want of any label, but, still today they use to carry the loads and palanquins for people as well as for the deities in their villages and in neighbouring villages also. Therefore, *Gauḍas* were the *Pālakī-Āpaṭa* (palanquin bearer) of the deities during festivals at Bhubaneswar since olden days. They are also called *Bāhika* as referred to earlier.

Chāsā: Literally they are the cultivators, the word corrupted from *Sanskrit 'karshaka'* or *'krishaka'*. As they were doing physical labour they would also have shared jobs involving physical labour in the temple festivals. In present setup they were assigned the duty of carrying the *vimāna* (float) of the deities in the Liṅgarāja temple during different festivals (*yātrās*). We are well aware of the peasantry settlements in the neighbouring areas through inscriptions as well as antiquities.[140] They would have sponsored voluntary services initially but later on crystalized for particular services. Though, agriculture as a main resource of livelihood is professed by various castes all over India, however, mostly in all parts of the India some caste people are specially recognised as agrarian class, e.g. in western Uttara Pradesh a people-community call themselves '*kisāna*' (peasant).

Keuta (the Fishermen / the Ferrymen): They are very ancient inhabitant of *Ekāmra*-Bhubaneswar. The vast water-lakes and rivers in this area would have led them to settle here for fishing and navigating the ships of traders (of ancient settlement of Sisuplagarh). However, as soon as *Ekāmra*-Bhubaneswar emerged as leading religious settlement they also involved themselves with the activities related to the temples. Now, they row the barge of Liṅgarāja during the *Chandana Yātrā* (Sandal festival). They have their own hamlet in the *Ekāmra*-Bhubaneswar called Keuṭasāhaī.

Ratha Paika (Chariot Attendant) : They belong to the village Nuapalli and are *'khaṅdaita'* (warrior) by caste. Their surnames are: 'santara' (a corrupt form of *Saṅskrit* *'sāmaṅtarāya'*, *'paika'* (a soldier of infantry or foot-warrior), *'jenā'* (a commander), and *'routrā'* (corrupt form of *Saṅskrit* *'routarai')*, The only sacred duty allotted to them was/is to guard the ratha (car) of the deity Liṅgarāja during the car-festival. The *khaṅdatis* are the eminent class of people who were warriors by profession.

Guḍiā (the Confectioner): Literally it means a man working with *'gura'* (molasses). The *Guḍiās* are the businessmen whose traditional profession is parching the paddy as well as rice, in Orissa. They parch the paddy and separate the husk then mix molasses or sugar with the parched grains and thus prepare a dry-*bhoga* (dry-offering-food for the deity) i.e. called '*khoī*' or *ukhuḍā'* . This *'bhoga'* is very much popular in all over Orissa and offered to the deities in the temples and religious festivals. They also prepare a raw sugar-candy called *'korā'*. The duty of a Guḍiā is to supply these preparations for the morning-*bhoga* of Liṅgarāja. At present they also sell this preparation as morning-*bhoga* to the pilgrims. It is offered to the deity and carried by pilgrims as *prasāda* for distribution. In the Oriya folk-songs the *korā-khoī* of Bhubaneswar is highly praised and treated as special and tasty preparation of the place which suggests that the Guḍiās of Bhubaneswar mastered in the art of confectionary from very old tradition. Nowadays, a number of stalls can be seen in front of the Liṅgarāja temple selling this famous item.

The parching of cereals and other grains is a very ancient tradition. The parched paddy was called *lājā* in Vedic literature.[141] Therefore, the parchers' community would have developed and served the people since very early times. In Uttar Pradesh they are called '*Bhuñjavā*' (lit. parcher) and produce similar preparation after parching the various grains. Due to emergence of *Ekāmra*-Bhubaneswar as a *tīrtha* of frequent public visit the *Guḍiās* of the neighbouring area would have got a high selling market. Initially they would have come on festivals and auspicious days but later they became residents of the sacred site.

Puṭulī Baniā (Itinerant Spice Sellar): They supply the betel and spices in the service of Liṅgarāja. Literally '*puṭulī*' means a bag of cloth or canvas the mouth of which is tied with rope or thread and *Baniā* is a trader, thus *Puṭulī Baniā* means a trader who used to carry his goods in bags. In the bi-weekly *hāto* (a day long camp market) of Old Town of Bhubaneswar on the southern side of Liṅgarāja temple I have seen two such traders selling the various items of spices such as turmeric, coriander , cumin-seed , clove, cardamom and several other like articles. They keep these articles in the bags of various sizes .The hawker-sellers of spices and other related articles are in this profession by old tradition. They can also be compared with the *paṅsārī* people of Uttar Pradesh, but the Paṅsārīs deal in wide range of herbs, salts, and other various articles of medicinal uses; they also keep permanent shops and do not wander with their goods. Regarding the antiquity of the association of *Puṭulī Baniās* with the temple of Liṅgarāja, we can site a dedicatory record in the Liṅgarāja temple which records about the arrangements of *nitya-tāmbula* [daily betels] in the services of lord Kṛittivāsas [Liṅgarāja].[142] The inscription is dated in the reign of the Gaṅga- king Narasiṁha I, thus the services of the Puṭuli Baniās from very olden times are well-established.

Telī [the Oil-seller]: In the present setup they supply the oil [oil of *Polaṅga*-seeds or Colophyllum] for lighting the perpetual lamp and carry the traditional torch i.e. called *triśākakha* in the night. Their association with the holy-*tīrtha* of *Ekāmra* is very old and well-established . The offering of the lamp to the deity was very common in old days; and in most of the dedicatory records the arrangement for providing perpetual lamp *[akhaṅḍadīpa or akhaṅḍavartti]* are made. In this arrangement the oil- producers *[telī]* were granted the land or money in currency for regular supply of measured quantity of oil. For example, the Gaṅga king Anaṅtavarman Choḍagaṅgadeva had granted the village Devadharmasri for the maintenance of perpetual lamp[by the income of that village], this arrangement was made with the particular caste people [*sapāṭake: sa* means with and *pāṭaka* is for the particular caste for oil supply i.e. *telī*] for the supply of 24 kuñchāgra [an oil-mesure] oil for burning the perpetual lamp till the sun, moon and stars[143] last. In other dedications by public or royal persons the money in currency is deposited with trustees [and if it is a land-grant the management is left on the trustees] and they were asked to purchase the required quantity of oil from the *telīs* [the oil producer] and to supply in the temple for the said purpose. The origin of the *Telī* or *Tailī* community goes to the hoary past. We find the references of their guilds in inscriptions and literature which were spread in different parts of ancient India. At *Ekāmra* they had been an integral part of local population and temple-functionaries since its emergence as great *tīrtha.*

Kumbhāra (saṅs. Kumbhakāra, the potter): The potter was an essential part of human society in all over the world.. We find the use of earthen pots since the very dawn of human civilization. Therefore, a number of people would have crystallized as a community related with this craft. Infact there is no need of proving the potters habitation at Ekāmra-Bhubaneswar in ancient days by any evidence. But, still there are some important records witnessing the potters' services in the temple-town. For example in a dedication by Govinda Senāpati (dated 13th January 1239 AD) the potter *(kumbhakāra)* was gifted two *vāṭīs (vāṭī* is a land measure) of land at the village Lakshmisagara for the supply of[144] pots. There is no doubt that the potters' community had got a good market for their various kinds of pottery products and settled near the temple area. At present their hamlet is called Kumbhārasāhī and located near Keuṭasāhī.

Kammāra and Baḍheī (the Blacksmith and the Carpenter): We have already stated them as workers in temple constructions alongwith the *stapatis*_(the sculptors). The *kammāra* and the *Baḍheī* also bear the title of *mahārāṇā.* The *Baḍheīs* are associated with Liṅgarāja by constructing the *ratha* (car and the *chāpa (barge).* In addition, to the traditional services they attend to all the major and minor and minor repairs of the temple. The *kammāras* are for the help of Baḍheīs in construction of the car and the *chāpa* both the craftsmen existed in the society from very early times.

Bhaṅdārī (the barber): The barbers are associated with the priestly class for performance of various menial jobs in *Yajña* (sacrifices) and other ceremonial rituals. With the *tīrtha* or sacred places their association is long cherished as they do *kshaura-karma* (shaving the hairs of pilgrims).

Darjī (the Tailor): The traditional sacred duty of this group was to do all sewing for the deity of Liṅgarāja. For example, the royal insignias such as *chhatra* (umbrella), *trāsā* (a leaf-shaped object fitted with a long stick, a royal insigia carried with the deity in the procession) and *meghadambaru* (a cloth canopy) etc. are made by the tailors. They also stich the garments of the deity as well as the *patākā* (flags), *chāṅduā* (sun-screen) and also fix the coloured cloths round the ratha (car). For these services they receive a monthly remuneration and new *dhotī* after the *rathayātrā* every year. In addition to the sacred services, they supplied stitched garments to the local residents.

The tradition of wearing stitched clothes in India is of course of very late origin. We get only some stray references of stitching in the ancient literatures.[145] However, the umbrella and the royal insignias such as *trāsā* and *patākā* had prolonged association with the royalty as well as divinity. Making of cloth-ceiling for temporary shed and for decorative purpose variously called *Chāṅduā* and *vitāna,* is also an old tradition. Therefore, we may accept the existence of the *Daraji-mahārānā* in the old society of Ekāmra. Pipli a nearby village, which is said to have been once the Mughal headquarters in Orissa, is famous for such traditional crafts. It is not improbable that earlier they would have been working for the requirements of various temples scattered in the whole coastal belt later on they shifted to the Mughal court where they had much scope for their craft.

Samartha: This is an interesting surname as well as caste in the temple-town of Bhubaneswar. The *sevakas* (servants) of this group are allotted the following duties:

(1) Grinding of pulses and spices etc for the preparation of *bhoga*.

(2) Beat the drum and the bell during food-offerings.

(3) Supply water in the kitchen.

(4) To carry the great lamp (*mahādīpa)* to the top of the temple-spire on every fourteenth day of each fortnight.

(5) To tie the flag at the temple spire whenever required. The duties of the *samartha* are of laborious and acrobatic nature.

The very characteristic surname and the caste title '*samartha*' (lit. the capable one) of this group would have been result of performing the last two duties. In olden days the priestly class would have felt those jobs very difficult and when some adventurous person would have done this job he was given the title of 'samartha' (the capable one) and his services were regulated in the temple. In due course this job has become patriarchal and professional one.

Kahaliā: They play a wind instrument called *kahalai.* For their origin O' Malley writes, "They are believed to be descendant from the illegitimate children of the dancing girls attached to the great temples but they themselves not unnaturally deny this and confess to have come from the banks of Ganges."[146] In many sculptured scenes at Liṅgarāja temple itself the persons playing such wind instruments welcoming some royal personage have been depicted. Therefore, the people playing the instrument called 'kahali' existed in the tenth century AD also.

Dhoba (the Washerman): Though, notionally they are well known community engaged in washing of clothes all over India. But, in Orissa, in addition to the above job they are also linked with the job of wood-cutter. Usually, if the other Hindus cut the trees they will be committing the great sin but the *dhobās* and *mahārāṇās* are exempted from these sins. He is also one of the *sevakas (*servants) of Liṅgarāja.

The *Bāurīs* and the *Hāḍīs* (sweeper) once untouchables are also among the sevakas (servants) of Liṅgarāja. The *bāurīs* are already mentioned. They served in the following capacities: cutting the first long for the deity's car repairing of the road for the car, acting as brakesman of the car and cleaning the temple compound.[147] The *hāḍī's* services are also of ancient origin. They have their hamlet on the north-western bank of Bindusagara. The nearness of their settlement to the sacred site also confirms that the *hāḍīs* were the much earlier settler of *Ekāmra*-Bhubaneswar.

Some other people with various crafts and trades which would have been resided or sold their goods in the festive and auspicious days at *Ekāmra*-Bhubaneswar are as follows:

(1) *Chūrṇakāra*[148] (the lime maker from shells). 2) *Tāmilī* (*Tāmbulī:* betel dealer), (3) Mālī (the florist) (4) *Śaṅkhari* (shell-

bangle maker), (5) *Kachāra* (glass-bangle maker) (6) *Paṭāra* (dealing with coloured yarns, fillets etc.) (7) *Kharuḍa* (the bell-metal maker, the word *kharuḍa* is like the Hindi word '*kharāda*' which stands for peeling and polishing the bell-metal, therefore, the Kharuḍas would have been a ancillary unit of the *Ṭhaṭaras),* (8) *Ṭhaṭarī* (the bell-metal maker; in Uttar Pradesh *Ṭhaṭherā:* they are maker as well as the vendors) (9) *Kaṅsārī (*brass and bronze maker), (10) *Tambārī* (copper - dealer), (11) *Baisya Baniā* (established business class which are termed in literature as well as inscriptions as *śhreshṭhin),* (12) *Rañgani* (dyer), (13) *Luniā* (the salt maker), (14) *Buna (*the weaver) etc.

(v) Pilgrims

The pilgrims (*tīrtha-yatrī*) who visited the sacred land though for few hours or few days, had always increased the grandeur of this place and kept it lively. The sacred land of *Ekāmra* was visited and favored by many imperial kings, regional chiefs, royal families, officers of high rank and file, rich people, scholars, sages as well as by common folks. The pilgrims were the recurring feeding source for the various classes of people dependent for their livelihood through their services in temples as well as gaining income through business professions.

SUMMARY

The above discussions show that the sacred land of Ekāmra was gradually occupied by various classes of people. Not only the priestly people but the peasants, warriors, traders, craftsmen and laborers also equally shared the various duties fitted for their professions and ultimately became a part of that society which was based on religious economy. However, it is also a true fact that the demography of this settlement would have been always changing. The place had seen many ups and down. We know that under royal patronage during the various kings and dynasties, viz: the king Śaśāṅka, the Śailodhavas, the Somavaṁśī-king Yayāti II and Uddyotakesarī and during the Gaṅgas and the Sūryavaṁśīs, there were more people. It is also true that many times new stocks of priestly people and Brāhmaṇs migrated to this land and abandoned also. In the early sixth century AD the Śaivas came from Mathura or Kayarohana, the Somavaṁśīs brought the *Kānyakubja* brāhmaṇs. During the Gaṅgas the Śaivites also came from south India. The Sūryavavṁśī-king Kapileṅdra when constructed the Kapileśvara temple brought the *Malihā* priests and settled them in that area which is now called Kapileśwar village. It is also true that due to various temple jobs some new castes were also formed. The temples' concerned business helped in localisation and specialization of some particular craftsmanship. Due to foundation of military forts around this place under the Gaṅgas subsequently creation of Kṛittivās-Kaṭaka a number of royal and military people had come here and stayed here. But as soon as the court patronage lacked the settlement also suffered from partial abandonment though never deserted fully.

4. A SCENARION OF BHUBANESWAR AS RECONSTRUCTED FROM THE RELIEF ART, MONUMENTS AND INSCRIPTIONS

The majestic temples and pleasing tanks are the chief grandeur of Bhubaneswar which would have been attractive in the past when various modern construction and encroachments making clumsy appearance of the settlement were not in existence. The land was animated with ascetics' practices & amazing activities, religious teachers, students busy in discourses and chanting *mantras,* priests engaged in worship rituals and *devadāsīs* performing dances before the deities: ascetics, architects, masons and sculptors busy in the temple constructions; and crowding pilgrims taking dips in holy water then rushing to the temples for making offerings.

At present a number of religious festivals and fairs are celebrated at Bhubaneswar. These are categorized as following:[149]

(i) Weekly and Fortnightly festivals

(ii) Occasional Rituals

(iii) *Caturdaśa Yātrās (*Fourteen Major Festivals), and

(iv) *Upayātrās* (Minor festivals)

The weekly festival is called *Somabāra Beḍhā* (monday festival). We found that most of the dedications were recorded on Mondays and this day is auspicious for Śiva. Thought not listed in weekly festivals yet Thursday is also very important day at Bhubaneswar; and on these days a market is held near Liṅgarāja temple. No doubt that these festivals are very old celebration and had given rise to the bi-weekly market days *(hāto)* at Bhubaneswar in weekly cycle. The fortnightly festivals are celebrated on each *chaturdaśī* which are succeeded by *Amāvasyā* and *Pūrṇimā.* On these occasions large number of pilgrims assemble for offerings in the temple. We can guess that in past these days would have attracted large number of pilgrims.

The occasional rituals are *Deula Sodhā* (cleaning of temple). *Dhokāla* (temple purification), Special pūjā and fire sacrifice.

The fourteenth major festivals are: (1) *Prathamashṭamī*, (2) *Prāvaraṇa Sashṭī*, (3) *Makara Sakrānti*, (4) *Māgha Saptamī*, (5) *Śivarātri*, (6) *Asokāsṭamī*, (7) *Basanta Pañchamī*, (8) *Damanka chaturdaśī*, (9) *Akshaya Tṛitīyā*, (10) *Paraśurāma Ashṭamī*, (11) *Śayana Chaturdaśī*, (12) *Pabitrāropaṇa*, (13) *Yama-dvitiyā* and (14) *Uttahāpana Chaturdaśī*.

The *upayatras* are seven in number, viz: (1) *Sarasvatī Pujā*, (2) *Bhaumī Ekādasī*, (3) *Kapila Yātrā*, (4) *Dola Purnimā*, *(5) Śiva Vivāha*, (6) *Janmāshṭamī* and (7) *Durgā Ashṭamī*.

These are traditional festivals and some of them are of much early origin. The Brahampuraṇa [150] refers to the Māgha Ashṭamī, Soler and Lunar eclipses, *Saṅkrāntis* of *Tula* and *Mesha* (the crossings of the Libra and Aries), the *Sankrantis* of *Karka* and *Makara* (the crossings of the Cancer and the Capricorn), *Akshaya Tṛitīya; Mithuna, kanyā* and *mīna rāśis* (the zodiacs of the Gemini, Virgo and Pisces) as auspicious days for ablutions, homage and offerings. Further, the *Devaśāyanī* and the *Devotthāna ekādaśīs*, *Asokashṭamī* and *Pavitrāropana* ceremony are also referred to as very auspicious, A dedicatory record in the wall of Liṅgarāja temple, datable in AD 1239 of thee reign of Aniaṅkabhīmadeva, informs us that arrangements were made for the celebration of *Vivāhotsava* (marriage festival of Śiva) by Goviṅda Senāpati by making the land-grants in several respects.[151] In this record the marriage festival is said to be the greatest festival (*Parvostarvamahotsavam Vivāhotsava*). On this occasion the donor had made arrangements for the roofing of the pavilion, and for the supply of earthen pots, lime and water. Śiva vivāha-scenes are frequently depicted in the narrative art of the temples of Bhubaneswar. We get it in the early temple like Paraśurāmeśvara as well as in the Liṅgarāja temple which is the climax of developed phase These evidences show that the festival was very much popular and was celebrated with great pomp and show.

During the festival-processions the *chalaṅtī pratimā* (proxi deity of Liṅgarāja altogether with Kapilanātha and Anaṅtavāsudeva are taken to different temple institutions in different occasions in a procession accompanied with the priests, servants and music band consisting of trumpets and drums. The deities are placed in beautiful palanquins carried by the traditional temple servants. During the *Asokāshtamī* the deities travel to the *Mausī Mā maṅdira* (Rameśvara Temple) being seated in big *ratha* (car). Thus the deity of Liṅgarāja visits the Maitreśvara on *Pṛathāmashṭamī* to Bhashkareśvara and Kedāra-Gaurī on *Māghasaptamī* to *Tīrtheśvara* on *Damanaka-Chaturdaśī* to Yameśvara on *Yama-dvitiyā,* to Bhīmeśvara on *Bhaumī-ekādaśī,* to Kapileśvara on Kapilayātrā and in *Śiva-vivāha* the deity goes as groom to Mukteśvara temple.

During the celebration of some of these festivals large fairs are being held. Thus, on the occasion of Asokāshṭami and *Śivarātri* large fairs are held around the temple of Liṅgarāja. Similarly, during *Māha-saptamī* a large fair is organized at Khaṅdagiri and one the occasion of *Damanaka-chaturdaśī* a such fair is celebrated at Bhīmeśvara area. These are the traditional fairs; and though city of Bhubaneswar, being the capital of the State of Orissa, is fully modernised the nature and scope of the fairs as well as the craze of the public for these fairs hardly witness any change.

Looking in to the celebrations of the above festivals and fairs one can directly peep in to the past that was the life at Bhubaneswar. Further, some animals like monkeys and aquatic creature like tortoise were also the attraction for pilgrims; sometimes these animals presented very interesting and amazing scenes. One such scene is described in the Anaṁta Vāsudeva Temple inscription of Chaṁdrikā devī while narrating the grandeur of Bhubaneswar, the composer of the inscription writes, "See how the female monkeys jump on the back of old turtles while swimming on the surface of its water and make them into pleasure boats, when the turtles dive in, the monkeys become submerged and jump on the back after their monkey like diving the water, beholding all these the crowd on the bank (of Bindusagara) are thrown into consternation."[152]

The sacred tīrtha of Bhubaneswar was visited by many kings and members of royal houses as well as by the high ranking officers, military commandors and various other officers serving in different capacities.[153] As per traditions the Gauḍa-king Śaśāṅka, the Somavaṁśī-king Yayāti and the Gaṅga-king Narasiṁha had very intimate association with Bhubaneswar. Further, the epigraphic records inform us about the activities or visits of a number of rulers, their kith and kins and officers viz: Śāṅtikara of Yamagartta maṅdala, Kolāvatī, the queen-mother of Somavaṁśī-king Uddyotakesarī, Gaṅga-king Choḍagaṅga, rājaṇ Pramādi, the younger brother of king Choḍagaṅga, the son of rājan Pramādi, Vīrachoḍa, the son of Kulottuṅga Rājeṅdra Choḍa (he was also the father-in-law of Choḍagaṅga), Bhīmadeva, (the Chālukya Bhīma with whom the Gangas had matrimonial relations), Bavila Barīshu (a ruler of the *aṅtara*-branch of Choḍagaṅga) Svapneśvaradeva, the Commander of the Gaṅga-armies, Goviṅda Senāpati, the Commander of the Gaṅga-armies in the reign of Aṅgabhīmadeva, Chaṅdrikādevī, the daughter of Bhānudeva and Kapileṅdra. No doubt, there would have visited many more such royal personalties of their age though the evidences of their visits are yet to be traced. The visits of these royal and rich persons would have proved very special and great occasion and charm. A depiction of such royal visit available in the wall of Liṅgarāja temple. In this scene the king is shown being received by religious teachers/priests and public with great reverence and welcome by the drummers and musicians while crowd is shown overwhelming to get the glimpse of their king. Thus, it is evident that such royal visits were not less than any festival and made lively the temple city.

Bhubaneswar was in a near approach of the Śailobhavas as well as the headquarters of the Bhaumakaras and the Somavaṁśīs i.e. Jajpur. During the time of the Gaṅgas and the Suryavamsis the royal headquarter shifted to Cuttack and this nearness provided the opportunity of more frequented visits of royal people and their associates. On the other hand military establishments of the Gaṅgas and the Suryavaṁśīs, viz: Chudagagarha (near modern Baranga Railway station, Bualigarha (at Chhandaka forests) and Kṛittvās-Kaṭaka (at Bhubaneswar) all located at or near Bhubaneswar contributed special charm and a new colour to the social atmosphere of Bhubaneswar. During those days the temple-city was pulsating with the movements of the civil and military officers of the kingdom.

NOTES AND REFERENCES

1. A.K. Rath, "A note On the Vigraha Dynasty", *Orissa Historical Research Journal* Vol. X1, op. cit., pp. 54-9.
2. S.N. Rajaguru, *Inscriptions of Orissa*, Vol.I, pt. ii, pp. 113-6.
3. ibid., pp. 54-6 and 63.
4. ibid., pp. 120-3.
5. ibid., 117-9.
6. ibid., pp. 124-7.
7. ibid., p. 264.
8. ibid., pp. 157-61. See also, E. Hultzsch, "Plates of the Time of Śaśāṅkaraja, Gupta Samvat 300", *Epigraphica Indica* Vol. VI (1900-1901), Calcutta, pp. 143-6.
9. ibid., pp. 141-3.
10. ibid., pp.136-40. See also, N.G. Mazumdar, "Four Copper Plates From Soro", *Epigraphica Indica* Vol. XXIII (1936), Delhi, 1938, pp. 203-3.
11. ibid., pp. 178-85.
12. R.D. Banarji, *History Of Orissa* Vol. I, p. 141.
13. S.N. Rajaguru, *Inscriptions Of Orissa, Vol. I*, pt. ii, pp. 178-85.
14. ibid., pp. 162-5.
15. ibid., Vol. II, p. 332.
16. Thomas Watters, On *Yuan Chwang's Travel In India,* Vol. II, (original London), Indian edition Delhi, 1961, pp. 196-7.

17. S.N. Rajaguru, *History Of the Gangas,* Vol.I, Bhubaneswar, 1968, p. 61.

18. D.K. Ganguly, *Historical Geography And Dynastic History of Orissa*, Calcutta, 1975, pp. 10-2 and 62.

19. S.N. Rajaguru, *Inscriptions of Orissa,* Vol. I, pt. ii, pp. 186-92.

20. ibid., pp. 199-205.

21. ibid., pp. 229-34.

22. ibid., Vol. IV, pp. 39-42. See also, Balachandra Jain, "Adhabhara Plates of Maha- Nannaraja", *Epigraphica Indica* Vol. XXXI (1956), Delhi, 1957, pp. 219-22.

23. ibid., Vol. I, pt. ii, pp. 206-12.

24. K.C. Panigrahi, *Archaeological Remains At Bhubaneswar,* op. cit., p. 229.

25. ibid., p. 11.

26. ibid., p. 230.

27. S.N. Rajaguru, "The Dasapalla Copper Plate Grant of Satrubhan jadeva", *Orissa Historical Research Journal,* Vol. I Bhubaneswar, 1952, pp. 208-13.

28. S.N. Rajaguru, *Inscriptions of Orissa*, Vol. II, pp. 258-63 (Badakhimidi Plates of Jayavarman).

29. K.C. Panigrahai, *Chronology of the Bhaumakaras And the Somavaṁśīs of Orissa,* Madras, 1961, pp. 8-9; and History of Orissa, op. cit., p. 77.

30. T.E. Donaldson, *Hindu Temple Art of Orissa, Vol.* I op. cit., p. 70.

31. D.R. Bhandarkar, "Sanjan Plates of Amoghavarsha I: Saka Samvat 793", *Epigraphica Indica,* Vol. XVIII (1926), p. 245, verse 24: *....sakosala-Kaliṅga-Veṅgī-Dahalaudrakan Mālvāṁ vilabhya ni jasevakaih.....*; and translation on p. 253.

32. K.C. Panigrahi, *Chronology Of the Bhaumakaras And the Somavaṁśīs Of Orissa*, op. cit., pp. 27-8.

33. ibid., p. 29.

34. ibid., p. 7.

35. D.C. Sircar, "Śantirgrama Grant of Dandimahadevi", *Epigraphica Indica*, Vol. XXIX (1951-52), Delhi, 1957, pp. 79-89; and F. Kielhorn, "Two Grants Of Dandimahadevi", *Epigraphica Indica*, Vol. VI (1900-1), pp. 133-42.

36. F. Kielhorn, "Bilhari stone Inscription of The Rulers Chedi" *Epigraphica Indica*, Vol. I, Calcutta, 1892, p. 256.

37. S.N. Rajaguru, "*Chronology of the Somavaṁśī kings of Orissa*" Orissa Historical Research Journal, Vol . VII (1959), pp. 103-4.

38. J.F. Fleet, "Records of the Somavaṁśī Kings of Kataka", *Epigraphica Indica*, Vol. III (1894-95), p. 352.

39. T.E. Donalldson, *Hindu Temple Art of Orissa*, Vol. I, op. cit., p. 196.

40. S.N. Rajaguru, *Inscriptions of Orissa*, Vol . IV, op. cit., pp. 374-5.

41. ibid., pp. 376-8.

42. K.C. Panigrahi, *Archaeological Remains at Bhubaneswar*, op. cit., p. 47.

43. S.N. Rajaguru, *Inscriptions of Orissa*, Vol. IV op. cit., pp. 227 and 232.

44. ibid., p. 389.

45. ibid., Vol. II, p. 341.

46. ibid.

47. ibid., p. 345.

48. ibid., p. 332.

49. ibid.

50. ibid.

51. ibid.

52. ibid.

53. ibid.

54. ibid.

55. ibid., pp. 337-8.

56. ibid., p. 338.

57. ibid.

58. ibid. pp. 338-41.

59. ibid., Vol. III, pt. ii, p. 365.

60. ibid., p. 366.

61. ibid.

62. ibid., p. 368.

63. ibid., p. 385; and pp. 19-20 in the pt. i of the same vol.

64. ibid., pt. i, p. 6.

65. T.E. Donaldson, *Hindu Temple Art of Orissa,* Vol. I, op. cit., p. 365.

66. K.C. Panigrahi, "Temple Inscriptions From Bhubaneswar", *Orissa Historical Research Journal,* Vol. I, No. 2. (1952), pp. 1-7.

67. S.N. Rajaguru, *Inscription of Orissa,* Vol. III, pt, i, pp. 66: the inscription is dated in Śaka year 1035 i.e. A.D. 1113 when king Anantavarmadeva (Choḍagaṅga) was ruling in Kaliṅga.

68. ibid., pp. 185-6; dated corresponds A.D. 1136.

69. ibid., p. 135, dated in Saka era 1053 i.e. A.D. 1131.

70. K.C. Panigrahi, *Archaeological Remains at Bhubaneswar,* op. cit,. p. 253.

71. S.N. Rajaguru, *Inscriptions of Orissa,* Vol. III, pt, ii, op. cit, pp. 394-6; and History of the Gangas, Vol. II, Bhubaneswar, 1972, p. 31.

72. R.D. Banarji, *History of Orissa,* Vol. V. pp. 258-9.

73. S.N. Rajaguru, *Inscriptions of Orissa,* Vol. V, pt, i, op. cit., pp. 60-1; see also *Epigraphica Indica,* Vol. XXIX, pp. 121-33.

74. ibid.

75. ibid., (Ananta Vasudeva Temple at Bhubaneswar) pp. 211-20; and K.C. Panigrahi, "Sonepur Stone Inscription of Bhanudeva", *Orissa Historical Research Journal,* Vol, X (1962), pp. 9-10.

76. R.D. Banarji, *Histoty of Orissa,* Vol. I, op, cit., pp. 264-5.

77. T.E. Donaldson, *Hindu Temple Art of Orissa,* Vol. I, op. cit., pp. 710-11.

78. ibid.

79. ibid.

80. ibid.

81. ibid.

82. ibid., p. 712.

83. ibid.

84. ibid.

85. ibid.

86. ibid.

87. S.N. Rajaguru, *Inscriptions of Orissa,* Vol, I, pt. ii, op. cit. See the copper plate grants of the Matharas, the Vigrahas and the Sailodbhavas.

88. Thomas Watters, *On Yuan Chwang's Travels In India,* Vol. II, op. cit, pp. 193-8. and see also, R.D. Banarji, *History Of Orissa,* Vol. I, op. cit. p. 141.

89. S.N. Rajaguru, *Inscriptions of Orissa,* Vol. I, pt. ii, op. cit., pp. 224-8.

90. R.D. Banarji, "Inscriptions In The Udaygiri And Khandagiri Caves", *Epigraphica Indica,* Vol., XIII, op. cit., p. 167.

91. ibid., p. 166.

92. D. Mitra, Udaygiri and Khandagiri, op. cit. p. 6.

93. ibid.

94. D. Mitra, "Buddhist Vestiges In Orissa: A Survey", *Orissa Review,* (Monument Special), Bhubaneswar, 1976, p. 32.

95. K.C. Panigrahi, *History of Orissa,* op. cit. pp. 313-4.

96. ibid., pp. 53-9.

97. D. Mitra, *Bhubanesawar,* op. cit, p. 42.

98. K.C. Panigrahi, *History Of Orissa,* p. 321.

99. ibid., p. 320.

100. ibid., p. 131.

101. ibid., p. 331.

102. ibid.

103. ibid.

104. ibid.

105. ibid.

106. ibid., p. 340.

107. S.N. Rajaguru, *Inscriptions of Orissa,* Vol. V, pt, i., op. cit. pp. 211-20.

108. K.C.Panigrahi, *History of Orissa,* op, cit, p. 343.

109. ibid., pp. 342-3.

110. ibid., p. 343.

111. ibid.

112. ibid., p. 342.

113. ibid.

114. ibid., pp. 343-7.

115. The cumulative effects of the influences of the Jagannatha cult on the Lingaraja temple have been that it has lost its distinctive character of a Śiva shrine and has adopted the cosmopolitism of the Jagannatha at Pur, The Sudra priests, known as Vadus, who have been described in the sixty-second chapter of the *Ekāmra Purāṇa* as the descendants of the Savara mother by a Saiva saint known as Siddhabhuti, are like the Daipatis of the Jaggantha temple, still the custodians of the Lingaraja shrine, although the Brahmins have also taken a share in the worship in recent times. Notwithstanding the well-known orthodox dictum that the offerings given to Śiva are not to be partaken of by any Hindu, even the cooked rice offered to Brahmins. The custom is certainly analogous to the one prevalent in the Jagannatha temple where, while partaking of the cooked rice offered to the god, no caste distinction is observed.

     That these influences of the Jagannatha of the Jagannatha cult are not of recent origin in the Lingaraja temple, is proved by the *Ekāmra Purāṇa* and the allied works. They are unanimous in their assertion that no real distinction exists between the Saiva and Vaishṇava and that it is only superficial people who find a distinction in them. Thus says the *Ekāmra-Purāṇa*:

     "There is not distinction between Vishnụ and Śiva. This is the eternal Dharma and the man who observes this Dharmạ attains merit."

In another place the same authority seeks to synthesize the four main cult, Vaishṇavism, Śaivism, Saktism and Saurism:

"O *Goddess* (*Pārvatī*), you are *Vishṇu*; and *Sūrya* and Achyuta are my emanations. The intelligent people say that there is no distinction between them. (i.e. between *Vishnu* and *Sūrya*), So *Vishṇu* and *Śiva* split the one body into two. The god (thus became) Ardhanarisvara and (in him) the female portion is really *Janārdana.* O goddess (*Pārvatī*), do not see any distinction, (for) he who is *Vishṇu,* is also *Maheśvara.* The vicious fool who makes a distinction goes to hell."

Instances can be multiplied from the same work to show that it takes all the main cults to be one and the same cult. It also utters warnings against the followers of Vishnu who was to vilify Saivism. It says:

"In the *Kali* age those persons who have been Vaishnavas, become the Villifiers of Śiva, will certainly go to hell. There is no doubt about it."

On the ground that the Liṅgarāja is the combination of *Vishṇu* and *Śiva,* it also justifies the partaking of the offerings made to him;

"*Bhubanśvara* is no *lingam;* (he) is the image of *Parama Brahma.* (So) the partaking of the offerings made to him leads to the merits that accrue from a great sacrifice."

The Svarnadri-Mahodaya in its sixth chapter further explain the reasons why the offerings given to *Bhubanesvara* are not to be tabooed. It says:

"This *lingam* is neither god, nor *Rudra,* nor *Mādhava.* Since half of each exists to a lingam are not to be partaken of, (but) *Bhubaneśvara* is not a *lingam,*. So. O son, eat the offerings with the gods and demons."

Culturally the Suryavamsi period merges into Ganga epoch, but towards the end of this period Vaishnavism became predominant in Orissa on account of the visit of Sri Chaitanya to this country his long so journ at Puri and his great influence on Prataparudradeva (A.D. 1497-1541), the great king of the dynasty. So the ascendancy of the Jagannatha cult popularly affiliated to the *Vaishṇava* cult, was more in evidence at Bhubaneswar. The earlier work *Ekāmra Purāṇa*, while describing in its 18th and 22nd chapters the procedure of visiting the main shrine enjoins that, a pilgrim after having a dip in the Vindusarovara, must first see Ananta-Vasudeva and then go to the shrine of Lingaraja. Though the procedure gave the first precedence to a Vaishnava shrine, nonetheless Ananta-Vasudeva was never conceived as the supreme deity of the place. On the other hand, the work has consistently maintained that at the request of Śiva, Vasudeva with his brother Ananta Agreed to fix his abode at Ekmra, and to serve as kshetrapal or the protector of the place. It thus maintains that Bhubaneswar was mainly a Śiva shrine and Vashnavism was allowed to exist here on equal terms. But this position was soon to change alongwith the growing popularity of Vaishnavism as a result of Sri Chaitanya's activities and the patronage of Vaishnavism by Prataparudradeva under the former's spiritual guidence. The *kapila-saṁhitā* gives an account of Śiva's coming to Ekamra which , on account of its historical significance, is worth quoting here in its main outline :

In the age of Treta Śiva once told Narada that, since Baranasi had become overcrowd, he would not like to stay there any longer and would choose a solitary place for his abode. Narada told him that there was a beautiful place known as Ekamra situated to the north of Nilachala on the sea-coast and that Vasudeva with his brother had been living there (at *Ekāmra*). If he wanted to fix his abode there he should first go to

Vasudeva, practise pennance to propitiate him and obtain his permission to stay there. According to the advice of Narada Śiva went to Ekamra and at his first meeting with Vasudeva, fell at his feet and propitiated him with various prayers. *Vasudeva* atleast agreed to assign him a place at Ekamra, but not without a condition. The condition was that he must not try to go Banarsi again. Siva agreed to the condition and was therefore given a place in *Ekāmra.*

The inner significance of the story is that Ekamra was originally a Vaishnava shrine and that Saivism was allowed to exist there after a compromise and certain conditions. This, however, is just contrary to the archaeological evidence that we find with regard to its origin.........

115. S.N. Rajaguru, *Inscriptions of Orissa*, Vol. I, pt, ii, op. cit., pp. 191-8. verses -14-15 :

केचिद वन्य मृगेन सार्द्धम चर(न) स्तातां

स्थितिलीलया केचिच्चोर्द्धमुखा सहस्रकिरण

ज्वालावली प्रेषिण: केचिद वल्कलिनस्थिजिनधरा:

केचिद जटाधारिण: नाना रूपधारास्तपन्ति

मुनयो दिव्यास्पदा कांक्षिण: ।। 14 ।।

केचिच्छैल गुहोदरेषु नियता धूमावलर पायिन:

अन्यवायु फलाम्बुभक्ष निरता केचिन्निराहारका: ।

इन्थं योग जुगो विहाय बसतिं ध्यायन्ति

दिव्यं पदं चित्तं मध्यमराज देवगुण धृग्रा ज्येति तत्प्राप्तबान् ।। 15 ।।

116. ibid., pp. 199-205.

117. ibid., pp. 218-40.

118. D.C. Sircar, "Two Plates From Kanas", *Epigraphica Indica*, Vol. XXVIII (1949-50), Delhi, 1958, pp. 329-31: *nana-gotra maitrayaniya-chhatra-brahmanam..........*

119. S.N. Das, "The Inscription of Prapannacharya In The Temple Of Parasuramesvara', Orissa Historical Research Journal, Vol. II, No. 384, (1953), pp. 46-8.

120. S.N. Rajaguru, "The Lingaraja Temple Inscription of The Time Of Rajarajadeva (s. 1094)", *O.H.R.J.* Vol I X, No. 1, pp. 8-9; see also *Inscriptions of Orissa*, Vol. III, pt, ii, pp. 309.

121. ibid., Vol. V, pt. i, pp. 314.

122. ibid., Vol. III, pt. i i, pp. 336-7.

123. ibid.

124. ibid, pt. i, p. 35.

125. M.M. Mahapatra, *Traditional Structure And Change In An Orissan Temple,* op. cit., pp. 92-8.

In the present setup of Liṅgarāja temple the priestly class of Bhubaneswar is divided into four major categories; viz. (i) *Kshetrabāsī,* (ii) *Pūjāpaṇḍā,* (iii) *Mahāsua* or *suar* and (iv) *Baḍu.*

(i) *Kshetrabāsī* (lit, sacred city-dwellers): They are vedapāṭhin brahmans and their duty is to read Veda daily in the audience hall of the temple. The *Kshetrabāsīs* are/were entrusted with all the priestly services in the ritual cycle. In addition to the temple duties, they are privileged to use the temple-tank Bindusagar to perform to sacred rituals for and on behalf on the pilgrims and clients. These rituals include: libation of water (*tarpaṇa*), food-offerings to forbearers (*piṇḍa*), *saṁkalpa* etc. Obviously this category is comparable to our previous mentioned brahmanas who devoted in learnings of *Vedas* and *Śāstras*. They were the *āchāryas* and well-versed in the *Śastrīya-Karmakaṇḍa* and Vishṇu Āchārya who conscrated the Megheśvara temple.

(ii) *Pūjāpaṇḍā*: They are another class of priests whose main traditional sacred duties are to worship and present the food to the deity and to act as supervisor of the temple. He also sprinkles water and flowers on the car during *chherā-panharā* ritual on the occasion of Car-festival. Obviously the job of Pūjāpaṇḍā involves more physical work than mental, his duties show that he was a person who observed the daily customary worship. Thus he was below the learned brahmans. The old term *Pujārī* which is also more common than the Pūjāpaṇḍā stands for these people.

(iii) *Mahāsuara or Suara* (the great cook): Their main job was to cook the food offerings to the deity and they are also privileged to sell the *Mahāprasāda.* However, they are also assigned to sell the *Mahaprasada.* However, they are also assigned other different services in daily rituals, viz. decoration of the place of offerings, supply of water to Pūjāpaṇḍā at the time of offerings, providing wooden seat (*pīdhā*) to *Pūjāpaṇḍā*, keeping guard to the main entrance to the deity during the offerings, supplying washed clothes in a special folded manner, holding of umbrella and *trasa* (insignia) of deity so on and so forth.

(iv) *Baḍus*: It is cited earlier that they originated from a *Śaiva* saint named Śivabhuti and a Śavara woman (see, ref. 114). They are called assimilated brahmans and originally assigned the five different duties, namely : attend the bath of the deity, clothing and giving of Bilvapatra to the deity, guarding the inner sanctum of the temple, arranging the bed of the deity, (*khāṭaseja*), cleaning of the floor of the cella and to screen the main entrance of the cella at the time of food-offering.

The duties of the Suara and the Suara and the Badus are of similar nature and they may called as the assistants of Pūjāpaṇḍā. Though their duties are of menial nature in comparison to *Kshetrabāsī* Brahmans and *Pūjāpaṇḍās* but their performances in the temple are of priestly nature.

Crystallization of such a interwoven class with particular assignment would have been a result of a long patriarchal tradition. In the initial stage we do not think that such elaborate rituals and their performers would have been in existence. The daily rituals would have been of simple nature and more stress was given in the studies of Vedas, *Śāstras* and religious discourses as evident from the epigraphical narrations and *Sikshādāna* sculptures. Socio-economic conditions of the latter period would also caused a long list of priestly people, who have no other way to earn their livelihood than serving in the temples.

126. S.N. Rajaguru, *Inscriptions of Orissa,* Vol. IV, op. cit, p. 248, verser-15 :

    रत्नालंकृत भूषिताङ्गसुषमा देदीप्यमानादि (व क्रीडन्य)

    स्तड़ित: स्थिरा इव कुचश्रोणीभरव्याकुला: (सुन्दर्योक्षि)

    कनीनिकां इव दृशमन्त: प्रविष्टा नृणस्मै

    चंञ्चलकञ्जाभ्र नयना दत्तास्तया दारिका: ।। 15 ।।

127. ibid., Vol. III, pt. ii, p. 335, verses

    यासां नेत्रांचल तरलिमा विश्ववश्यैकमन्त्र:

    पादन्यासस्त्रिभुवनगनिस्तम्भनं संबिधते ।

    नृत्यारम्भे बलयमणिभिर्निर्मिताऽयत्न दीपास्तस्मै

    दत्तस्त्रिपुरबिजयिने तेन तास्ता मृगाक्ष: ।। 25 ।।

128. S.N. Das, "The Inscription Of Prapannacharya In The Temple Of Parasuramesvara", *Orissa Historical Research Journal,* Vol. II, op. cit., pp. 46-8.

129. D.C. Sircar and J.C. Ghosh, "Three Lingaraja Temple

Inscriptions", *Indian Culture*, Vol. III, po. i, (A Journal of Indian Research Institute), (reprint) Patan, 1984, p. 126.

130. ibid.

131. S.N. Rajaguru, *Inscriptions of Orissa,* Vol. V, pt, i. op. cit., pp. 306-14.

132. ibid., Vol. III, pt. i, p. 309. See his article, "The Lingarja Temple Inscription of The Time of Rajarajadeva (s. 1064)", *O.H.R.J.* Vol. IX, No. 1, pp. 8-9.

133. ibid., pt. i. p. 218.

134. ibid, pt. ii, p. 294.

135. We notice in the dedicatory record at Parśurāmeśvara temple (see, *Orissa Historical Research Journal* , Vol. II, op, cit., pp. 46-8) that the arrangements of the supply of food-stuffs were made with the *Kshetrapālā* or the person who was looking after the field. This suggests that though the land belonged to the temple but it was maintained and cultivated by a farmer. The monks attached with the temple were getting the land-produce in a fix quantity.

136. B.V. Nath, "The Parvati Temple At Bhubaneswar", *Orissa, Historical Research Journal* Vol. III, no. 3. 1954, pp. 146-9.

137. S.N. Rajaguru, *Inscriptions of Orissa,* Vol. V, pt. 2. Bhubaneswar, 1975, p. 610.

138. M.M. Mahapatra, *Traditional Structure And Change In An Orrisan Temple,* op. cit. pp. 98-114, 174-6 and 190-1.

139. D.C. Sircar, and K.G. Krishanan "Bhubaneswar Inscription of Ganga Narasimha", *Epigraphica Indica* Vol. XXXII, (1958), Delhi, 1959, p. 232; and see also S.N. Rajaguru, *Inscriptions Orissa*, Vol. V pt. i, op. cit., p. 314.

140. S.N. Rajaguru, *Inscription of Orissa,* Vol. III, pt. 2. op. cit., p. 309 and Vol. V. pt. i, op. cit., pp. 118-9.

141. R.C. Mazumdar, (ed.) *The Vedic Age* (Bhartiya Vidya Bhavan), London, 1951, pp. 393 and 519-20.

142. S.N. Rajaguru, *Inscriptions Orissa* Vol. V, pt. i, op. cit., pp. 124-5; and see also, D.C. Sircar and J.C. Ghosh, "Three Liṅgarāja Temple Inscriptions", *Indian Culture* Vol. III, pt. i, op. cit., pp. 124-5.

143. K.C. Panigrahi, "Three Temple Inscriptions From Bhubaneswar", *Orissa Historical Research Journal* Vol, I. No. 2, op. cit., p. 7., fn. 20 and 21.

144. S.N. Rajaguru, *Inscriptions of Orissa,* Vol. V, pt, i, pt. i, op. cit., pp. 118-9.

145. Moti Chandra, *Costumes, Textiles, Cosmetics and Coiffure In Ancient India,* Delhi, 1973,

The archaeological evidences of stitching tool i.e. a needle with an eye on the other end, are reported from the sites namely; Bur jahom (J & K), Bagor (Rajasthan) and Chirand (Bihar) from the Neolithic levels. These are made out of bone. From Harappa and Lothal the similar needles are found in copper. From Lothal a needle is found with an eye hole in the pointed end. See, A. Ghosh, An Encyclopedia of Indian Archaeological, Vol II, New Delhi, 1989, pp.

146. M.M. Mahapatra, *Traditional Structure And Changes In An Orrisan Temple*, op. cit., p. 110.

147. S.N. Rajaguru, *Inscriptions of Orissa,* Vol. V. pt. i, op, cit., pp. 118-9.

148. ibid.

149. M.M. Mahapatra, *Traditional Structure And Changes In An Orissan Temple*, op, cit,. pp. 53-90.

150. *Brahmapurana*, Ch, 41, op. cit,. pp. 257-8. verser-54, 55, 57, 58 and 72.

चकार ऋषिभिः सार्धतेन बिन्दुसरः स्मृतम् ।
अष्टम्यां बहुले पक्षे मार्गशीर्षे द्विजोत्तमाः ॥ 54 ॥
यस्तत्र यात्रां कुरूते विषुवे विजितेन्द्रियः ।
विधिवाद्बिन्दुसरसि स्नात्वा श्रद्धासमन्वितः ॥ 55 ॥
स्नात्वैवं विधिवत्तत्र सोऽश्वमेधफलं लभेत् ।
ग्रहोपरागे विषुवे संक्रान्त्यामयने तथा ॥ 57 ॥
युगाषु षडशीत्यां तथा न्यत्र शुभे तिथौ ।
ये तत्र दानं विप्रेभ्यः प्रयच्छिन्ति धनादिकम् ॥ 58 ॥
शयनोत्थापने चैव संक्रान्त्यामयने तथा ।
अशोकाख्यां तथा ष्टम्यां पवित्रारोपणे तथा ॥ 72 ॥

151. S.N. Rajaguru, *Inscriptions of Orissa,* Vol. V, pt. i., op. cit,. pp. 118-9.

152. ibid., pp. 211-20, see, verses-12 :
यस्यान्तः प्लवमान वृद्धकमठी पृष्ठन्थत्न स्थायुक
प्रेङ्खन्दानर कामिनीभिरकलि क्रीडेडुप प्रपिता।
मग्न यामनुज्जता (ती) भिरिह तत कापेय परिप्लवा
दुतप्त्नूत्यभिमुख प्रतीर लोलोकः समुत्र (त्रा) स्यते॥

153. Information about some distinguished visitors is available from the dedicatory records and euologies, most of these records are compiled in the *Inscriptions of Orissa* Series, Vol. III, IV and V, edited By S.N. Rajaguru.

6

End of Hindu Rule and Aftermath

After the death of Pratāparudradeva, and the murder of his tow sons the minister Govinda Vidyādhara himself ascended the throne. It was throne. It was a crucial period in the history of Orissa as her independence was threatened not only from the north-east and north by the Muslim-rulers of Bengal and Delhi, but also from the south by the kings of Golkunda. Vidyādhara's reign (AD 1540-49) was marked by not only incessant attacks of the Muslims but also by internal rebellion. The absence of effective and recognised central authority continued until the emergence of Mukundadeva Harichandana, a Telanga by birth who ascended the throne in circa AD 1559-60. Though northern orissa and fallen to Sultan Ghiyasuddin Jalalshah of Bengal, Mukundadeva succeded in driving the Muslims out and reoccupied the northern districts up to the Bhagirathi river. In the eighth year of his reign he even advanced as far as the Ganges and concluded an alliance with emperor Akbar.

But Mukundadeva was attacked by Sulaiman Karrani the Afghan Sultan of Bengal in AD 1568, who sent an expedition under the command of his son Bayzid. Mukundadeva sent two chosen officers to resist the march of the invaders. But they sided with enemy and even instigated them to attack their own master, Mukuṅdadeva took refuge in the fort of Kotsimha. The Muslims forces under Bayzid and Kalapahāra advanced up to Cuttack and laid seize to it. In the battle the warden of Cuttack fort was killed. The royal palace was stormed and the iconoclastic activities were conducted throughout the country by the Muslims. At this time the commandant of Sarangagarh fort Rāmachandra Bhaṅja revolted and proclaimed himself the king of Orissa. In this civil war Mukundadev was killed by Rāmachandra near Jajpur and the latter was killed the same day by the forces of Bayzid.

Two independent kingdoms were founded upon the remains of the empire of Pratāparudradeva, one in the south and the other in the north, both of which claimed to be the (Sūryavaṁśī) Gajapatis. In the north the chiefs of Khurda were given the rank of Rājā of Orissa by the Mughal emperor Akbar at the intercession of Mansingh who settled the ensuing dispute between Rāmachandra and the heirs of Mukumdadeva in favour of the former. Orissa, thus, lost her independence with the administration at Khurda passing in to the hands of the Mughals and the southern dominions being in control of the Sultans of Bahmani.

After the defeat of Mukuṅdadeva Harichandan in Afghan invasion of 1568 the sovereign Hindu kingdom ended in Orissa and Afgan rulers at Bengal occupied the northern Orissa and made their headquarters at Cuttack and Jaipur. The Afgans did not bring any change in administrative system that prevailed previously. They even failed to exercise control over all parts of Orissa. Political conditions were remained unsettled. The Afgan rule of a couple of years was not conducive to the people's interests and rather very adversely terrorized by their much indulgence in the iconoclastic activities. They frequently raided many Hindu temples, more particularly the temple of Jagannatha and won the unpopularity amongst the Hindus who were always remaining succeptible to harassment on religious ground.

On the other hand Ramaī Routray who was the son of Danaī Vidyādhara, the Prime Minister of Gajapati-king Govinda Vidyādhara, established a kingdom at Khurda and ascended the throne as Ramchandra I at a time when the Mughals were fighting with the Afgans for the conquest of Orissa. His kingdom extended upto river Mahanadi on the north on the north and to the borders of Khemedy on the south. On the east it touched the present Jagannātha road and the Chilika lake; while on the west it did not go beyond the borders of Nayagarh, Khandapara and Daspalla region. The dynasty founded by Rāmchandradeva I was known as Bhoi-Gajapati and the rulers of the dynasty had their headquarter at Khurda till the British conquest of Orissa when they retired to Puri merely as the pensioners of the British Government.

Orissa was annexed to the Mughal in AD 1576 when sultan Daud, the son of Sulaiman Karrani was defeated and killed in the battle of Rajmahal. But it took long to put down Afgan opposition in Orissa and it was in 1593 when rājā Mansingh on behalf of Akbar got evicted all the Afgan rebellions, suppressed Rāmchandradeva, the rājā Khurda and subsequently reinstated and recognised him as rājā.

Thus, the established Hindu-kingdom became a dependency of Mughal-court. Rājā Ramchandradeva I was assigned the principality of Khurda with Purushottama Chatter of Puri and certain mahals free of tribute as Zamindari tenure. He was conferred the title of Maharājā and status of manasabadāra of 3500 in the Mughal court. It was settled that he would pay revenue to the imperial treasury instead of paying it to the Afgans. He was also allowed to exercise sovereignty over and to collect tribute from 129 killas or jurisdictions of the *Khatris* of Orissa, including all the present Cuttack tributary mahals south of Mahanadi and the estate of Ghumsur reaching as far as the borders of Khimedy in Ganjam. The remaining Orissa was controlled by the governors and the officers of the Mughals being stationed at Cuttack.

The Gajapatis of Khurda tried to maintain peace and patronized Hindu culture in their territory. The Gajapatis were recognized as the superintendent of Jagannath temple at Puri and they were called *Thākura-Rājā* and favoured by the peoples sentiments due to their association with the temple. The Gajapati-kings generally patronized the Vishṇu temples, though Śiva-worship in the Śaiva temple was not neglected. Brhāmaṇa-śāsana (villages for the Brahmans) were founded by the kings and their officers. The Sisupalgarh and Dhauligarh were the important Zamindari settlements during this time it appears that the Mahābhoiśāsana near Sisupalgarh, the Pūrvaśāsana and Uttarasasana near Dhauli were also founded during this time.

But, inspite of all their noble efforts the Gajapatis proved helpless against the frequent incursions of the Mughal-officers after the time of emperor Akbar. Akbar's rule in Orissa, based on the principles of tolerance and that of non-interference in religious matters, helped to build an atmosphere of social security and thus encouraged Hindus to freely worship their gods and practise their faith. However, the impact of liberal Islam on Orissa was practically over with the end to the reign of Akbar. The religious policy of both Jahangir and Shah Jahan gave a twist towards conservatism and orthodoxy though a regular persecution was not sanctioned. During their reign the royal house of Khurda as well as the temple of Jagannātha at Puri were frequently raided by the Mughal-officers. Several Mughal-governors and officers are noted for their hostile attitudes towards Hindus and their iconoclastic activities in the history of Orissa. During Jahangir's reign, the Mughal-Governor Hashim Khan and Mukarram Khan frightened the priests and disturbed the worship of lord Jagannatha. The Muslim occupation of Khurda, the capital, drove out Purushottamdeva from his palace and he alongwith his family was forced to reside at Manitri, on the border of Ranpur. In the reign of Shah Jahan, Mutaquad Khan invaded Puri, plundered the store of Jagannatha and slew Narasimhadeva, the *raja* of Khurda. With the beginning of reign of Aurangzeb, Orissa had to feel even more the impact of orthodox Islam than before. The reign of Aurangzeb was most disturbing and humiliating to the worship of Jagannātha and to the *rājās* of Khurda. Apart from his emphasis on Orthodox Islam, what had been much perturbing and regarded most oppressive by the Hindus of Orissa, was Aurangzeb's order for the destruction of temples. In the post-Aurangzeb period, too, the policy pursued by the governors of Orissa never had any change. Though, generally not so conservative or fanatical as that of Aurangzeb, it could never go back to the days of liberal Islam of the reign of Akbar. During the administration of Taqui Khan, the frequent Muslim invasion on Khurda affected the worship of Jagannātha as well as the peace of the whole territory. The Mughal-troops stormed the various zamindari villages including Sisupalgarh and Dhauligarh lying near Bhubaneswar and disturbed the public life.

Thus in the Mughal period, except during the reign of Akbar and Murshid Quli II, the religious sentiments of Hindus in Orissa were hurt, the royal house of Khurda was frequently subjugated, the rājās were killed or captivated or sometimes compelled to flee from the palace. The worship at the temple of Jagannātha was very often disturbed. The lives of the priests were always unsafe and people remained always panic-striken. The whole strength of rājās of Khurda was exhausted in safety and repairs of Jagannātha temple and they could not pay much attention in patronizing the religious institutions of Bhubaneswar. The huge amount of taxes collected from the people of Orissa was going to the Mughal-court and therefore, Orissa was becoming economically weaker. The political and economic conditions adversally affected the trade and commerce in the religious places. All these circumstances led to the abandonment of Bhubaneswar by peoples depending upon trade and commerce as well as Brahmaṇs and other people of priestly class. Puri, though it was the main target of Muslim-incursions, thrived as the main city which became the main identity of Hindus in Orissa and therefore, equal resistance was rendered by the cumulative Hindu-forces for the safety of that place. The most of the Brāhmaṇs migrated to the new *śāsanas* founded by the Bhoi-Gajapatis. The military settlement of Krittivās-Kaṭaka would have deserted soon after the fall of the Sūryavaṁśīs or at the time of the raids of Kālāpahār, the General of Afgans. The civil as well as the religious population moved due to economic reasons and Bhubaneswar was reduced to small village of poor priests and petty traders and craftsmen. The pilgrim-tax was imposed in Orissa sometimes during the reigns of Mughals which had minimized the number of common pilgrims. Though the pilgrim centre of Puri was not affected much due to its greater popularity however, it adversely affected the pilgrim centre of Bhubaneswar which was already falling into background due to lesser interest of the royal house and changes in public faith. The several draughts in Orissa also weakened the economic condition of the people and thus they could not afford the pilgrim-trips and charity -expenditures in various *tirthas*. The number of Bhubaneswar-oriented pilgrims badly decreased and it was visited only by the pilgrims on the way who were going to or coming back from Puri. Under these circumstances most of the priests and religious teachers would have deserted the temples and monasteries in search of livelihood. It appears that at last only a few temples viz: Liṅgarāja, Anaṅta Vāsudeva and Kapileśvara were getting regular worship, and Bhubaneswar was inhabited only by a few people of priestly class who solely depended upon the incomes coming from the landed properties of the temples and monasteries as is evident from the very poor description of this sacred place by the scholars of eighteenth and nineteenth centuries which is totally in contrast of the present condition as well as the glorious past. At present the temple city (Old Town) is always crowded with the pilgrims of the native area as well as from the whole India which again enables us to speculate about and to peep into its glorious past.

NOTES AND REFERENCES

This narrative is based on the following books:

1. P.K. Pattanaik, *A Forgotten Chapter of Orissan History,* Calcutta, 1979.

2. B.C. Roy, *Orissa Under The Mughals,* Calcutta, 1981.

3. N.K. Sahu (ed.), *A History of Orissa By W.W. Hunter, Sterling, John Beams And Sahu,* Calcutta, 1966.

7

Conclusion

Our survey of the habitation sites, archaeological remains, monuments, epigraphic records, historical and religious backgrounds and traditions helped us to have a knowledge about the various types of settlements in Bhubaneswar area and the contemporary society. It is Bhubaneswar area and the contemporary society. It is observed that the geomorphology and the natural resources such as the hills and forests; alluvial plains and the laterite tracts; rivers and lakes have been favourable factors to inspire the rise of various settlements of diverse nature in this area. Traces of human foot-prints are available since the pre-historic periods in this area and no doubt that convenient circumstances of hunting and fishing would have attracted human-settlings in this area in the initial stages. However, the well-watered alluvial plains rendered good scope for cultivation, and hence the growth of aggrarian settlements had been inspired in due course. The agrarian settlements with their stable economy and secure surroundings due to forests and hills were again the important factors for making this area as a place of royal court. The plentiful availability of building material; such as stone, laterite and timber made easy for the large scale constructions of civil, military and religious edifices. On the other hand ascetics and monks also felt convenient to settle here, getting safe and comfortable natural resorts as well as royal and public patronage for sustenance.

In due course various kinds of settlement-units developed in this land which have been discussed in Chapter II. The prominent habitation-sites were the (i) Dhauli Hills and the neighbouring area, (ii) Udayagiri and Khandarigi Hills, (iii) Sisupalgarh and its neighbouring area, (iv) Ekāmra-Bhubaneswar in the laterite tract and (v) other few sites inhabited by ascetics; and the peasantry villages witnessed by stray caves, antiquities and inscriptions.

In the Dhauli area the erstwhile existence of monastic dwellings is established on the basis of settlement remains and some corroborations. In the Asokan Rock-Edicts of Dhauli, mention of a city named Tosalī suggests the location of the city in the neighbourhood of this site. However, controversy prevailed among the scholars for the identification of the Tosalī with the fortified remains of Sisupalgarh or with scanty remains available at the foot of Dhauli hills in want of substantial evidences. The problem is discussed in detail considering its different aspects. Tosalī was an ancient city which existed in pre-Mauryan period. The origin of this pre-Mauryan city can be taken back to sixth century B.C. This period is well-known for the growth of urban settlements and referred to as the age of second urbanization in ancient India (the first being the Harappan).[1] While discussing the location of pre-Mauryan city of Tosalī in this dissertation, the Dhauli area is proposed as the site of the old city of Tosalī. On the other hand, the fortified-remains at Sisupalgarh have been identified as the new royal establishment founded by Asoka; on the basis of chronology, settlement pattern, canonical principles of layout and planning and the settlement-traditions. All these factors have been discussed elaborately and are self-explanatory. Similarly, the terms *'nagara* and *khibīra* mentioned in Hathigumpha inscription of Khāravela are also suitably explained by their corroboration with fortified remains at Sisupalgarh and in the adjoining area. The definition of settlement unit *nagara* and the various meanings of this word in Sanskrit lexicons and in regional usages have been surveyed. In this approach it is found that *nagara* stands for a fortified settlement. In Oriya its parallel *naara* (नँअर) denote the place of royal abode; and in ancient Tamil literature the word *nakar* was also used in a similar meaning. In the Hindi speaking areas also it is observed that the names of villages or towns, where the house of a zamindar or a chief is located, are traditionally coined as *nagara*. Simultaneously, the word *nagarā* or *nagalā,* is also used to denote an extension or new colony in the neighbourhood of an old settlement. Considering these definitions and other corroborations, it is concluded that fortification at the site of Sisupalgarh was founded by Asoka that was in the neighbourhood of old city of Tosalī and came to be called as *Nagara* for elaborate provisions of fortification, for its being a place of royal abode and for its isolated location from the old city. This was the place where the *Kumārāmātya* and the *Nagala-Viyohālaka* (*nagara-vyāvahārika*) resided and held their court alongwith other officials and their staff. This fortified settlement continued as the place of royal court in the succeeding period also as speak the stratigraphical chronology of the site yielded by excavations. On the other hand the Hathigumpha inscriptions of Khāravela located at Udayagiri hill and not far from the fortified site refers to Kaliṅga-nagarī with its variant *nagara* which was of fortified nature. Comparing the physical remains of Sisupalgarh with those characteristics of the nagara or city described in the Hathigumpha inscription we come to reasonable conclusion that both the evidences can be corroborated with each other. Our additional argument to affirm this corroboration is already referred to, which traces an earlier tradition of term *nagara* for fortification at Sisupalgarh that was coined in the pre-Mauryan period itself. Therefore, the *nagara* with its variant *Kaliṅga-nagarī* referred to in the Hathigumpha inscription stands for the existing fortification-remains at Sisupalgarh. Another cave inscription datable in the same period speaks about the *'nagara-akhaṅdhas'* or *nagardhayksha* translated as Town-Judge, this evidence again confirms the location of the said nagara in the neighbouring area.

Similarly, identification is proposed for a word *khibīra* occurring in the Hathigumpha inscription. In this proposition the etymological corruption of the word and the settlement pattern of the court-cities have been taken into consideration. Thus it is traced that word *khibīra* is a corrupt form of Sanskrit word *śhivira*. In the ancient canons of settlement-planning the establishing of an advance army-encampment near the royal abode or capital but outside the fortification, is emphatically prescribed. This army-encampment is called *Śivira.* The Saṅskrit lexicons also confirm similar meaning of this word. This practice of laying out a Sivira in the vicinity of the capital cities is also confirmed by various literary sources. At present also we find several place-names in the erstwhile capital cities, which coined after permanent encampment of armies, e.g., 'Lashkar' at Gwalior and 'Lashkarpur' at Agra. In the similar usage the word '*kaṭaka*' was popular in the mediaeval Orissa. Thus following the pattern of settlement a site of *Śivira* with essential commodities for soldiers and associated peoples and animals, such as water and sufficient camping ground, etc., cannot be ruled out in the neighbourhood of the fortified city of Sisupalgarh. In this context an extensive area, with a large tank having terraced embankments and wide steps, to the north of Sisupalgarh beyond the fortifications, can be proposed as the site of an advance army-encampment or *Śivira* in all respects. In this proposition a statement in Hathigumpha inscription regarding providing of embankments to the tank of *khibīra* and to other small tank, can be reasonably corroborated with the settlement remains of a large tank on the northern side and another smaller one on the south-western side. The *Śivira* was an essential part of settlement; and in this connection an evidence from *Harashacharit* traced by P.V. Begde is worth mentioning.[2] While reconstructing the layout of king Harsha's palace from the description in Harshcharit Begde has referred to an open space infront of the palace as *Śibīra*. This evidence of *Śibīra* or *Śivira* suggests that an open space adjoining to royal abode was traditionally being assigned for army-encampments and such an space even infront of main royal palaces coined as Śivira due to its use by patrolling guards and sentries. All the above arguments formed the basis for the proposed identification of *khibīra* with *Śivira*.

We find a good picture of the monastic settlement of *Jaina* ascetics at Udauagiri and Khandagiri hills; and of the sacred tīrtha of Śaivites all together. The sacred site of Ekāmra gradually evolved to a holy temple-city with a cluster of magnificent temples and monasteries. Apart from its religious sanctity it also achieved a political status in the process of development, and emerged as one of the important *kaṭakas* or royal military bases. Our survey also provides the norms of settlement-planning of civil and monastic establishments in this area. We get immense and sensitive defence arrangements in the fortification at Sisupalgagh, while in the monastic retreats of Udayagiri and Khandagiri, we find simple village-like dwelling caves provided with minimum commodities required for survival. Further, we notice that Saivite *tīrtha* of Bhubaneswar grew totally upon circumstantial requirements barring any particular settlement planning but depending upon public requirements and convenience. The *muhallās* or *sāhīs* of various classes grew up in these process of evolution; the priests, the caste-professionals and craftsmen and the outcasts settled in different quarters as per their convenience and availability of space. Though, in this study the existence of Buddhism in the given area is discussed on the basis of available sources, however, for detailed information about the Buddhist monastic settlement at Bhubaneswar, more evidences are still wanting. Archaeological excavations in the area near Brahmeśvara temple, Bhākareśvara temple and at Dhauli, may unfold the status of Buddism in the Bhubaneswar region. Similarly further excavations and Sisupalgaru also may help to retain our above propositions and identifications or open further grounds for discussion. But the merit of the findings of an archaeological excavation is always limited by the nature of deposit at the site. If the deposit is not much disturbed by the public activities it may yield very good and authentic evidences for reconstructing the past; otherwise if it is disturbed by robbing and quarying we may get only a vague idea about the past. In this context it is a bitter and disappointing fact that our most of the sites mentioned above are facing hard problem for their perpetuation as the repository of our cherished past. It is already mentioned that how the palacial remains of Krittivās-kaṭaka, which were existing in the early nineteenth century and witnessed by Andrew Sterling, have vanished completely due to public vandalism. The whole site of Sisupalgarh is under the private holdings and used for cultivation. The settlement remains underneath the cultivated lands have been quarried and still being quarried for the building materials to be used in the modern construction by the people which not only deprive us from an evidence of the past but haphazard digging of this type also disturb the stratigraphy of the deposits. Thus viewing the population influx, rapid growth of urban area archaeological remains hardly endure to provide some vivid picture of the past.

Likewise, the natural and geomorphological features have also been greatly disturbed by public activities and this process of mishandling of natural environment is further expedited by the rapid urban and industrial growth. The old water-course are vanishing due to filling up and levelling for modern constructions. Such activities may disturb the underground water-percolation and affect the survival of water-springs which were the part and partial of the ancient settlements here. The crowding shops and market-complexes have hidden the original surroundings and environment of the temples and other edifices of the past. In this context mention may be made of *Sharasra-Liṅga* think which lay infront of great Liṅgarāja temple and once increased the aesthetic beauty of the place, but unsightly clumsy growth of shops and houses has not only made it invisible but also deprived the most splendid and majestic temple of the place from its grand and picturesque frontage. Therefore, alongwith the conservation of individual monuments, it is also desirable that original environment of the place should also be preserved by restricting the adverse and unlike land-use in the Bhubaneswar area.

In this dissertation we have noticed that the whole occupational history of Bhubaneswar region in the period under review, can be divided into two broad settlement periods. The early period which begins with circa third century B.C. and ends in circa A.D. fourth century, is the period of Sisupalgarh settlement or political settlement. In this period this land was the place of royal court and the monastic and religious settlements depended upon this settlements. Thus the contemporary society was constituted by royal people, civil and military personnels, royal servants, soldiers, traders, craftsmen and so on and so forth as described in Chapter III. Being the place of royal court, this time Sisupalgarh was, no doubt, a flourishing centre of trade and commerce which also had trade links with the Roman-world as evident from the finds from archaeological excavations. In the period the land witnessed the presence of two great monarchs of their age; viz: Asoka and Khāravela. The inscriptions of both these monarchs and the narrative relief art of Udayagiri caves and the dedicatory records altogether therein provide a glimpse of contemporary society. These sources are utilized for reconstruction of the social history of this periods. The archaeological excavation at Sisupalgarh is another major source which throws a flood of light on contemporary art, crafts, dwelling patterns and architecture; and this source is also well utilized. However, we lack any such contemporary poetic work or itinerary account which can provide a detail description of the particular place. For the literary sources mention can be made of *Gaṇḍavyūha* in which a passage occurs about the city of Tosalī, and another literary source is *Brahmapurāṇa* which gives a graphic description of Ekāmra-Bhubaneswar; a few verses of this description can be taken for the Sisupalgarh period.

In about the fourth century A.D. the settlement of Sisupalgarh came to an end, the royal court deserted from the place and the site was totally abandoned in a very short period. We do not have any documentary evidence about the causes of desertion and abandonment of the place. In the Chapter IV, however, we have tried to explain the probable causes of this abandonment. The desertion of the royal camps and capitals is a very usual phenomenon in the course of history. The royal camps and the capital cities are laid-out in suitable places by the will of rulers and their counsellors. Such settlements are succeptible for desertion due to invasions by powerful enemy-kingdom. In many cases the defeated cities were restored to their grace under the victorious invader, but, in most cases the invaders converted into ruins the defeated cities as did Kahārvela himself in the case of the city of Pithuṇḍa. The abandonment of court cities was always liable upon the gravitation of the political power. In the course of history it is noticed that the capitals moved with the royal powers and royal camps were converted in capital cities within no time, the Sisupalgarh was a complete royal city and this was deserted as soon as the royal power abandoned this place.

The fourth-fifth century AD is a period of transition in the settlement history of Bhbaneswar. In the same chapter we have noticed the decline of the political city-settlement and the rise of Śaivite *tīrtha* that grew with the overwhelming religious sanctity and established as the celebrated pilgrimage centre of the Hindus with everlasting grace and existence. The antiquity and evolution of this Śaivite *tīrtha* is traced with available archaeological and literary evidences, However, to trace the root-source of the movement of Pāśupata-cult to Bhubaneswar, we have to wait for the new discoveries in that sphere.

In our broad periodisation of settlement-history of Bhubaneswar, the second period is the period of a great religious settlement i.e. *Bhubaneśvara* the Śaivite *tīrtha.* This period begins from sixth century AD (as we get *Śiva*-temple of this period). We have given a very long span to this period which ends with the Muslim conquest of Orissa; though, this span is still shorter for Bhubaneswar as we know that unlike the early settlement the religious settlement of Bhubaneswar still continues in its full grace and charm, however, as the Muslim conquest of Orissa was the significant event which brought many qualitative and quantitative changes in the political and social setups and also affected the religious activities of the people; and all these changes acted upon in the cultural and economic setup of Bhubaneswar. Therefore, we have taken this event as a landmark to conclude our studies about Bhubaneswar. In Chapter V, we have dealt with the political condition and the religious background and their impact on the development of

Bhubaneswar. This long period saw many ups and downs in political arena: however, Bhubaneswar was gradually becoming near to royal court and received court patronage. Not only this, the land became politically so sensitive that a royal military camp was also established here and at this royal-military camp one of the greatest Hindu monarchs Kapilendra, the founder of the Sūryavaṁśī-Gajapati dynasty, who usurped the throne from Gaṅga-king, was corronated.

Amalgamation of *Śaiva* and *Śākta* cults is a Characteristic feature in the early stage of this period, however, in the late phases a synthesis in between Śaivism and Vaishṇavism is discernible in the religious life. As discussed in same chapter the contemporary society of Bhubaneswar was constituted by various classes of people adhering different professions, trades and crafts, however, the *Brāhmaṇas* and the priestly class held the superb position. Pilgrimage was the main industry which fed all the classes, though inter-alia economy also developed. Due to this industry we find that the priestly class gradually crystalized in many sub-groups such as the *Badu* and the *Suara* or *Mahāsuara* are found only at Bhubaneswar and these titles arose from their particular type of job in temple rituals. Similarly, many fairs and festivals owe their very local religious origin and are being organised/celebrated in the honour of Lord Śiva. The temple of Lord Kṛittivāsas and other similar institutions became the centre of community life. Not only the priestly class depended on temples but other caste people also took adherence for their livelihood. The temples were having enough landed and monetory endowments which were the source of recurring income and that income was spent in performance of various temple rituals and celebration of festivals. The workers involved in these performances and celebrations were paid wages in cash and kind. Thus, many craft-professionals were permanently associated with temple services. This economic condition had given rise to a new social setup which can be called as the temple community. The contemporary dedicatory records help us to know a very important fact about the maintenance of the perpetual offerings in which land or cash currency were given/ deposited to some trustee, thus creating a recurring treasure *(nidhi)* in the presence of some important people, that trustee had to supply the desirable article in the temple. This system reveals that the endowments done in the name of the deity were managed by the trustees and not by the priests.

In our study we have observed the growth and development of two divergent settlements, (1) royal or political settlements (2) religious settlement. In this regard it is a noteworthy point that the royal-camps or capitals move with their owners and are not bound with the place. However, the religious settlements grow only due to sanctity of place and therefore such settlements can never be deserted. They can grow and flourish on court patronage but do not solely depend upon that. The political crisis or any other such calamity may prove an obstacle but it cannot desert such settlements forever. The religious faith of the people is such strong factor which always inspire to continue such religious centre of great sanctity. This phenomenon also happened with Bhubaneswar, as we notice that during the political disturbances and weakened economic atmosphere the Śaivite *tirtha* suffered for want of pilgrims, and with available resources the large population could not survive, therefore, people left the place in search of better scope. However, the place was not abandoned fully only the population was reduced. The remaining population maintained sanctity of the place and as soon as favourable conditions appeared this religious centre was again restored to its popularity and grace as evident from the present environment of Lingaraja temple and the whole of the Old Town or the Temple City which can be seen always crowded with pilgrims and other associated people.

NOTES AND REFERENCES

1. A. Ghosh, *The City In early Historical India,* op. cit., pp. 32-58 and 89-90. See also the discussions and comments of different scholars on Urbanization in India, *Puratattva, No. 6 & 7 (Bulletin of Indian Archaeological Society),* New Delhi, pp. 27-42 and pp. 87-91 respectively.

2. P.V. Begde, *Ancient and Medieval Town Planning in India* op. cit., p. 237.

BIBLIOGRAPHY

(Modern Books)

A. Ghosh, *The City In Early Historical India,* Simla, 1993.

A. Ghosh (ed.), *An Encyclopaedia of Indian Archaeology,* Vol. 1 & II, New Delhi, 1989.

A.K. Coomarswamy, *History of Indian and Indonesian Art,* London, 1927.

A.S. Altekar, *State and Government in India,* (first ed. Varanasi, 1949) Delhi, (reprint) 1977.

B.B. Dutt, *Town planning In Ancient India,* Delhi, 1977.

B.C. Bhattacharya, *The Jaina Iconography,* Lahore, 1939.

B.C. Chhabra (ed.), (revised by D.R. Bhandarkar), *Corpus Inscriptionum Indicarum,* Vol. III (Inscriptions of the Early Gupta kings) New Delhi, 1981.

B.C. Mazumdar, *Orissa In The Making,* (reprint) Patna, 1984.

B.C. Ray, (*Orissa Under the Mughals*), Calcutta, 1981.

Charls Fabri, *History of the Art of Orissa.* Bhubaneswar, 1974.

D. Das, *The Early History of Kalinga,* Calcutta, 1977.

D.K. Ganguly, *Historical Geography and Dynastic History of Orissa,* Calcutta, 1975.

Debla Mitra, *Bubaneswar,* A.S.I. New Delhi, (reprint) 1978.

Debla Mitra, *Udayagiri and Khadagiri,* A.S.I. New Delhi. (reprint) 1975.

D.C. Sircar, *Select Inscriptions, Vol.* I, Calcutta, 1965

E. Hultzsch, *Corpus Inscriptionum Indicarum,* Vol. I (Inscriptions of Asoka), Varanasi, 1967.

F.E. Pargiter, *Ancient Indian Historical Traditions,* Delhi, (reprint) 1972.

H.C. Raychaudhury, *Political History of Ancient India* Calcutta, (sixth ed.) 1953.

James Fergussion and Jas Burgess, *The cave Temples of India* Delhi, (reprint) 1969.

K.C. Misra. *The cult of Jagannath,* Calcutta, 1971.

K.C. Panigrahi, *Archaeological Remains At Bhubaneswar,* cuttack, (second ed.) 1981.

K.C. Panigrahi, *Chronology of the Bhaumakaras and the Somavamsis of Orissa*, Madras, 1961.

K.C. Panigrahi, *History of orissa,* cuttack, 1981

Kripa Shankar Mathur, *Manava Pragati Ki Kahani* (Hindi Translation of V. Gordon Childe's *Man Makes Himself),* Uttar Pradesh Hindi Grantha Academy, Lucknow, 1971.

K.V. Saundar Rajan, *Early Kalinga Art and Architecture,* Delhi, 1984.

Moti Chandra, *Costumes, Textiles, Cosmetics and Coiffure In Ancient India,* Delhi, 1973.

M.H. Kuraishi (revised by A. Ghosh), Rajgir, A.S.I. New Delhi, (reprint) 1975.

M.M. Ganguly, *Orissa and Her Remains,* Calcutta, 1912.

M.M. Mahapatra, *Traditional Structure And Changes In An Orissan Temple,* Calcutta, 1981.

N.K. Bose, *Canons of Orissan architecture.* (first ed. 1932) reprint, Delhi, 1982.

N.K. Sahu (ed.), *History of Orissa by W.W. Hunter, Andrew Streling, john Beams And N.K. Sahu,* Vol. I & II, Calcutta, 1966.

P. Banerjee, *Early Indian Religions*, Delhi, 1973.

P.K. Pattanaik, *A Forgotten Chapter or Orissan History,* Calcutta, 1979.

P.V. Bedge, *Ancient and Medieval Town Planning In India,* New Delhi, 1978.

R.C. Mazumdar (ed.), *The Vedic Age,* (Bharatiya Vidya Bhavan), London, 1951.

R.C. Mazumdar (ed.), *The Age of Imperial unity,* (Bharatiya vidya Bhavan) London, 1951.

R.C. Mazumdar, H.C. Raychaudhury and K.K. Dutta, *An Advanced History of India*, London, 1949.

R.D. Banerjee, *History of orissa,* Vol. I & II, 1931, (reprint) Delhi, 1980.

R.L. Mitra. *Antiquities of Orissa.* Vol. I & II, 1880, (reprint) Cuttack, 1963.

R.P. Mohapatra, *Udayagiri And Khadagiri Caves*, Delhi.

Romila Thapar, *Asoka and The Decline of the Mauryas,* Delhi, 1973.

S.K. Sarswati, *A Survey of Indian Sculpture,* New Delhi, (reprint) 1975.

S.N. Rajaguru, *Inscriptions of Orissa,* Vol. I. pt. ii, Bhubaneswar, 1958.

S.N. Rajaguru, *Inscriptions of Orissa,* Vol. II, Bhubaneswar, 1960.

S.N. Rajaguru, *Inscriptions of Orissa,* Vol. III, pt. i. Bhubaneswar, 1960

S.N. Rajaguru, Inscriptions of Orissa, Vol. III, pt. ii, Bhubaneswar, 1961.

S.N. Rajaguru, Inscriptions of Orissa, Vol. IV, Bhubaneswar, 1966.

S.N. Rajaguru, *Inscriptions of Orissa,* Vol. V, pt. i, Bhubaneswar, 1975.

S.N. Rajaguru, *Inscriptions of Orissa,* Vol. V, pt. iii, Bhubaneswar, 1976.

S.N. Rajaguru, *History of the Gangas*, Vol. I, Bhubaneswar, 1968.

S.N. Sarangi, *proposed Master Plan of Bhubaneswar,* Bhubaneswar, 1988.

Thomas E. Donaldson, *The Hindu Temple Art of Orissa,* Vol. I, II and III, Lyden, 1985-87.

Thomas watters, *On Yuan chwang's Travel in India*, Vol. II, (Original London) Delhi, 1961.

Vasudeva Saran Agrawal, *Bharatiya Kala,* Varanasi (reprint) 1987.

Vidya Dehijia, *Early Stone Temples of Orissa,* New Delhi, 1979.

V.A. Smith, *Asoka The Buddhist Emperor of India,* New Delhi, (reprint) 1970.

V. Gordon Childe, *What Happened in History,* penguin Books, 1942, (reprint) 1978.

(Articles in Research Journals, Felicitation Volumes and commemoration volumes)

A. Das, "Some New Facts about Sisupalgarh". *Orissa Historical Research Journal,* Vol. XI, Bhubaneswar, 1967.

A. Ghosh, "Parasuramesvara Temple inscription", *Epigraphica indica* Vol. XXVI (1941-42), Delhi, 1952

B.B. Lal "Sisupalgarh 1948: An Early Historical fort In eastern India", *Ancient India.* No. 5, A.S.I., New Delhi, 1949.

B.B. Lal, "Sisupalgarh: Some Reflections on its Layout", *Art And Archaeology* (K. Deva Felicitation volume), edited by C.P. Sinha and others, Delhi, 1990.

B.M. Barua, "Hathigumpha Inscription of Kharavela", Indian Historical 'Hathigumpha Inscription of Kharavela", *Indian Historical Quarterly,* vol. XIV, Delhi, (reprint) 1985.

B.M. Barua, "Minor Old Brahmi Inscriptions in the Udayagiri Khandagiri Caves", *Indian Historical Quarterly,* Vol. XIV.

B.N. Srivastava, 'Sasanka king of Gauda', *Bulletin of the U.P. Historical Society,* Vol. X, No. 3, 1965.

B.V. Nath, "The Parvati Temple Inscription At Bhubaneswar", *Orissa Historical Research journal,* Vol. III, No. 3, Bhubaneswar, 1954.

Debala Mitra, "Buddhist Vestiges In Orissa: A Survey", *Orissa Review* (Monument special), Bhubaneswar, 1976.

Debal Mitra, "Glimpses of Jaina Relics In Orissa", *Orissa Review* (Monument Special), Bhubaneswar, 1976.

D.C. Sircar, "Two plates from Kanas". *Epigraphica Indica* Vol. XXVIII (1949-50), Delhi, 1958.

D.C. Sircar, "Bhubaneswar Inscription of Anangabhimadeva III, Anka year 34" and "Bhubaneswar Inscription of Bhimadeva", *Epigraphica Indica* Vol. XXX (1953-54), Delhi, 1958.

D.C. Sircar, "Bhubaneswar Inscription Of Anangabhimadeva III, Anda Year 34" and "Bhubaneswar Inscription of Bhimadeva", *Epigraphica Indica,* Vol. XXX (1953-54), Delhi, 1958.

D.C. Sircar, "Santirgrama Grant of Dandi Mahadevi", *Epigraphica Indica,* Vol. XXIX (1951-52), Delhi, 1957.

D.C. Sircar, "Two Lingaraja Temple Inscriptions", Indian *Culture* (A Journal Of Indian Research Institute), Vol. IV, Patna, (reprint) 1984.

D.C. Sircar and J.C. Ghosh, "Three Lingaraja Temple Inscriptions", *Indian Culture* (A journal of Indian Research Institute), Vol. III, pt. i, Patna, (reprint) 1984.

D.C. Sircar and J.C. Ghosh, "Three Lingaraja Temple Inscriptions", Indian *Culture* (A journal of Indian Research Institute), Vol. III, pt. i, Patna, (reprint) 1984.

D.C. Sircar and K.G. Krishnan, "Bhubaneswar Inscription of Ganga Narasimha", *Epigraphica Indica,* Vol. XXXII, Delhi, 1958.

D.R. Bhandarkar, "Sanjan Plates of Amoghvarsha: Saka Samvat 793", *Epigraphica Indica,* Vol. XVIII (1926).

D.K. Chakravarti, "Concept of The Urban Revolution and the Indian Context", *Purātattva*, No. 6 (Bulletin of Indian Archaeological Society), New Delhi, 1972.

D.K. Chakravarti, "Some Traditional Aspects of Early Indian Urban Growth", *Puratattva* No. 7, New Delhi, 1974.

F. Kielhorn, "Two Grants Of Dandimahadevi" and "Two Bhubaneswar Inscriptions: A. Inscription of Svapnesvara of The Time of AniankBhima, B. Eulogy of Bhatta Bhavadeva", *Epigrahica Indica,* Vol. VI (1900-1).

F. Kielhorn, "Bilhari Stone Inscription of The Rulers of Chedi", *Epigrahica Indica,* Vol. I, Calcutta, 1892.

J.F. Fleet, "Records of the Somavamsi Kings of Kataka", *Epigraphica Indica*, Vol. III (1894-95).

J.Ph. Vogel, "Prakrit Inscriptions From A Buddhist Site at Nagarjunikonda", *Epigraphica Indica,* Vol. XX (1929-30), New Delhi (reprint) 1983.

J.W. Mc. Crindle, "Ptolemy's Geography of India and Southern Asia", *Indian Antiquary,* Vol. XIII, 1884.

K.A. Chadhury and S.S. Ghosh, "Wood Remains From Sisupalgarh", *Ancient India* No. 8, A.S.I. New Delhi, 1952.

K.C. Panigrahi, "Three Temple Inscriptions From Bhubaneswar" *Orissa Historical Research Journal*, Vol. I No. 2, Bhubaneswar, 1952.

K.P. Jayaswal and R.D. Banerji, "Hathigumpha Inscription of Kharavela", *Epigraphica Indica,* Vol. XX (1929-30), Delhi, (reprint), 1983.

K.S. Behera, "The Antiquity of Purushottam-Jagannatha at Puri", *The Cult And Culture of Lord Jagannatha,* (ed. Daitari Panda and S.C. Panigrahi), Cuttack, 1984.

Lionel D. Barnet, "Bhubaneswar Inscription In The Royal Asiatic Society", *Epigraphica Indica*, Vol. XIII.

Martin Brandtner, "Archaeological Remains At Dhauli", *KṚṢṆA PRATIBHĀ* (Prof. K.C. Panigrahi Commemoration Vol, ed. H. Das and others), Delhi, 1994.

M.C. Joshi, "Early Historical Urban Growth In India: Some Observations", *Purātattva,* No. 7, New Delhi, 1974.

N.K. Sahu, "Historical Geography of Orissa", *Souvenir, Orissa Sahitya Academy*, Silver Jubilee, 1982.

P.K. Ray, "The Oriya Inscription of Lingraja Temple", *Orissa Historical Research Journal*, Vol. XV, No. 1 & 2, Bhubaneswar, 1967.

R.D. Banarjee, "Inscriptions In The Udayagiri And Khandagiri Caves", *Epigraphica Indica*, Vol . XIX (1927-28).

S.C. De, "Three Ancient Naga Images From Bhubaneswar", *Orissa Historical Research Journal*, Vol. III, No. I, Bhubaneswar, 1954

S.N. Das, "The Inscription of Prapanncharya In The Temple of Parasuramesvara" O.H.R.J. Vol. II, No. 3 & 4. Bhubaneswar, 1953.

S.N. Das, "Votive Inscription in the Lingraja Temple (2) Stone Inscription of Govinda Senapati", O.H.R.J. Vol . II , No. 3 & 4.

S.N. Rajaguru, "Daspalla Copper Plate Grant of Satrubhanjdeva", O.H.R.J., Vol. I, Bhubaneswar, 1952.

S.N. Rajaguru, "Chronology of the Somavamsi Kings of Orissa", *O.H.R.J.* Vol. VII, Bhubaneswar, 1959.

S.N. Rajaguru, "Two Lingaraj Temple Inscriptions of the Time of Raghavadeva", *OH.R.J.* Vol. V, No. 4, Bhubaneswar, 1957.

S.N. Rajaguru, "The Lingaraj Temple Inscription of the Time of Rajarajadeva (s. 1094), *O.H.R.J.* Vol. IX, No. 1, Bhubaneswar, 1960.

S.P. Gupta, "Two Urbanizations in India : A Side Study in Their Social Structure", *Puratattva,* No. 7, New Delhi, 1974.

S.R. Balsubrahmanyan, "Bhubaneswar" *Orissa Historical Research Journal*, Vol. VI, No. 2., 1957.

S. Tripathy, "Ekamra-Bhubaneswar", *Orissa Historical Research Journal* Vol. XXXIII, No. 1 & 2.

V.S. Pathak, "Ekamra-Bhubaneswar In Inscriptions", *Orissa Historical Research Journal,* Vol. VII, No. 1, 1952.

(Annual Reports, Reviews of Archaeological Survey of India And Gazetteers)

J.D. Beglar, *Reports of Tours In South-Eastern Province In 1874-75 and 1875-76,* Calcutta, 1882, *Archaeological Survey of India Reports* (ed.) A. Cunningham, Vol. XIII.

Indian Archaeology 1958-59 - A Review.

Indian Archaeology 1961-62- A Review.

Indian Archaeology 1963-64 - A Review.

Indian Archaeology 1969-70 - A Review.

Indian Archaeology 1984-85- A Review.

Gazetteer of India Pakistan, Vol. I & II, Director of Military Survey GSGS. Army India, Delhi, 1950.

Puri District Gazetteer (ed.) N. Senapati, Cuttack, 1977.

(Sanskrit texts)

Arthasastra

Brahampuranam

Mahabharat

Skandapuranam

Sabdakalpadrumah

Svarnadri Mahodaya

Index

A

B

E

F

G

H

N

O

P

List of Illustrations

Pl. 1

Pl. 2

Pl. 3

Pl. 4

Pl. 5

Pl. 6

Pl. 7

Pl. 8

Pl. 9

Pl. 10

Pl. 11

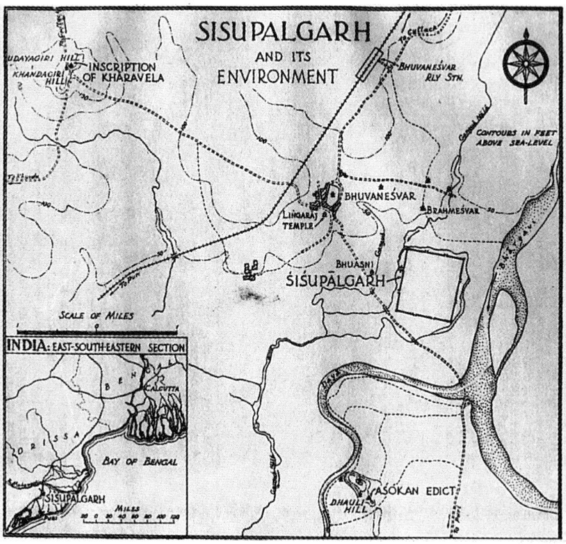
SISUPALGARH
AND ITS
ENVIRONMENT
UDAYAGIRI HILL
KHANDAGIRI HILL
INSCRIPTION OF KHARAVELA
BHUVANEŚVAR RLY STN.
CONTOURS IN FEET ABOVE SEA-LEVEL
BHUVANEŚVAR
LINGARAJ TEMPLE
BRAHMEŚVAR
BHUASNI
ŚIŚUPĀLGARH
SCALE OF MILES
INDIA: EAST-SOUTH-EASTERN SECTION
CALCUTTA
ORISSA
BAY OF BENGAL
SISUPALGARH
MILES
AŚOKAN EDICT
DHAULI HILL

Pl. 12

Pl. 13

Pl. 14

Pl. 15

Pl. 16

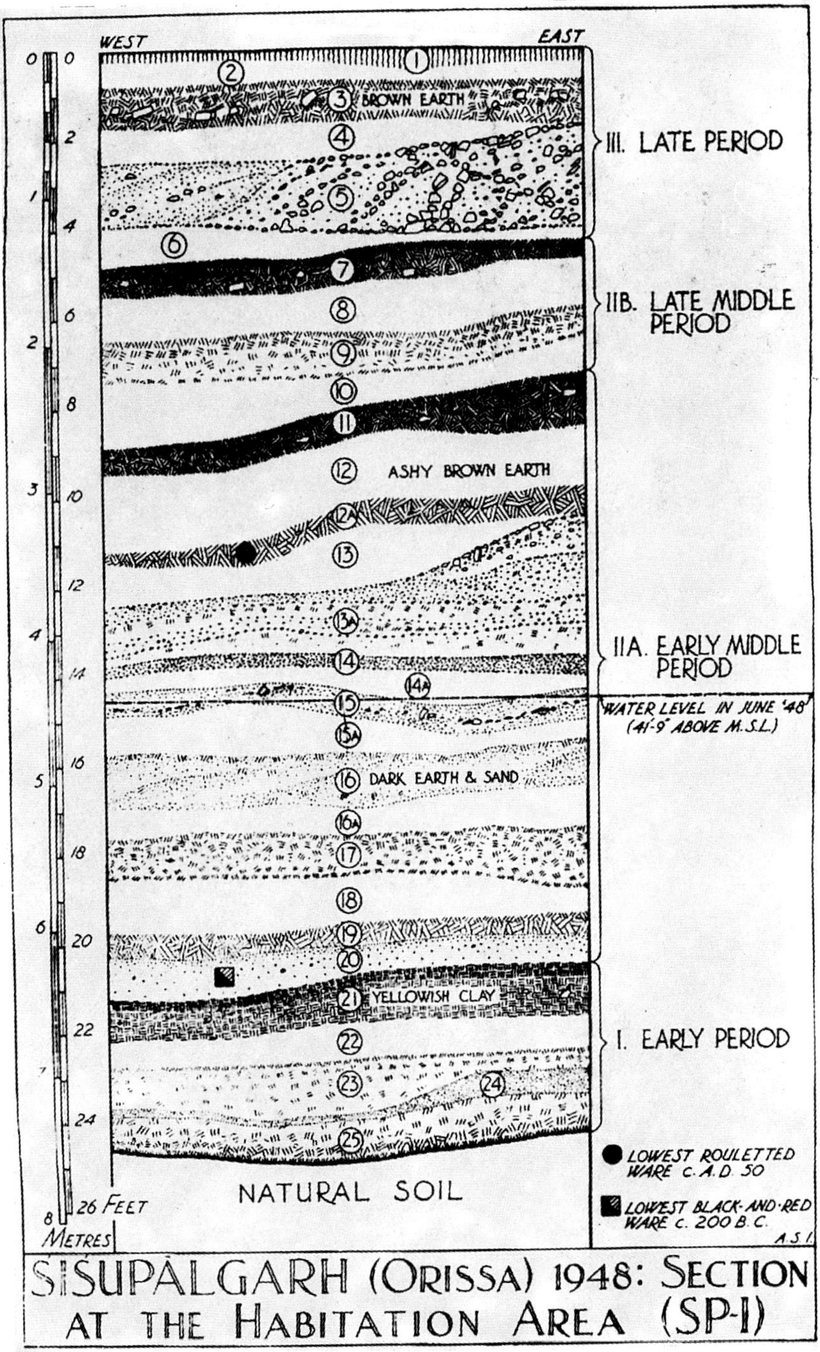

SISUPĀLGARH (ORISSA) 1948: SECTION AT THE HABITATION AREA (SP-I)

Pl. 17

Pl. 18

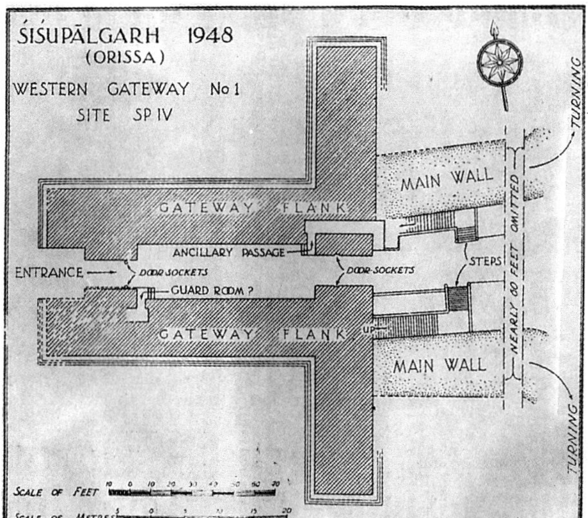
ŚIŚUPĀLGARH 1948
(ORISSA)
WESTERN GATEWAY No 1
SITE SP IV
MAIN WALL
GATEWAY FLANK
ANCILLARY PASSAGE
ENTRANCE
DOOR SOCKETS
DOOR-SOCKETS
STEPS
GUARD ROOM ?
GATEWAY FLANK
UP
MAIN WALL
NEARLY 60 FEET OMITTED
TURNING
TURNING
SCALE OF FEET
SCALE OF METRES
B. SARAN

Pl. 19

Pl. 20

Pl. 21

Pl. 22

Pl. 23

Pl. 24

Pl. 25

Pl. 26

Pl. 27

Pl. 28

13
14
15
16
17
18
19
INCHES
CMS

Pl. 29

20
21
22
23
24
25
26
27
28
29
30
31
32
INCHES
CMS

Pl. 30

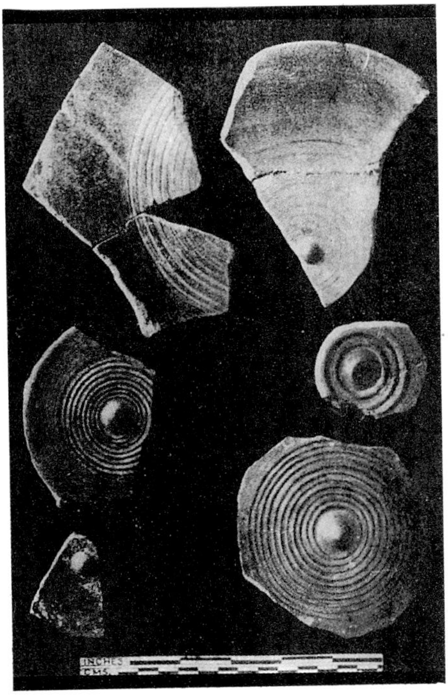

Pl. 31a

Pl. 31b

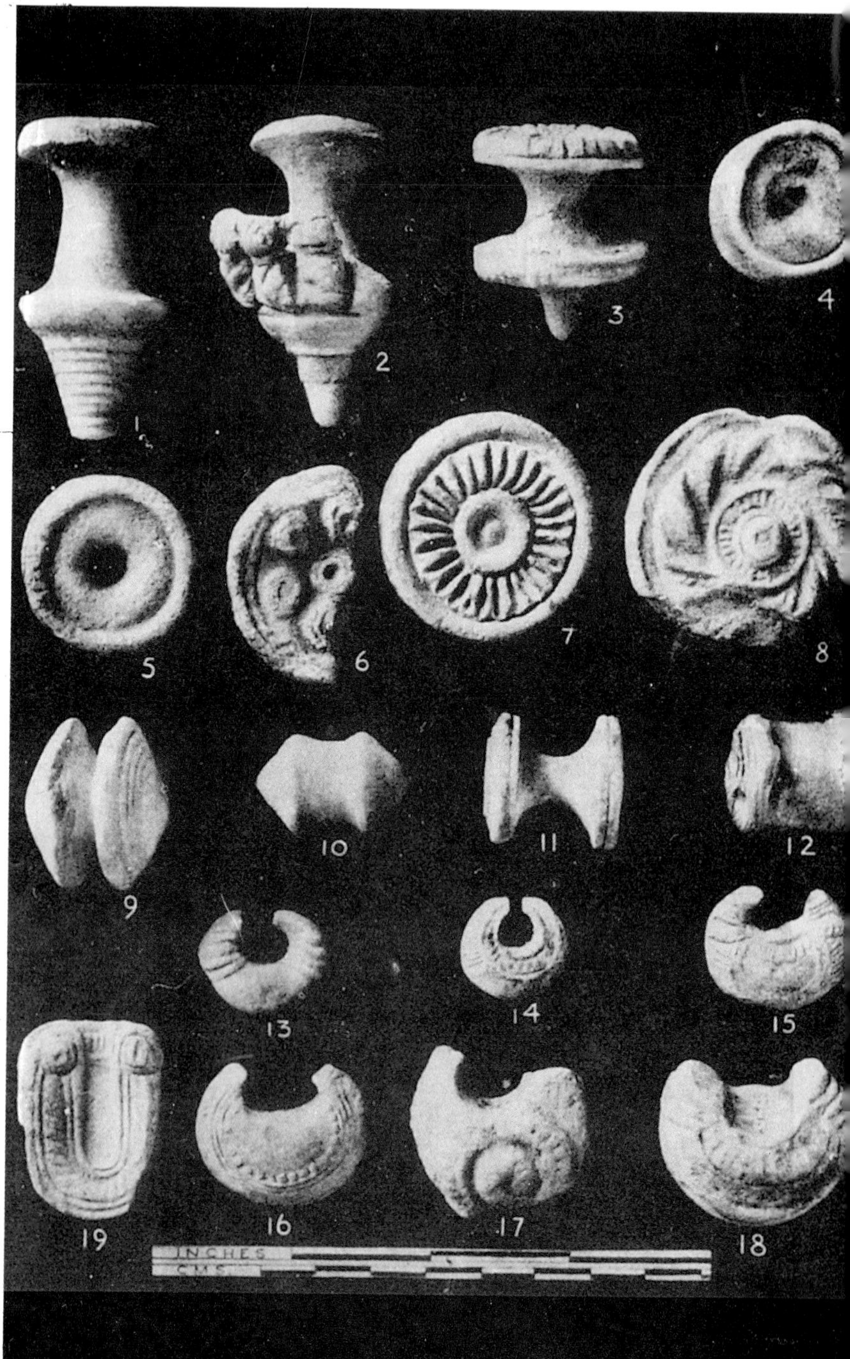

Pl. 32

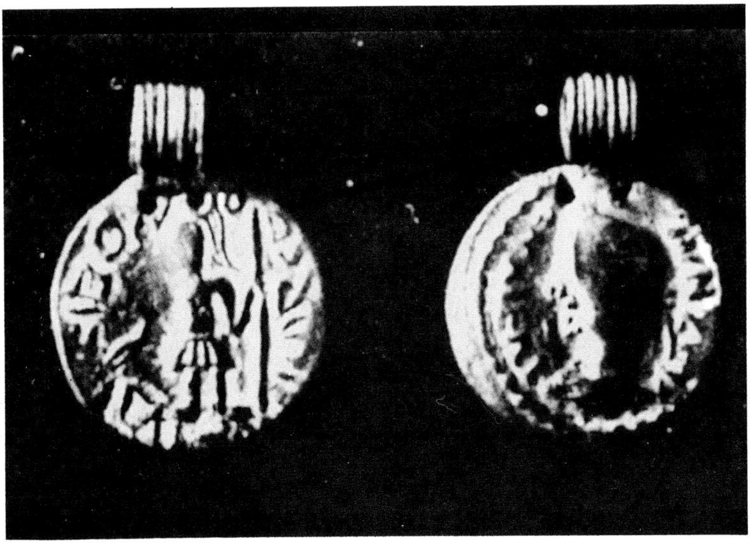

Pl. 33

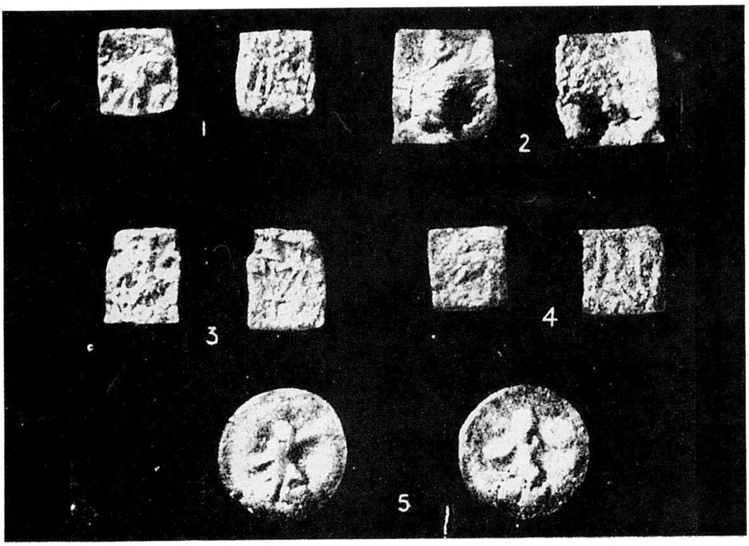

Pl. 34

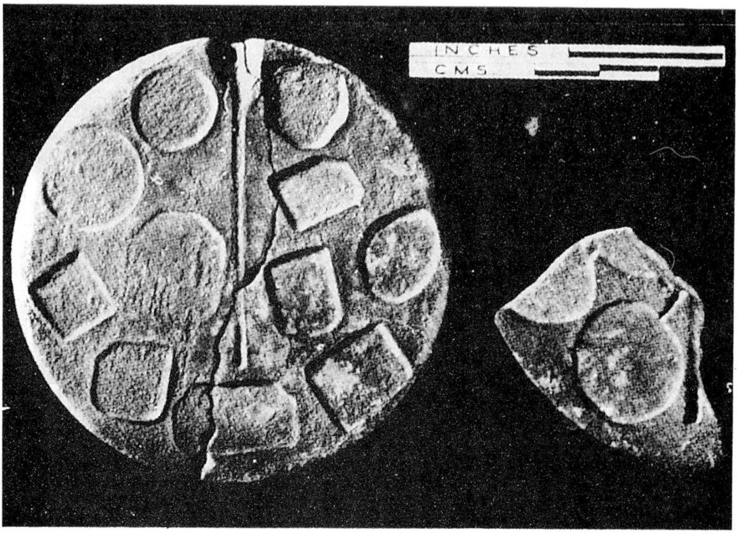

Pl. 35

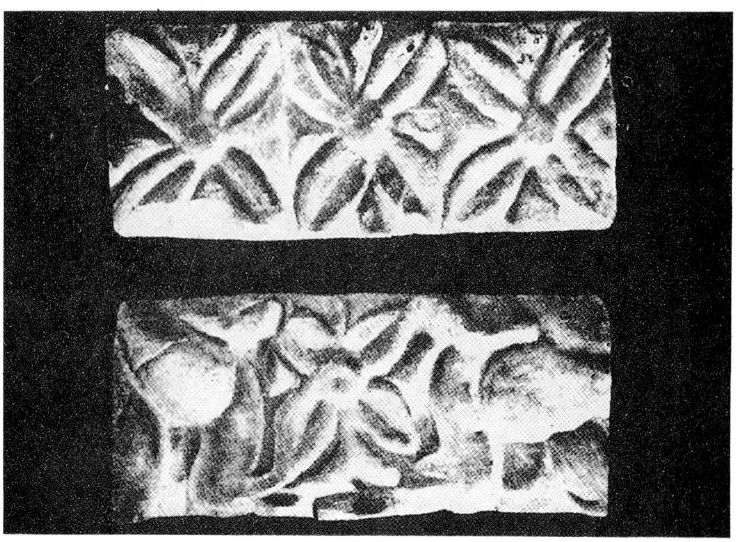

Pl. 36

Pl. 37

Pl. 38

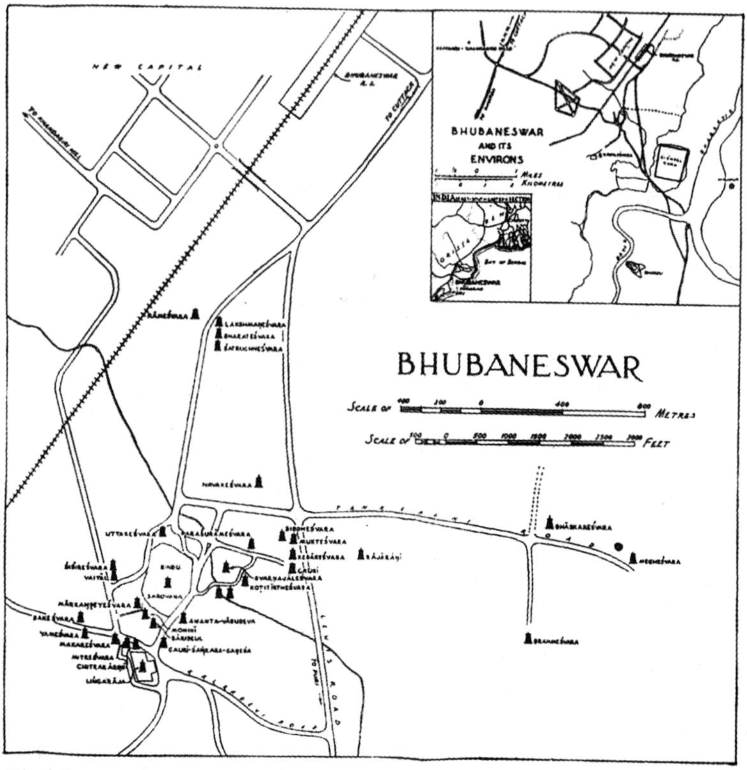
NEW CAPITAL
TO CUTTACK
BHUBANESWAR
AND ITS
ENVIRONS
MILES
KILOMETRES
BAY OF BENGAL
BHUBANESWAR
SCALE OF
400
200
0
400
800
METRES
SCALE OF
500
0
500
1000
1500
2000
2500
3000
FEET
LAKSHMAṆEŚVARA
BHARATEŚVARA
ŚATRUGHNEŚVARA
UTTAREŚVARA
PARAŚURĀMEŚVARA
SIDDHEŚVARA
MUKTEŚVARA
KEDĀREŚVARA
RĀJARĀṆI
GAURĪ
SVARṆAJALEŚVARA
KOṬITĪRTHEŚVARA
BINDU
SAROVARA
VAITĀL
MĀRKAṆḌEYEŚVARA
ANANTA-VĀSUDEVA
MOHINĪ
YAMEŚVARA
MAKAREŚVARA
MITREŚVARA
LIṄGARĀJA
BHĀSKAREŚVARA
MEGHEŚVARA
BRAHMEŚVARA
ROAD
TO PURI

Pl. 39

Pl. 40

Pl. 41

Pl. 42

Pl. 43